Navigating World History

Navigating World History
Historians Create a Global Past

Patrick Manning

First published in 2003 by
PALGRAVE MACMILLAN™
175 Fifth Avenue, New York, N.Y. 10010 and
Houndmills, Basingstoke, Hampshire, England RG21 6XS
Companies and representatives throughout the world

PALGRAVE MACMILLAN is the global academic imprint of the Palgrave Macmillan division of St. Martin's Press, LLC and of Palgrave Macmillan Ltd. Macmillan® is a registered trademark in the United States, United Kingdom and other countries. Palgrave is a registered trademark in the European Union and other countries.

1–4039–6119–0 hardback — 1–4039–6117–4 paperback

Library of Congress Cataloging-in-Publication Data Available from the Library of Congress

Manning, Patrick, 1941-
    Navigating world history : a guide for researchers and teachers / by Patrick Manning.
        p. cm.
    Includes bibliographical references.
    ISBN 1–4039–6119–0 (alk. paper) – ISBN 1–4039–6117–4 (pbk.)
        1. World history—Study and teaching. 2. World history—Research. I. Title.

D21.3 .M285 2003
907–dc21   2002030745

A catalogue record for this book is available from the British Library.

First edition: May, 2003
10   9   8   7   6   5   4   3   2   1

Transferred to digital printing 2005

Contents

Acknowledgments

The work that has led to this book began in 1990. It developed especially out of the graduate courses that I taught at Northeastern University on historical methodology and global historiography, along with research seminars in global social, political, and cultural history. What I thought would be an article on global historiography was politely declined by two major journals; as I revised it further, it kept growing at a rate that averaged a chapter or two each year.

I want to express special thanks to four M.A. students at Northeastern who, in the days before our doctoral program, gave me great assistance and important feedback in developing approaches to graduate teaching in world history: Suzanne McCain, Kelly Farrell, Rebecca Eads, and Perry Tapper. From 1994, when our doctoral program began, Pam Brooks, Yinghong Cheng, and Sarah Swedberg were the pioneers who first gave shape to the new program. They were followed in 1995 by Desirée Evans, James Beauchesne, and Whitney Howarth. From 1996, when a larger cohort of students arrived and worked together to create the *Migration in Modern World History* CD-ROM, the students were also able to critique my growing set of chapters. These included Eric Martin, David Kalivas, Jeff Sommers, Deborah Smith Johnston, Whitney Howarth and the late Hector Enrique Melo of the doctoral program, and Patty Whitney, Beimei Long, Dolly Wilson, Julie Gauthier, David Kinkela, and Athanasios Michaels of the master's program. From 1998 a group including Christopher Harris, Bin Yang, George Reklaitis, George Dehner, Tiffany Trimmer, Jeremy Neill, Stacy Tweedy, Josh Weiner, John Kelly, and Tiffany Olson helped me get through the later versions. Kristie Stark, an extraordinary graphic designer, along with doctoral students Parker James of Tufts University and Richard Rath of Brandeis University, helped clarify my ideas of how to make world history work.

I am thankful to all of my colleagues in the departments of History and African-American studies at Northeastern for the support they provided for the venture in world history. Adam McKeown and Clay McShane provided me with particular inspiration and with gentle but firm critique; Bill Fowler, Robert L. Hall and Jeffrey Burds provided steady cheer and support. Colleagues at other institutions who have debated and developed the ideas I have expressed in these pages include Peter Gran, Andre Gunder Frank, Alfred Crosby, Maghan Keita, Nicola di Cosmo, Ross Dunn, Heidi Roupp, Lynda Shaffer, Joseph Miller, and Kim Butler.

At the editorial stage of reorganizing once again, plus forcing general ideas into coherent sentences, I wish to thank Kim A. Pederson of Editheads, who once again has asked the tough questions and suggested sensible answers, Deborah Gershenowitz and Brendan O'Malley of Palgrave Macmillan, whose enthusiasm and guidance have been immensely helpful to me, and Enid Stubin for expert copyediting.

Let me also acknowledge in advance the colleagues who will identify the gaps and gaffes in this book—the many works I should have cited and the trends I should have elucidated. The errors and deficiencies of this work are my own, but I hope to continue with others in the rewarding work of advancing the academic discourse on world history.

The past dozen years of work on world history at the graduate level have provided me with all the academic excitement I could hope for. The frustrations and defeats could never match the exhilaration of opening new categories of connections or supporting the completion of successful dissertations. The world history program at Northeastern demonstrated, to my satisfaction and I think to the satisfaction of many others, that world history is indeed a legitimate and even important field of specialization at the doctoral level. The Northeastern program now teeters toward collapse for lack of fiscal or academic support from a university and a funding community that cannot envision world history as a legitimate research field. But programs collapse in one place and begin anew in others— that is an established pattern in American university life. The study of world history may soon, after many false starts, get past its long-term status as an amateur activity, and become a professional field of study, devoted to articulating the connected patterns of the human past.

My wife Susan joined me as we passed middle age, turning what might have been the beginning of quiet decline into a time of exciting new ventures in an atmosphere of warmth and comfort. I offer heartfelt thanks to her for support, to my thriving daughters Gina and Pamela, and to Susan's children, also thriving, Tania, Charles, and Amakca. I thank my brother Curt for years of conversations about systems; the scope of his work on cosmology far exceeds the spatial and temporal reach of this study.

I dedicate this book to my father, John V. Manning. In my early years he introduced me to the world, to appreciating its variety and the challenge of trying to change it. He continues to explore and enjoy this planet and struggles daily and critically to make it a better place. In so doing he provides a wonderful role model for his offspring and to others.

PATRICK MANNING
Boston, June 2002

Preface

This volume presents an overview and critique of world history as a field of scholarship and teaching. In it, I review the narratives and analyses of historians who focus on large processes and connections in the past, ranging from the time of the earth's formation up to our own day. For researchers, teachers, and general readers, I offer clues for navigating the literature on world history and propose guidelines for conducting research on global historical issues. I recommend courses of study for prospective and current graduate students, for practicing teachers and scholars seeking to add breadth and depth to their knowledge of world history, and for researchers in other disciplines who are looking for a broader view of their own field.

While I have put energy into defining the field of world history and its elements, I do not propose an authorized version of world history. Instead, I offer a review of the evolving patterns of study in order to display the range of possibilities in world history. I have organized this book around five principal objectives, devoting a section to each. My first goal involves defining world history in terms of the patterns of its current rapid development and its firm base in earlier writings. This discussion focuses on illustrating the long-term continuities in the conceptualization and study of world history and on demonstrating how the ideas of world historians and their institutions have influenced the current expansion of teaching and research in world history.

My second objective centers on showing how the current expansion of world history is part of a wider revolution in historical studies. The development of new theories and new data in the disciplines of social sciences, humanities, and natural sciences has pushed back the boundaries of historical studies in general and created an exciting set of "world-centric" insights. In turn, the development of area studies and the rise of several sorts of global studies have greatly expanded the world history field.

Third, I summarize recent advances in each of several sub-fields of world history and examine the current main debates among world historians. In the course of this review, I contrast the advances in global political and economic history (historically the strongest sub-fields) with recent developments in social history, with the interplay of technological, ecological, and health history, and with the many and varied aspects of cultural history.

My fourth objective is to enunciate guiding rules for conducting a logical analysis in world history. This task involves summarizing the historian's choices in selecting a geographic scale, a time period, and a topical focus, outlining strategies for researching and interpreting the global past and delineating techniques for verifying interpretations. Overall, these guidelines restate in a world-historical context the discipline of history as the art, craft, and science of collecting evidence and then using these data to characterize and interpret the past.

Finally, I offer a set of program and curriculum recommendations designed specifically for promoting successful study and research in world history. These guidelines address in particular the needs of those participating in organized world history programs such as graduate or undergraduate schools, or professional development workshops for teachers. They should also prove useful, however, to readers engaged in individual studies of world history.

I have written this volume as a review of the field and as a guide for those seeking to advance their skills in navigating world history. To aid in these efforts, I offer readers dependable techniques for exploring world history's literature and resources. Navigating world history is an ambitious but limited goal, one quite distinct from the unattainable aim of "mastering" the topic. No one can learn all of world history. Anyone who pursues such a goal is sure to become lost. To strike an analogy, all those who have attempted to conquer the world have failed, but many of those who have traveled the globe have gained pleasure and expanded their understanding. Similarly, my purpose is to identify techniques that will help readers find their way from one place to another in their voyage toward understanding world history.

The researchers perusing this book may include professional historians writing monographs, graduate students getting ready for general exams, and advanced undergraduate students looking to add to their knowledge and experience. The teachers among these groups may be preparing classes for high school or middle school classes, or for graduate or undergraduate students. One of their primary concerns is how to convey the lessons of world history in their classrooms. General readers seeking an introduction to world history may be students of national history exploring wider connections or scholars in fields such as economics or biology who seek to set their work in the context of change over time.

The members of these groups need to locate works that answer their specific questions about the global past and to learn techniques for reading world history works from a critical perspective. Many will look for an overview of world history. Some require help in designing and conducting research, whether it be as extensive as writing a book or as restricted as preparing materials for an hour-long class. World history teachers—both practicing instructors and those preparing to teach— face a heavy responsibility: they must be able to convey the global past in a manner that addresses every region of the world, that spans two millennia or more, and that examines the full range of human actions and ideas.

For each of these overlapping groups of readers, I offer advice on executing some key tasks in world history. For researchers, I recommend ways of setting up a logical analysis, including procedures for defining and conducting research. I also illustrate the importance of libraries, history departments, and funding agencies, with an emphasis on how these institutions provide resources and help set directions for study. For teachers, I point out the availability of a wide range of resources, give tips on maintaining a balance between continuity and innovation in world history, and offer techniques for identifying opportunities and making decisions regarding approaches to connections, comparisons, time periods, and disciplines. I also encourage teachers to stay in touch with research developments in world history and to convey their questions and those of their students to researchers. For readers from other fields, my advice includes demonstrating how to link many fields of knowledge to world history's processes and patterns and suggesting ways to practice making these historical connections.

In addition to those five major goals, I have interwoven several other themes throughout the book. The first of these I introduced earlier: students of world history can develop "navigational" techniques that will provide a useful, if never definitive, understanding of the field.

Second, the work in world history to this point has not simply summarized the past. It has also made remarkable discoveries of patterns, including those related to comparative population history, the patterns of major civilizations, early connections in the global economy, ecological transformations of the earth, and the location of major interchanges among the world's major religions.

Third, the study of world history faces major difficulties and dilemmas. Teachers confront immense obstacles in their efforts to convey to students a sense of the past addressing over two thousand years, seven continents, and topics from gender to ecology. For researchers, finding the resources for topics such as comparing trade in the twelfth and eighteenth centuries is no easy task.

Fourth, I have given particular prominence to African examples in my portrayal of world history. I take this approach partly because of my own role as a historian of Africa. More precisely, however, I emphasize African examples because they emphasize interconnection rather than dominance in world history. While it lost its ascendancy over human population and human history thousands of years ago, Africa has since remained home to a substantial portion of human population, has been a significant center of innovation, and is a region connected to most other world areas. Interpretations of world history that leave out Africa are open, I find, to suspicion. Stories of the world focusing on the search for the leading power, or on struggles between paired great powers (Britain and France in the eighteenth century, the United States and the Soviet Union in the mid-twentieth century), focus on dominance rather than system in world history. I find more interesting and more representative the tales of more complex interactions. Even for stories of popular culture in our own century, I prefer versions that stress the interplay of musicians in the Caribbean, Central Africa, and South Asia with those in Paris, London, and Los Angeles, rather than assume that a U.S.-based juggernaut of MTV is carrying all before it. I am not proposing the equality or entitlement of every region in world history, but I do propose that world historians should go beyond dominance to focus on interaction.

Fifth, while world history has made great advances through an accumulation of individual efforts, its study is nearing a great bottleneck. If world historians do not gain substantial funding and other institutional support for their efforts, they will be unable to sustain the very promising work in teaching and research that has emerged in recent years. For all its achievements and advances, world history remains an arena of amateur activity. Only if it can attract the backing necessary to create substantial centers of research and graduate training will it become a field of professional study.

The chapters of this book address a broad range of issues. Consequently, some readers may wish to read them in a different order, skip some, or mark some for later reference. The section and chapter introductions have been written to guide those who wish to read the chapters in order and those who wish to read more selectively. However, at this point a general overview of the book's contents and organization is perhaps more useful for giving readers an initial idea of the scope and progression of ideas in this volume, which is why I offer it here.

The bibliography at the end of the book includes over a thousand citations of studies in world history and works relevant to world history. They are organized by author within the four periods analyzed in chapters 2 through 5. Over half of the citations are for works published since 1989. In the footnotes throughout the book, a full citation is given for the first reference to each work, and a short-form citation is given subsequently.

Part I, "The Evolution of World History," traces the ancestry of world history in ancient times and its development to the present. The opening chapter defines world history in the present day and assesses the field's direction in research and teaching. Chapter 2 reviews world history studies from the Renaissance to 1900, focusing first on European writers and then moving on to historical traditions in other areas of the world. Chapter 3 moves to the twentieth century and centers on the macrohistorical syntheses of Oswald Spengler, Arnold J. Toynbee, and William H. McNeill. The fourth chapter, which addresses the years 1965 to 1990, centers on thematic analyses, giving primary attention to the work of Philip D. Curtin, Immanuel Wallerstein, and Alfred W. Crosby. The fifth chapter explores the development of new institutions for world history such as journals and graduate study programs, along with the accompanying, accelerating research production. Chapter 6 reviews global historiography through a narrative of the changing interpretation of world history.

Part II, "Revolution in Historical Studies," describes the dramatic transformation in historical studies through changes in each and every area within the field. Chapter 7 focuses on new theories and methods in the disciplines of social science, humanities, and natural sciences. Chapter 8 traces the rise of area-studies scholarship, with an emphasis on its interdisciplinary approaches. Chapter 9 maps out the emergence, somewhat later, of global studies, with its dimensions of environmental studies, economic analysis, and the movement for global-studies teaching encouraged especially by geographers. It poses the dilemma of the place of world history in global studies.

In part III, "Results of Recent Research," I review the impact of the revolution in method on research results in world history, organizing the chapters according to the disciplinary sub-fields of world history. Chapter 10 explores new work in political and economic history; chapter 11 considers social history; chapter 12 addresses the combination of history of technology, ecology, and health; and chapter 13 reviews cultural history. The section concludes, in chapter 14, with an assessment of current debates in world history.

Part IV, "Logic of Analysis in World History," explores the impact of the revolution in method on the techniques of research and analysis in world history. Chapter 15 explores the scale of world history in geographical scope, in time frame, and also in the range of themes addressed in various studies. Chapter 16 offers guidelines for several aspects of research design, including research agenda, frameworks, strategies, and models in historical analysis. The verification and presentation of research results is the subject of chapter 17, and chapter 18 ends the section with a recap of methods for world historical analysis.

Part V, "Study and Research in World History," reviews and proposes programs of graduate study and independent study for world historians. Chapter 19 reviews the development of graduate studies. Chapter 20 proposes study programs for three different levels of intensity in graduate study. Resources for world history study is the topic of chapter 21, and chapter 22 provides additional tips for conducting research projects in world history. Chapter 23 concludes the book with a set of keynotes for study and research in world history.

World history has come a long way, but world historians are not yet able to debate effectively with national historians, for lack of an organized body of global research. The dearth of research by young scholars is in turn related to the lack of programs of graduate study and the absence of world historians from such centers of priority-setting as the Social Science Research Council in the United States. I hope that more historians will take up specialization in the study of large-scale patterns and interactions in the past. And I hope that universities and funding agencies will recognize the potential of world history for new discoveries and improved interpretations on the past and will commit energy and resources to the support of study by world historians.

Part I

The Evolution of World History

The past is over and will not change. Yet history—the portrayal and interpretation of what has gone before—shifts with every generation. Such revisions come because historians and their audiences develop new outlooks and because new or different information about the past becomes available. Historians of the present generation, more than at any previous time, have become more interested in and given greater preference to world history.

In the six chapters of part I, I examine the recent flowering of attention being lavished on world history. I also trace the long-term evolution of world history as a field of study. These chapters summarize the discourse of world history—that is, the knowledge and beliefs about the world gathered, interpreted, and recorded by historical writers.

I begin by outlining the present configuration of world history and the directions in which change is taking it. I see world history developing rapidly along two distinct though mutually reinforcing avenues: the historians' path and the scientific-cultural path. My attempt to define world history comprises a concise exploration of both of these channels. I review the development of world historical thinking in four periods: the writings of historical philosophers from ancient times to the end of the nineteenth century, the works of global synthesis from 1900 to 1964, the thematic analyses of world history from 1965 to 1989, and the analyses accompanying the organizing of professional study in world history from 1990. My rapid review of two millennia of changing narrative in world history emphasizes the periodic increase in information available on the world and the evolution of social and philosophical perspectives on the global past.

Overall, the chapters of part I stress historiography—the summary and critique of historical interpretation. For the world historical literature—which I define to include works directly addressing or relevant to connections and broad patterns in world history—I highlight the development of knowledge about the world, the shifts in perspective over time, and the recurring tension between an emphasis on span and depth.

Within this broad view of historiography, I also focus on three more specific issues, each of which addresses the direction of change and the rate of change in the literature on world history. One is the evolution and accelerating exploration of the historians' path to world history. Within the established realms of historical studies—politics, trade, social conflict, and literary culture—ideas have developed and are developing on the nature of civilizations, global economic patterns, revolutionary upheavals, and the aesthetics of high culture. The second subject is the emergence in the nineteenth century of the scientific-cultural path to world history. This path appeared as the collected efforts of specialists in geography, botany, linguistics, anthropology, and other fields began to change in scope and have an influence on interpretation of history. (I trace this second course in more detail in part II of this book, where I explore the development and interaction of the two paths under the rubric of method.) The third topic is the recurrence of major issues and dilemmas in the interpretation of world history. New ideas abound in global history, yet the discipline is equally and productively full of venerable, tested concepts. Past authors wrote effectively on the dynamics of great states, the development of major religions, the encounters among peoples, and the roles of human agency, providence, and destiny in determining events. Thus, world history is developing rapidly in the study of established topics and for new topics, yet world historians often find themselves to be reconsidering issues that have been previously debated.

Chapter 1

Defining World History

To put it simply, world history is the story of connections within the global human community. The world historian's work is to portray the crossing of boundaries and the linking of systems in the human past. The source material ranges in scale from individual family tales to migrations of peoples to narratives encompassing all humanity. World history is far less than the sum total of all history. Nevertheless, it adds to our accumulated knowledge of the past through its focus on connections among historical localities, time periods, and themes of study.

World history is new to most historians. Since it is not a small or simple terrain, the question "What is world history?" gets asked repeatedly. Some may pose the question out of suspicion or even hostility, but most who ask simply want to get beyond a vague sense of a field dealing with large expanses of time and space. They also want to learn a more specific way of dealing with immense topics in an orderly fashion. Defining world history requires clarifying such terms as *connections* and human *community*, and addresses the dilemma of depth and span in analyzing the past.[1]

Two Paths to World History

The study of world history has formed in part out of new ways of looking at the established materials of history, particularly in politics, trade, and culture. Historians have known for hundreds of years about the near-simultaneous rise of great empires around the world in the sixteenth century and about the global flow of silver in the

1. I wish to acknowledge Alfred Crosby's effectiveness in demonstrating the power and utility of the term *connection,* in the course of personal conversation from 1994 to 1996. After further thought, I have concluded that it is helpful to be more explicit and say that world history is "the study of connections among subsystems in history." The argument in succeeding chapters develops the meaning and application of this definition. On "depth" and "span," see Philip D. Curtin, "Depth, Span, and Relevance," *American Historical Review,* No. 89 (1984), 1–9.

sixteenth and seventeenth centuries. But only recently have historians sought to understand connections among such events as the imperial expansions of Habsburg Spain, Mughal India, and Russia. Only recently have scholars systematically traced silver flows from mines in Peru, Mexico, and Japan to markets in Europe, South Asia, and especially China.[2] In this work, historians have found that some historical patterns can be explained better through global linkages than through localized case studies. Expanding the scale of analysis helps locate interconnections that explain the patterns. This path takes an "internal" route to world history.

The other path to world history is "external." This one involves the emergence of immense quantities of new information about change over time from outside the traditional bounds of history. For example, in recent decades we have learned much about environmental changes, the history of disease, and the stages of human evolution. The disciplines of linguistics, archaeology, and chemistry have revealed important historical information. As this information has worked its way into history, the boundaries of historical studies have expanded. Environmental scientists began giving historical interpretation to their findings, and some historians responded by studying changes in the environment. As specialists in various fields have developed global insights into change over time, their work has been instrumental in fostering the incorporation of previously excluded fields of study into history.[3]

The events and thought involved in each of the two expanding channels to world history—the internal historians' path and the external scientific-cultural path—have helped fuel the growth and define the character of world history. Historians now examine old and new topics, using old and new approaches to discover many new patterns in the past. "History," as a result, now addresses a wider range of areas, a longer time period, and a greater range of topics than ever before. At the same time, the patterns now being uncovered in our past help make sense of the enormous amount of new evidence.

Along the internal path—that is, within the traditional arena of historical studies—the history of slavery, freedom, and racial discrimination in the Atlantic

2. The Spanish, Portuguese, Ottoman, Safavid, Mughal, and Russian empires all expanded dramatically within less than a century. David Abernethy, *The Dynamics of Global Dominance: European Overseas Empires, 1415–1980* (New Haven, 2000). On silver trade, see Dennis O. Flynn and Arturo Giráldez, "Born with a 'Silver Spoon': The Origin of World Trade in 1571," *Journal of World History* 6 (1995), 201–21; Flynn and Giráldez, "Cycles of Silver: Global Economic Unity through the Mid-eighteenth Century," *Journal of World History* 13 (2002), 391–429.

3. The impact of new scientific work on interpretations of world history has already been substantial. The work of new cultural studies (with more variables to account for and less financial support for research) is moving more slowly, yet its implications for world history are sure to be profound. Although scientific and cultural studies differ sharply from each other, they share the experience of finding that their results have substantial historical implications, but that they are outside the areas of principal historical emphasis. For a study that reveals details of these dynamics in the case of environmental history and that does not neglect cultural issues, see John R. McNeill, *Something New Under the Sun: An Environmental History of the Twentieth-century World* (New York, 2000).

world provides an instructive example of world historical connections. Over the years, writers in the United States developed a national history of the slavery experience. So had historians in Brazil, the British West Indies, and other areas. In each case, scholars sought to explain slavery, abolition, and racial discrimination within the boundaries of a single national territory or imperial system.[4] But when a view of slavery from a world historical perspective arose, particularly through the study of slave trade volume and distribution, a series of new lessons emerged.[5] The United States, for all the importance of slavery there, received just over 5 percent of the captives brought across the Atlantic. Racial discrimination developed not in a single place and time, but evolved out of interplay all around the Atlantic. Just when Jim Crow laws emerged in the wake of the U.S. Civil War, racial polarization emerged in other societies where slaves gained their freedom. In Jamaica, for instance, the efforts of ex-slaves to buy land brought a vicious repression in the Morant Bay events of 1865; thereafter the British eliminated representative government and replaced it with Crown Colony rule.[6] The U.S. and British experiences thus had remarkable parallels.

This same connected logic of world history has also led historians of slavery to examine the place of African societies in the worldwide system of slavery and its consequences. Indeed, the racial polarization of the Americas in the late nineteenth century was also felt in Africa. For instance, Samuel Crowther, the Nigerian-born ex-slave who rose to be the first Anglican bishop of West Africa, lost his position in 1890 through the action of younger, white clerics. After being ruled by the Portuguese for over three hundred years, Angola experienced sudden changes at the end of the nineteenth century that brought residential segregation to the capital city of Luanda and removed blacks from positions of influence in the colonial administration. In French-ruled Senegal, an outbreak of bubonic plague at the turn of the twentieth century prompted residential segregation in the growing port town of Dakar.[7]

In sum, the rise of racial discrimination and racial segregation that began in the 1890s all around the Atlantic in apparently independent situations suggests

---

4. Stanley M. Elkins, *Slavery; a problem in American institutional and intellectual life* (Chicago, 1959); Gilberto Freyre, *The masters and the slaves: A study in the development of Brazilian civilization,* trans. Samuel Putnam ([1933] New York, 1946); Michael Craton, *Searching for the Invisible Man: Slaves and plantation life in Jamaica* (Cambridge, Mass., 1978).

5. Philip D. Curtin, *The Atlantic Slave Trade: A Census* (Madison, 1969); Joseph C. Miller, *Way of Death: Merchant Capitalism and the Angolan Slave Trade, 1730–1830* (Madison, 1988).

6. Philip D. Curtin, *Two Jamaicas: The role of ideas in a tropical colony 1830–1865* (Cambridge, Mass., 1955); Thomas C. Holt, *The Problem of Freedom: Race, Labor, and Politics in Jamaica and Britain, 1832–1938* (Baltimore, 1992).

7. E. A. Ayandele, *The Missionary Impact on Modern Nigeria, 1842–1914* (London, 1966); Douglas L. Wheeler, " 'Angola is Whose House?' Early stirrings of Angolan nationalism and protest, 1822–1910," *African Historical Studies* 2 (1969), 1–22; Elikia M'Bokolo, "Peste et société urbaine à Dakar: l'épidémie de 1914," *Cahiers d'Etudes Africaines* No. 85–86 (1982), 13–46.

that some underlying common cause affected all these situations. A regional or national narrative does not explain the global timing of events.

Similarly, the old stories about the history of industrialization tend to change when the topic is viewed as world history. The English spinners and weavers of wool and cotton retain their place, along with the entrepreneurs who organized the division of labor in early pin factories. Yet they must share their pages in the history of industrialization with the entrepreneurs and workers who mechanized sugar production in the Americas, and with cotton producers in the United States, India, Brazil, and Egypt. The established story of the evolution of European social structure, with its transformation through the rise of a class of wage workers, must now make room for the emerging stories of work forces and the industrial revolution overseas. In the Americas, both free and slave labor forces expanded; in Africa, the population declined but slave labor expanded; and in India, the hand-loom textile industry was crippled by mechanization.[8] Industrialization, we can now see, has been a global phenomenon since its earliest stages.

World history also addresses past connections in areas new to the work of historians. One example of change along this second, scientific-cultural path to world history involves the histories of agriculture and disease. Jared Diamond, a physiologist by training, did much to clarify this linkage. Diamond summarized available archaeological and botanical information on the main centers of agricultural innovation: the Fertile Crescent, China, North and South America, West Africa, Ethiopia, and New Guinea. Each of these agricultural expansions, beginning some ten thousand years ago, led to denser populations. They also led to the development of new infectious diseases. Diamond also noted that in Eurasia, the domestication of large animals led to sharing of diseases among humans and their animals. As a result, the populations of Eurasia (and Africa) not only were dense but also carried microbes and immunities for a wide range of diseases. By contrast, in the Americas, Australia, and the Pacific, large animals had been eliminated for the most part by the sudden appearance of *Homo sapiens* from ten to sixty thousand years ago, so that few remained (llamas are one exception) for domestication. From the sixteenth century, contact between Eurasian populations and those elsewhere resulted, therefore, in horrendous mortality rates in the regions previously lacking large animals.[9]

8. M. Flinn, *The Origins of the Industrial Revolution* (New York, 1966); E. P. Thompson, *The Making of the English Working Class* (Harmondsworth, 1964); Gavin Wright, *The Political Economy of the Cotton South* (New York, 1978); Patrick Manning, *Slavery and African Life: Occidental, Oriental, and African Slave Trades* (Cambridge, 1990); Prasannan Parthasarathi, "Rethinking Wages and Competitiveness in the Eighteenth Century: Britain and South Asia," *Past and Present* 158 (1988), 79–109.

9. Jared Diamond, *Guns, Germs, and Steel: The Fate of Human Societies* (New York, 1997). Area-studies knowledge remains important for the world historian. Diamond brought to bear a particular strength on New Guinea and Australia, thus improving his argument overall as well as the details for that region. But his relative lack of knowledge on Africa caused him to underrate substantially the significance of agriculture and domestic animals in East, West, and North Africa, and thereby to add a racial tinge to his analysis.

In each of these cases, a world historian looking further afield brought another story into focus. World history, in other words, expands the study of traditional topics in history. It also broadens the scope of historical study to include a range of issues regarding the past first germinated in other disciplines.[10] World history has gained more prominence and practice now because of the great expansion of historical data, the expanded range of issues to be addressed, and the greater attention being given to interconnections in historical processes. At the most expansive level, I could claim that all historical studies have now become world history, since all historians are now expected to pay attention to interdisciplinary approaches and historical connections. On a more modest and practical plane and coming full circle to the start of this chapter, however, I define world history as a field of study focusing on the historical connections among entities and systems often thought to be distinct.

The Ancestry and Evolution of the Historians' Path

Historians are those who assemble knowledge about the community. They have been active in almost every society, small or large. In part out of long practice and in part to sustain themselves, historians have honed the art of collecting information and presenting it in ways tailored specifically to the interests of their audiences. Over the centuries, the works of good historians continue to be read because human communities reproduce common patterns of behavior. For example, Herodotus, Thucydides, and Sima Qian, the early giants of written history, wrote analyses of personal characteristics and social situations that still ring true today.[11] In general, the tales of individual ambitions and group interactions preserve much that is similar over the eons. Even so, there are distinct genres of history that relate specifically to particular audiences, and there have been times when history has changed in response to new knowledge and new ways of organizing and prioritizing knowledge. In fact the list of genres and audiences is rather long.

Out of the family came *genealogy*, from the begettings of the Bible to the family tree of lineages both humble and lordly. Out of the village came the *local history*, tales of key local events that included biographies of outstanding local figures. From the traveler and compiler came *geography*, which accounted for distant lands and peoples. From the dynasty came the *dynastic chronology* of successive rulers and their exploits. From the warrior came the *war story*, which recounted heroism in victory or defeat. The *universal history* of priests accounted for the relations of man and God over time. From these groupings emerged various themes in history: the dynasty created *cultural history*, based on the work of palace poets and sculptors, but so did the masses, based on village songs and dances. Philosophers produced *reasoned histories* of the world that struggled with the question of where

10. Following dictionary definitions, I distinguish between *method*, the practical ways of conducting historical analysis, and *methodology*, the science or logic of applying reasoning to the analysis of history.
11. Herodotus, *History* ([ca. 450 B.C.E.] Chicago, 1987); Thucydides, *History of the Peloponnesian* War ([ca. 420 B.C.E.] Chicago, 1989); Ssu-ma Chien [Sima Qian] *Historical Records* ([ca. 100 B.C.E.] Oxford, 1994).

it had come from and where it was going.[12] All these types of history were born out of their various social settings, and these locales continue to produce new histories for their audiences. The genealogies and local histories were parochial, while the geographies and universal histories were cosmopolitan. All were about the world, but more importantly, all were about different portions of it.

These histories maintained their discrete character. They also became intertwined, in the hands of good storytellers, into myths combining such elements as genealogy, biography, dynastic chronology, geography, and universal history. Oral histories survived in detail for several generations and then were lost or transformed, yet their basic social message (for example, the truism behind a fable or parable) often survived much longer. Written histories could be passed on with greater precision. Such languages as Chinese, Greek, Latin, Persian, Arabic, and Sanskrit have sustained long traditions of historical interpretation, and out of these written traditions came *civilizational* histories.[13]

World history too persisted through all these times, in the minds of those few who strained vigorously against the limits on their knowledge of past and present. World history emerged from the dreams of prophets, generals, emperors, and perhaps also inventors, as they strove to explore and master the furthest dimensions of their environments and the forces at work within them. Many people might have been interested in the world, its meaning, and its fate, but only a few were interested in world history. Nevertheless, the tiny, tenuously connected series of world historians persevered, developing their interpretations step by step.

Each of the major written languages preserves contributions to the understanding of world history, but those writing in the modern European languages took the lead in creating the modern vision of world history.[14] Among European writers, studies on the history of the ancient world developed out of the tradition of classical studies that began in the Renaissance and thrived thereafter. Parallel studies of the history of Christianity and of medieval Europe were gradually

---

12. Ibn Khaldun, *An Introduction to History: The Muqaddimah,* trans. Franz Rosenthal, ed. N. J. Dawood ([1377] London, 1967); Jacques-Benigne Bossuet, *Discourse on Universal History,* trans. Elborg Forster, ed. Orest Ranum ([1681] Chicago, 1976); G. W. F. Hegel, *Lectures on the Philosophy of World History,* trans. H. B. Nisbet ([1830] Cambridge, 1975).

13. D. T. Niane, *Sundiata: An Epic of Old Mali,* trans. G. D. Pickett ([1960] New York, 1965); Ibn Khaldun [1377].

14. In the literature on world history, it is important to distinguish between materials that have proven useful for world history and materials that focus on interpreting world history. The vast majority of writings that are useful for interpreting the world were not themselves created with a global perspective.

    For key analyses in the eighteenth century that did address global issues directly, see Giambattista Vico, *The New Science of Giambattista Vico,* trans. Thomas Goddard Bergin and Max Harold Fisch ([1725] Ithaca, 1984); Voltaire, *The General History* [Essai sur les Moeurs et l'Esprit des Nations] trans. William Fleming ([1754–1757] Akron, 1901–1904); Voltaire, *La Philosophie de l'histoire,* ed. J. H. Brumfitt ([1753–1754] Toronto, 1969); Edward Gibbon, *The History of the Decline and Fall of the Roman Empire,* 6 vols. (London, 1776–1778); Stephen K. Sanderson, ed., *Civilizations and World Systems: Studying World-Historical Change* (Walnut Creek, 1995).

drawn into these classical studies and included in the analysis of politics, elite culture, and ideas. During the nineteenth century, studies of world history in early times broadened fundamentally as new fields of study emerged, especially archaeology, linguistics, Orientalism, and Sinology.[15]

The frameworks and the questions of scholars in the eighteenth and nineteenth centuries continue to dominate the study of world history in early times. This work is centered on "civilization"—the emergence of civilization, the course of civilizational histories, and interactions among civilizations. This focus on large states and on world religions means that the literature on early world history tended to neglect peoples outside the bounds of "civilization."[16]

For the world history of times since about 1500, the framework of analysis was formed out of each era's social analysis. Enlightenment scholars debated whether the Renaissance or the Reformation was key to the rise of modern society. Political history centered on the competition of states for hegemony and on the development of political theory harking back to Plato but focusing mainly on Locke and Montesquieu. In economic history, some visions of modern world history have developed out of studies of world trade and industrialization.[17] A number of interpretations of modern world history, for instance, have focused on Europe's role in creating a global community, on the process of incorporating various regions into that community, and ultimately on the global hegemony of Western Civilization. All these patterns of world historical thinking settled into place during the nineteenth century—in other words, before the grand syntheses of Spengler, Wells, and Toynbee and long before world history became a field of widespread interest.[18]

Then, in the nineteenth century, came the nation. The ascendancy and resilience of the nation as the primary form of political and social organization during the past two centuries changed the nature of history. This marked above

15. In Renaissance studies, the founding text was Jacob Burckhardt, *The Civilization of the Renaissance in Italy,* trans. S. Middlemore ([1860] New York, 1958). For two revealing (though conflicting) analyses of linguistics and archaeology in the development of Indo-European studies, see Colin Renfrew, *Archaeology and Language: The Puzzle of Indo-European Origins* (New York, 1988); and J. P. Mallory, *In Search of the Indo-Europeans: Language, Archaeology, and Myth* (London, 1989).

16. In one attempt to address this omission, William McNeill adopted an ancient terminology to include, in the course of the history of civilizations, not only periodic "closure of the ecumene" among civilizations, but also interactions among civilizations and "barbarians." McNeill, *The Rise of the West: A History of the Human Community* (Chicago, 1963).

17. John Locke, *Two treatises of government,* ed. Peter Laslett ([1690] Cambridge, 1988); Charles de Secondat, baron de Montesquieu, *The spirit of the laws,* trans. Thomas Nugent ([1748] New York, 1949); Montesquieu, *Lettres persanes,* ed. Gonzague Truc ([1721] Paris, 1946); Adam Smith, *An Inquiry into the Nature and Causes of the Wealth of Nations,* eds. R. H. Campbell and A. S. Skinner ([1776] Oxford, 1976).

18. Oswald Spengler, *The Decline of the West,* 2 vols., trans. Charles Francis Atkinson ([1918–1922] London, 1926–1928); H. G. Wells, *The Outline of History, Being a Plain History of Life and Mankind* (London, 1920); Arnold J. Toynbee, *A Study of History,* 12 vols. (Oxford, 1933–1961).

all a new definition of community. The leaders of emerging national communities and polities redefined and reaffirmed the priority of political history, directing historians to tell the tale of each nation's emergence and claim to destiny. Historians specialized in the study of state archives to the exclusion of most other sources. They developed national specializations and doctoral programs. They also formed professional associations such as the American Historical Association (founded in 1884). These national historians stigmatized historians outside the university and the discipline as amateurs. They treated local historians as regional chauvinists and world historians as philosophical gadflies. At the same time, these historians documented their claims of national destiny with clear references to each nation's place in the developing global community.

In the twentieth century, changing priorities within the nation created a need for a *social history* to recount the story of groups struggling for position in the national society. Social history helped bring social peace. With it, national history expanded to accommodate various ethnic and religious groups, laborers and entrepreneurs, and women and men, but this wider range of interaction remained firmly bound within the national borders. Here again, historians relied heavily on government documents to study social history. (This social history, indirectly and after a time, became linked to the expansion of world history.)[19]

Thus, for most of the time between 1850 and 2000, the most prestigious history was the study of national states. Studies of ancient history were labeled as "classics" or "archaeology" and set outside of history; studies of medieval history were left in limbo or treated as preconditions for modern nations. The national historians raised their study to an unprecedented level of skill and precision but at the cost of an extreme narrowing of their field and a neglect of most other types of history.

World history continued to develop in the nineteenth and early twentieth centuries, but it grew as a marginal field. As the discipline of professional history developed in the expanding universities, none of its professorships, scholarly journals, study programs, or prizes went to world history. World history was still considered an amateur pursuit. Yet the discussion and interpretation of global connections continued quietly, as it had in earlier centuries. As a result, when world history underwent its great expansion in the late twentieth century, it was not invented *de novo* but rather continued to expand along the lines of its previous development.

The Emergence of the Scientific-Cultural Path

As national history began to dominate the emerging historical profession, changes in the way humans were investigating and thinking about their world would ultimately give the expansion of world history a boost. In the eighteenth century

---

19. Social history, viewed on its own terms, is far more complex than I have allowed here. The study of families, ethnic groups, and laboring classes arose in substantial degree as a contestation of the hegemony of politics and the national state in historical studies. Nonetheless, social historians accepted and worked within the national framework, and exploration of such social historical issues as family and gender has been slow to develop in world history.

Thomas Malthus began speculating on the changing sizes of human populations, and Sir William Jones devised a common linguistic grouping that encompassed languages from the Celtic tongues of the British Isles to the Sanskrit of ancient India. In the nineteenth century Charles Darwin ruminated on humanity's origins in Africa, home of our closest animal relatives, while Henry Maine and Lewis Henry Morgan contemplated early systems of kinship. At the turn of the twentieth century Alfred Wegener studied geological patterns and wondered about continental drift. Meanwhile Karl Marx and Friedrich Engels sought to synthesize all this new knowledge into a grand picture of a transformed human society, focusing on economic organization.[20]

As of the mid-twentieth century, some of these speculations had been confirmed. The excavations of paleontologists established the patterns of human evolution, the laboratory work of radiocarbon and potassium-argon dating confirmed the span of social and geological time, and field geology determined the outlines of continental drift. The long-term patterns of social change were not so easy to verify, but enough was learned to disprove some overly simplistic interpretations of human society. Through the contributions of these and other disciplines, the discipline of history underwent significant change and expansion.

Can one speak of a "revolution" in historical studies? Since history has long been thought of as a rather stodgy, dusty field, it may seem strange to apply such a dashing term to it. Yet such is the argument I advance in these pages: historical study is indeed undergoing a revolution, with world history currently in the lead. At the turn of the twenty-first century, our understanding of the past and the tools of our analysis have advanced dramatically, not least because of the expansion of source material along the scientific-cultural path. Historical "documents" are no longer restricted to diplomatic correspondence and census records. They now include oral tradition, language patterns, blood types, geological and archaeological remains, musical scores, tree rings, and astronomical observations. The repositories of this information now include electronic databases along with the more traditional paper archives. While historical "analysis" still relies on reading and assessing paper documents, it also includes applying theories derived from biology, demography, and literary studies and using quantitative and statistical techniques.

Two Paths Interacting

To further illustrate the revolutionary changes in the study of history that have helped define and "legitimize" world history, I return to the two main drivers of this

20. Thomas Malthus, *Essay on the Principle of Population*, 2nd ed. ([1803] Cambridge, 1992); Charles Darwin, *The Origin of the Species by means of natural selection* (London, 1856); Henry Maine, *Ancient Law: Its connection with the early history of society and its relation to modern times* (London, 1861); Lewis Henry Morgan, *Ancient Society, or, Researches in the lines of human progress from savagery, through barbarism to civilization* (New York, 1877); Karl Marx, *Capital*, vol. 1 ([1867] Moscow, 1971); Friedrich Engels, *The Origin of the Family, Private Property, and the State* ([1884] London, 1968); Alfred Wegener, *Die Entstehung der Kontinente und Ozeane* (Braunschweig, 1915).

change: the internal and external changes in data and perspectives on the past. Internally, new knowledge has appeared in fields traditionally studied by historians. Historians extended their analytical and interpretive techniques socially and geographically during the last half of the twentieth century. Economists became active in economic history and supplemented the earlier description of the economic past with analysis and attempts to verify interpretations. In social history, historians have learned to study families and communities in a more systematic fashion, extending their study to include women and ethnic groups previously neglected. The collection of new survey data on contemporary populations—and the development of a quantitative methodology for analyzing earlier data from censuses and other registers—created an immense field of study. Analytical frameworks emphasizing gender, class, race, and ethnicity helped develop new fields of study in working-class history, ethnic history, migration studies, and popular culture. This marked a move beyond the focus on the state in national histories to a wider range of social interactions. In area studies, professional historians applied the techniques of history, as studied in Europe and North America, to the various regions of Asia, Africa, and Latin America and consequently developed thriving historical literatures for each of those regions. The growth of area studies led inexorably to comparisons among the regions and thus to explorations of global history.[21] These are just a few examples of how newly accessible data, new analytical methods, and changing social perspectives encouraged innovative work in established fields of history.

Externally, researchers in disciplines other than history turned increasingly to analyzing their data in historical terms. The rise of environmental science led some of its practitioners to historical inquiry; the study of language drew some linguists into historical analysis; and geologists, zoologists, and plant physiologists began cooperating in historical studies. Literary theorists and cultural anthropologists applied the insights of new theories to change over time, and students of folklore and popular culture introduced new topics of historical discussion. At the extreme edge of investigation, physicists and astronomers debated and documented the history of time. All these scholars were amateurs in history but professionals in their own fields.

---

21. I have placed area studies along the first rather than the second path to world history, though it is a borderline case. Latin American history became established in U.S. universities in the 1920s, and area-studies programs from the 1950s included the study of "history" for each region. Area studies ultimately overcame the skepticism of historians of Europe and the United States—they won in the argument that they were doing standard history in different regions, and they were rewarded with an allocation of area-studies positions in history departments. On the other hand, to the degree that they introduced anthropology, oral tradition, linguistics, and other disciplines into their work, area-studies historians were situated outside "history" and thus on the second path to world history. In any case the connections and comparisons among historians of Africa, Asia, and Latin America during the 1990s provided substantial momentum for world history. See, for instance, Frederick Cooper, Florencia E. Mallon, Steve J. Stern, Allen F. Isaacman, and William Roseberry, *Confronting Historical Paradigms: Peasants, Labor, and the Capitalist World System in Africa and Latin America* (Madison, 1993).

In the natural and social sciences, inundated by new knowledge, the term *inter-disciplinary* arose: it referred to the cooperation of professionals with training in different specializations as they sought to explain their subject matter. The terms *interaction* and *connection* gained wider use: they referred at once to interactions among researchers, among their disciplines, and among the elements of each issue under study. The emphasis on interaction also brought emphasis on "relativity," as it became clear that a single phenomenon appears different when viewed from different perspectives. The philosophical and methodological assumptions of interdisciplinary investigators changed as knowledge expanded. In their theories, researchers continued to use nineteenth-century positivist analysis, with its distinct categories and discrete analysis, but they also added a theoretical and practical emphasis on connections and on relativity. For instance, continental drift is now taken to explain the distinctive fauna of Madagascar, as the separation of Madagascar from Africa allowed for the separate development of lemurs in Madagascar and monkeys in Africa. To go back further in time, scientists now argue that flowering plants and flying insects developed in tandem, though the details are still obscure.[22]

Most of the new information about the past came from the efforts of social scientists along with physical and biological scientists, medical researchers, engineers, and even government officials.[23] Yet historians are an omnivorous group, one that eventually consumes the data and the methods of every other investigative group. For example, while a graduate student in history at Yale, William Cronon digested and integrated the results of several disciplines into an attractive little book that forever expanded the frontiers of history into environmental analysis. In *Changes in the Land*, Cronon documented the impact of human activity on the New England landscape, showing the astonishing succession of plant species and land uses as Native Americans and waves of immigrants struggled with each other and with nature to sustain themselves.[24] The book is a model of the way in which new

22. Newton's laws of motion posited that the distance moved by an object is its velocity multiplied by the elapsed time. Einstein added the condition that time is different for places moving at different velocities. At very high velocities, the differences between Newtonian and Einsteinian calculations of distance become significant, and Einstein's calculations fit the experimental result. Einstein did not refute Newton, but he replaced an invariant time with an interactive term, which in certain circumstances becomes highly significant. I find this to provide a useful metaphor for the relations between old knowledge and new knowledge in history. See Walter C. Mih, *The Fascinating Life and Theory of Albert Einstein* (Commack, N.Y., 2000), 81–88. On Madagascar, see Nick Garbutt, *Mammals of Madagascar* (New Haven, 1999), 17–18.

23. The rapid improvement in technology for storing and transmitting information has done much to expand the amount and availability of knowledge about the past. Computer simulations and other types of modeling have enabled analysts, including historians, to improve greatly in their ability to extract knowledge from such disparate raw materials as words, rocks, trees, the atmosphere, and the ocean.

24. William Cronon, *Changes in the Land: Indians, colonists, and the ecology of New England* (New York, 1983). For this work, Cronon was able to draw upon an earlier model, cast at a broader scale, in Alfred W. Crosby, *The Columbian Exchange: Biological and Cultural Consequences of 1492* (Westport, Conn., 1972).

scientific information becomes new historical information and the way in which the discipline of history is transformed and broadened as a result.

While historians could assimilate outside changes into their field, they struggled greatly over the structure of historical knowledge. This was the problem of perspectives on history. Some extraordinary shocks to the world in the mid-twentieth century helped provoke a reconsideration of how we define our community. The experience of global war, the threat of nuclear destruction, the advent of decolonization, the worldwide critique of racial discrimination, the petroleum crisis of the 1970s, the rise of international organizations, the reopening of international migration, and the expansion of multiculturalism—these experiences upset old ideas. The debates over these issues provoked many people to define their community in broader terms than before. Challenges to the maintenance of separate legal status for women or for racial or religious groups within nations, the creation of pan-Arab or pan-African identities, the rise of such categories as "the West" and "the Third World" were not smooth transformations or additions to history. Rather, they arose from angry accusations and agonizing reappraisals.

All of the changes in knowledge and analysis I have listed contribute in some way to the expansion of world history. To sort out and rank the various causes of the development of world history, I find it helpful to adopt the social-science terminology distinguishing proximate causes from ultimate causes—the most immediate factors associated with a change, as compared with the most basic and underlying causes of change. I believe that the ultimate cause for the expansion of world history is the external changes, the accumulation of new work along the scientific-cultural path to world history. This work analyzed and continues to analyze the explosively growing information about every aspect of the human past.[25] The expanded information on the natural sciences is now being supplemented by new information and new theories on human culture. Eventually, this new information will benefit from historical analysis, regardless of whether current historians participate in the effort. The growth in information also explains why world history has been growing so rapidly as a field of study in schools and colleges. People across the political and social spectrum recognize increasingly the need to make sense of our expanded knowledge of the human past.

The proximate cause of the expansion of world history, however, is that people have been changing their definitions of community. While masses of new knowledge have accumulated in every field, the debates within the traditional

---

25. Bruce Mazlish has argued for a distinction between "world history" and "global history," using a somewhat similar argument. World history, in his view, is an extension of universal history, remains primarily concerned with comparisons and interactions of civilizations, and is somewhat backward looking. Global history, in his terms, has been provoked by the dramatic changes of globalization, focuses on dynamics at the planetary level, and is forward-looking. Mazlish has thus identified two paths to world history, one from inside and one from outside the established study of history. Where he has located the key change in the new perspectives arising out of recent events, I acknowledge that factor but subordinate it to the new information available in all the fields of knowledge. Bruce Mazlish, "An Introduction to Global History," in Mazlish and Ralph Buultjens, eds., *Conceptualizing Global History* (Boulder, 1993), 1–24.

boundaries of history have determined what "world history" would become and when it would be taken seriously. These debates featured struggles over nationhood, civilization, race, gender, and class, as well as the national and global frameworks in history. As always in periods of major transformations in knowledge and concepts, some are attracted to the new framework and some prefer to hold on to the old knowledge and the old framework. National history retains its strength and its audience, though it has now entered into discourse with world history and various local histories. Those same global shocks that make some people more cosmopolitan in outlook cause others to adopt a more exclusive view of themselves and to affirm a patriotism based on locality.[26] Some critics of world history doubt that proper study can be carried out on a field as large as the globe, and others are certain that studying each corner of the world may divert attention from the really significant developments in history.

Conclusion: Exploring Connections within the Human Community

The definition of world history is open to debate. If the past is any guide, each future generation will redefine and rewrite its world history. But I can state the basic nature of the world historical beast with some confidence: it is the story of past connections in the human community. World history presumes the acceptance of a human community—one riven sometimes by divisions and hatreds but unified nonetheless by the nature of our species and our common experience. It is the study of connections between communities and between communities and their environments. This vision of a common, connected humanity is not new, though it is far more broadly shared at this time than in the past. We are ready, however, to discern and compare connections among the peoples of Africa, East Asia, and Europe. Local histories remain as relevant as ever, but now, thanks to the concepts and practices of world history, we can combine them in new ways to produce a global picture of history that grows clearer each day.

---

26. On these "culture wars," see Gary B. Nash, Charlotte Crabtree, and Ross E. Dunn, *History on Trial: Culture Wars and the Teaching of the Past* (New York, 1997).

Chapter 2

Historical Philosophy to 1900

Professional historians were few in number before the twentieth century. Professional historians focusing on the broad patterns and connections of world history could hardly be found at all. In fact, the professional study of world history did not begin until one hundred years after the nineteenth-century creation of modern universities. Yet many thinkers before the twentieth century searched for broad patterns in human history, and their ideas and terminology continue to influence those who have come after. In this review of global historical thinking, I begin with the European Renaissance and trace historical thinking from that time to the opening of the twentieth century. Then I cast the historiographical net more widely, considering how world historical analyses from regions outside Europe and from earlier times fit into current understandings of world history.[1]

I use the term *historical philosophy* to describe early interpretations of world history because the authors, lacking today's libraries, relied more heavily on philosophical presumption than on historical documentation. Their works are speculative, yet they succeeded in addressing many key issues in interpreting global patterns of the past, relying on their reading, experience, and reasoning powers. The accomplishments of early interpreters of world history, usually done in the absence of adequate information, remind us that historical writing requires

---

1. For a general survey of the historiography in the Western tradition, including a substantial treatment of world and universal history, see Ernst Breisach, *Historiography: Ancient, Medieval, and Modern* (Chicago, 1983); see also Georg G. Iggers, *Historiography in the Twentieth Century: From Scientific Objectivity to the Postmodern Challenge* (Hanover, N.H., 1997); Joyce Appleby, Lynn Hunt, and Margaret Jacob, *Telling the Truth about History* (New York, 1994); Bonnie G. Smith, *The Gender of History: Men, Women, and the Historical Practice* (Cambridge, Mass., 1998); and R. G. Collingwood, *The Idea of History*, ed. Jan van der Duesen ([1946] Oxford, 1993). For recent reviews of global historiography, see Jerry H. Bentley, *Shapes of World History in Twentieth-Century Scholarship* (Washington, 1995), and Daniel Segal, "'Western Civ' and the Staging of History in American Higher Education," *American Historical Review* 105 (2000), 770–805.

documentation of the past and an interpretive framework to organize it—each supplements the other. While evidence has become more available and more important in historical studies over time, the philosophy and logical structure of historical writing have retained equal importance.

Renaissance and Enlightenment

The European discovery of the Americas is an arbitrary but nonetheless interesting moment for beginning a review of global historiography. It reveals, among Renaissance writers, the coexistence of contradictory frameworks for considering the world—frameworks that remain with us today. A good example involves the views of two major figures of the Florentine Renaissance: Niccolo Machiavelli, the political consultant, and the historian Francesco Guicciardini. Guicciardini observed that the voyages to the Americas had proved that people could live south of the equator and near the poles. They had also shown that people existed who had not heard the Gospel from the apostles. For him, the expeditions to the Americas had revealed important new knowledge about the world.[2] For Machiavelli, in contrast, the history of the world stretched back in time rather than across its geographical span and centered on the issues and values of the classical era. He focused on reconsidering the established knowledge of human society rather than opening up new categories. Living in the same city as Amerigo Vespucci, Machiavelli took no notice of the Americas. As a humanist, he was concerned with individual character and with the ancients. For Guicciardini, a different sort of humanist, contact with the Americas held possibilities for great changes in the future but also prompted a reconceptualization of the past.

These two highly educated writers, maturing under similar circumstances, maintained sharply different perspectives of key issues in their understanding of the world. As noted, their differences showed up especially with regard to the Americas. In a sense, Machiavelli took the Western Civilization approach to the world, and Guicciardini took the world historians' approach. Jacob Burckhardt, the nineteenth-century Swiss historian whose *Civilization of the Renaissance in Italy* crystallized the historical vision of the Renaissance (and also the canons of modern cultural history), adopted the same humanistic vision as Machiavelli. In other words, he was willing to propagate a vision of the Renaissance that linked the story of Italian cultural flowering entirely to the Old World and the discourse with the classical era, rather than contemporary times and the New World. Consequently, we are left to puzzle out whether the parochialism of sixteenth-century Tuscany

---

2. This contrast is explored skillfully by Kenneth R. Bartlett, who notes that Guicciardini's *History of Italy* "is remarkable because of its clear recognition that neither the ancients nor Scripture held all knowledge." Bartlett, "Burckhardt's Humanist Myopia: Machiavelli, Guicciardini and the Wider World," *Scripta Mediterranea* 16–17 (1995–1996).

was a reality of the time or a figment of Burckhardt's imagination three centuries later.[3]

Again, the choice facing Machiavelli and Guicciardini was in part whether to explore the universality of the human condition by looking far back in time or far afield. A generation later, French historian Jean Bodin sought to do both in his 1566 introduction to history. Bodin proposed to divide the previous five thousand years of history in this manner. The first two millennia were dominated by states south and east of the Mediterranean, with an emphasis on religion. Thereafter, the peoples of the Mediterranean (the "middle" of the earth) took the lead for two millennia, relying on their practical acumen. In turn, inhabitants of the north conquered Rome and opened a new era centered on their skills in warfare and invention. Turning his scrutiny to the space of the globe, Bodin viewed the citizens of the world, in all their variety, as having developed gifts that would contribute to the good of the whole.[4] As interpreted by J. B. Bury, the noted historian of the idea of progress, Bodin's vision of the past challenged the common belief of medieval writers that humankind was degenerating from its earlier golden era. Bodin's understanding of history was humanistic, that is, he focused on the free will of humans as an active agent of change. However, he also sought to fit human history into the wider story of astrological influences on the world.[5] His was a broad and eclectic universal history, in which Providence and a divine plan ruled over a range of interacting phenomena.

A century later, universal historians had to take a narrower approach in order to maintain Providence as the central force in human history. To the discoveries of the Americas had now been added Francis Bacon's campaign to expand knowledge and René Descartes's demonstration of the power of deductive reasoning. The range of discoveries of this "new science" suggested the world might operate as much by natural laws as by Providence. In response, Jacques-Benigne Bossuet published in 1681 a general statement that reaffirmed the centrality of Providence in history and became a widely read manual. Bossuet, a Catholic bishop devoted to the service of Louis XIV of France, focused on historical depth rather than geographical or social span. His universal history responded implicitly to the philosophical debates of his time by restating the providential interpretation developed over a thousand years earlier by St. Augustine. It also served as an apology for the French regime.[6] Bossuet concentrated especially on refuting the philosophy of

3. Bartlett labels this as Burckhardt's "humanist myopia." Burckhardt was a student of Leopold von Ranke. Ibid.; Burckhardt [1860].

4. Jean Bodin, *Method for the Easy Comprehension of* History, trans. Beatrice Reynolds ([1566] New York, 1960); and Bodin, *De la vicissitude ou variété des choses on l'univers* (Paris, 1577); J. B. Bury, *The Idea of Progress: An Inquiry into its Growth and Origin* (New York, 1932), 37–38, 43–44.

5. In the words of Bury, "Both here and in his astrological creed, Bodin is crudely attempting to bring human history into close connection with the rest of the universe, and to establish the view that the whole world is built on a divine plan by which all the parts are intimately interrelated. He is careful, however, to avoid fatalism." See Bury 1932: 43. For a later and somewhat conflicting history of the idea of progress, see Robert Nisbet, *History of the Idea of Progress* (New York, 1980).

6. Bossuet [1681].

Baruch Spinoza and the views of dissident priest Richard Simon on biblical history. In a concession to Simon, Bossuet did accept the historicity of the Bible— that is, he considered it to be a historically constructed, though divinely inspired, set of documents. Accepting Simon's position provided Bossuet with an additional advantage, in that he distanced himself from the Vatican's position, which gave the Church primacy in interpreting doctrine. By creating this distance from the Vatican, Bossuet was able to underscore the legitimacy of the Gallican church, thus reaffirming the power of the king rather than the pope to appoint bishops in the French realm. The details of the book are restricted to Roman and early Christian times, and Bossuet devoted himself to sustaining the view that his history was consistent with the will of God and the Bible. That is, he carried through the seventeenth century the previous habit of using earlier history as a way of talking about one's own times.[7] "Universal history," as Bossuet used the term, referred to Biblical history of the Old and New Testaments and to the implications of that history for later times.

As Bossuet was writing, a debate linking such varied areas as human history, natural history, and the arts came to a head. This argument was known as "the quarrel of the ancients and the moderns."[8] For centuries, tentative challenges had arisen to the Medieval- and Renaissance-era consensus that the world was degenerating from its peak in the Classical and early Christian eras. The dispute raged during the seventeenth century in the form of a comparison of ancient and modern literary achievements.[9] Eventually, the quarrel became linked to the natural sciences as some writers began to ask whether nature had lost its vitality. An offshoot of this question was whether humans now lacked the intellectual powers of their predecessors.

The most effective proponent of the moderns was Bernard le Bouvier de Fontenelle. In 1688, the young Fontenelle published a compact pamphlet, *Digression on the Ancients and Moderns*, which argued explicitly that knowledge was advancing and implicitly that progress in history was inevitable.[10] Fontenelle began with the question of whether trees were greater in ancient times than today. He concluded that nature's works were the same over time (although they varied with climate). He also concluded that the passage of time allowed for the rejection of false theories. As to whether the collective mind of humanity was aging and degenerating, he argued that aging in the individual life cycle could not be generalized to humankind in

---

7. Over a century later, as Napoleon reestablished French relations with the Church, Bossuet's book again became popular among readers seeking reassurance of the existence of a divinely directed order.

8. The distinction between "history" and "natural history" remained as an artifact of the classical era. The *Natural History* of Pliny the Elder, widely read in Latin and in translation in early modern Europe, ranged across the cosmos, geography, medicine, animals, plants, and minerals, but also included painting and architecture. See Pliny, *Natural History*, 10 vols. ([ca. 70 C.E.] Cambridge, Mass., 1949). Two fine general treatments of the dispute of the ancients and the moderns are Bury 1932: 37–126, and Hugh Kearney, *Science and Change 1500–1700* (New York, 1971), 216–35.

9. Bossuet's approach to universal history, emphasizing the centrality of the early Christian era, put him on the side of the ancients.

10. Bury 1932: 98–126.

general and that humankind would never degenerate. The literary debates would go on for some time, but as others affirmed Fontenelle's points, the idea of human knowledge progressing cumulatively seemed established. Aside from this achievement, however, the question remained of whether the social condition of humanity would move forward over time. The debate continues on the latter issue.

Echoes of Francis Bacon's call for accumulating new knowledge sounded in human as well as natural sciences. The "descriptions" of Portuguese and Dutch travelers and compilers in the sixteenth and seventeenth centuries gave way to a new use of the term *universal history* that became common in the eighteenth century. This happened most notably in England, where *universal history* referred to an accumulation of histories and narratives drawn from all possible times and places. The most extensive such collection was published in sixty-five volumes appearing between 1736 and 1765. Other collections of voyages and narratives fit into the same framework.[11] The authors and editors were more concerned with collecting additional information than with classifying it, more interested in exploring the limits of the human condition than in defining their own ancestry as its essence.

In contrast to the eclectic, empirical, and largely contemporary approach of the English universal historians, an Italian writer of the same era sought to develop a logically rigorous approach to the human past. In the early eighteenth century, Giambattista Vico, a Neapolitan scholar, wrote a complex and obscure work that would inspire social theorists of the later eighteenth and nineteenth centuries. Vico's "new science" attempted to establish a method encompassing natural and human sciences. His analysis led him to speculate on history from earliest to contemporary times. He focused on processes of change and development, centering on philology and the logic of changes in language. His belief that alterations in language transformed society was at the base of his wide-ranging though often enigmatic writings on social change.[12]

The era now known as the Enlightenment, from the mid-eighteenth century, brought new breadth in historical studies as in other arenas of intellectual effort.

11. *Universal History*, 65 vols. (1736–1765), published by George Sale, George Psalmanzar, Archivald Bower, and others. See also Thomas Astley (Pub.), *A New General Collection of Voyages and Travels*, 4 vols. (London, 1745–1747); for an analogous publication in French, see J.-P. Labat, *Voyage du Chevalier des Marchais*, 3 vols. (Paris, 1728). For a discussion of the impact of this literature on later British thought, see Philip D. Curtin, *The Image of Africa* (Madison, 1964). For examples of earlier narratives and compilations on travel, see G. R. Crone, trans. and ed., *The Voyages of Cadamosto and Other Documents on Western Africa in the Second Half of the Fifteenth Century* (London, 1937); Pieter de Marees, *Description and Historical Account of the Gold Kingdom of Guinea*, trans. and ed. Albert van Dantzig and Adam Jones ([1602] Oxford, 1987); Olfert Dapper, *Beschreibung von Afrika* ([1670] New York, 1967).

12. Vico [1725]; Giorgio Tagliacozzo and Hayden V. White, eds., *Giambattista Vico: An International Symposium* (Baltimore, 1969); Peter Munz, "The Idea of 'New Science' in Vico and Marx," in G. Tagliacozzo, ed., *Vico and Marx* (Atlantic Highlands, N.J., 1983), 5–10; Isaiah Berlin, *Vico and Herder: Two Studies in the History of Ideas* (London, 1976); Breisach 1983: 210–13.

The indefatigable Voltaire, much more a man of letters than of science, wrote in history, literature, and ethics. Despite the ironies and cynicism he emphasized and the prejudices he revealed, he sought to demonstrate and bring change. His most popular work, the tale of *Candide*, serves as a sort of contemporary history of the Mediterranean and Atlantic basins in the mid-eighteenth century.[13] In it Voltaire makes clear the campaign for religious toleration that he carried forth on many fronts. He also wrote a multivolume global history, which began with concise sections on China and the Muslim world and then continued on with a European narrative from Charlemagne to Louis XIV. His interpretive summary focuses on the carnage and wasted effort involved in campaigns, never successful, to conquer the world. As a successor to this series, Voltaire wrote *The Philosophy of History*, which went beyond the European political narrative of his history to include speculations on race, religion, and social life, with numerous pointed critiques of Bossuet.[14] Voltaire thus remained within the humanistic tradition, yet he did so with an avid interest in a wide range of human societies.

Other writers of the French Enlightenment showed interest in a wider range of social processes than Voltaire. Denis Diderot, at once an eclectic and influential writer, addressed issues from art history to mechanical technology. He led in gathering numerous writers to summarize their knowledge in the *Encyclopédie*. While it proposed to divide all knowledge into history, philosophy, and poetry (or memory, reason, and imagination), in practice it published successive entries alphabetically and thus privileged empiricism over system. Wide-ranging as this effort was, the authors had to be wary of royal censors; consequently, their output concentrated heavily on technology and on Europe, so that a twenty-first-century world historian can only be disappointed with the limited nature of these articles on history.[15] Meanwhile, the encylopedist *abbé* Raynal published a global history of the East and West Indies that included a critique of French colonial policy.[16]

Edward Gibbon's *Decline and Fall of the Roman Empire* was a work of such sweep and literary power that it clearly invoked the notion of world history.[17] It appeared as Britain was nearing the peak of her powers but also losing the American colonies, which appeared to threaten England's dominion. Gibbon reexamined the ground St. Augustine had covered thirteen centuries earlier and came up with a contradictory view. Augustine, in *City of God*, refuted claims that the spread of Christianity was responsible for the barbarian sack of Rome and the

---

13. Voltaire, *Candide, or Optimism*, ed. and trans. Robert M. Adams ([1756] New York, 1991).

14. Voltaire [1753–1754]; Voltaire [1754–1757].

15. Denis Diderot, ed., *Encyclopédie, ou dictionnaire raisonné des sciences, des arts et des métiers*, 17 vols. (1751–1765); P. N. Furbank, *Diderot: A Critical Biography* (New York, 1992).

16. Guillaume-Thomas-François Raynal, *Histoire des deux Indes*, 3rd ed. (Amsterdam, 1781). Diderot contributed significant sections to the second and especially the third editions of this work.

17. Gibbon 1776–1778; St. Augustine, *The City of God against the Pagans*, 7 vols., trans. George E. McCracken ([426] Cambridge, Mass., 1957).

threatened collapse of the empire. In the same work, he also developed the theology of predestination, which proposed that man's only act of free will could be to turn away from God. Gibbon's detailed and imaginative portraits of individuals underscored the power and influence of their wills, and he asserted that the decisions of Christian churchmen and Christian rulers felled the Roman Empire. His argument was made for the deists of the eighteenth century, but Gibbon's premise and his narrative as a whole have remained influential since then. Gibbon, though encyclopedic in his knowledge of Rome, opted for depth rather than span in his view of the world. Yet he created out of Rome a metaphor for recent times that has been used from the eighteenth century forward.

As the eighteenth century proceeded, European philosophers became more explicit and detailed in their analysis of history. The German scholar J. G. von Herder emphasized the genetic unity of mankind over the differentiating effects of environment, yet the latter gave him lease to focus on "national genius" as the proximate cause of historical change.[18] He sought the origins of modern Europe in developments of the medieval period: the rise of trade and chivalry and the cultivation of reason. He treated Catholicism as beneficial to Europe's development at that stage, yet he saw the subsequent Protestant Reformation as the force that truly unleashed the European nations' potential for greatness. In general, he sought to explain the national states that were crystallizing before him and did so by focusing on general historical principles and on specific turning points in human development.

Herder's French contemporary, the Marquis de Condorcet, was even more explicit in concentrating on the progress of the human spirit. Inspired by the French Revolution that would soon take his life, Condorcet proposed a scheme of ten stages in the evolution of science and philosophy. More explicitly than Herder, Kant, or Hegel, he restricted his analysis to Europe and its adopted classical ancestors. (At the same time, Condorcet was a prominent antislavery propagandist and member of the Société des Amis des Noirs.) The doctrine of progress, ridiculed three decades earlier by Voltaire in *Candide*, was here affirmed in strong terms.[19]

In another thirty years, G. W. F. Hegel began delivering lectures on the philosophy of history that would be codified in 1830 into a widely read book. Hegel's general principles of history as recorded in these notes comprise one of the stronger statements of his philosophical idealism. He categorized historical works into original history (such sources as the narrative of Thucydides), reflective history (interpretive works, including universal history as a sub-category), and

18. Johann Gottfried von Herder, *On World History*, eds. Hans Adler and Ernest A. Menze, trans. Ernest A. Menze and Pichael Palma ([1784–1791] Armonk, N.Y., 1997).
19. The ten stages in Condorcet's scheme are divided by the following turning points: the formation of peoples, the development of agriculture, the development of the alphabet, the height of Greek sciences in the age of Alexander, the decadence of sciences in the late Hellenistic era, the rise of medieval scholarship in the era of the Crusades, the development of printing, the rise of philosophy shaking authority (Descartes), the French Revolution, and the future. Marie Jean Antoine Nicolas Carstat, Marquis de Condorcet, *Tableau historique des progrès de l'esprit humain* ([1795] Paris, 1900). On Condorcet's antislavery activism, see David Brion Davis, *The Problem of Slavery in the Age of Revolution, 1770–1823* (Ithaca, 1975), 97, 328.

philosophic history. It was in the latter, all-encompassing category that Hegel asserted that the history of the world is the progress of the consciousness of freedom. In his view, the distinct national spirits then emerging were but way stations on the path toward development of a universal human spirit.[20] In his more practical section on modern times, however, Hegel showed that he was capable of being drawn into the details of national prejudices. He viewed the Reformation as a necessary step in the development of freedom, and while he saw the French Revolution as a world historical event, he concluded that the German nation would be the main beneficiary of its advances.[21]

Positivism and Materialism

Hegel's philosophy, while it grasped the spirit of a romantic age in which intellectuals and activists sought to change the world, was ultimately more important in generating responses against it than in creating a lasting mindset. His work punctuated the transition from the encyclopedists of the eighteenth century to the systematic thinkers of the nineteenth century. Hegel's idealism, in which the vision of freedom is seen as determining the world, soon gave way to philosophical systems in which laws of nature and society were seen to propel human history. Of these systems, positivism arose as the scientific study of all and the logical ordering of society, and materialism developed as the critique of social orders. Positivism lent itself better to microlevel analysis; materialism lent itself better to macrolevel analysis.

As Hegel completed his lectures, the young Leopold von Ranke began applying philosophical idealism to the study of modern history in a career that would span most of the nineteenth century. In 1833, Ranke wrote a general synthesis on the European powers since Louis XIV; at his death in 1886, he was halfway through a multivolume history of the world. The range of his work, and particularly of his last project, suggests that his mind sustained at least three distinct frameworks: national, continental, and civilizational. Ranke was above all a historian of diplomatic relations: even in discussing the French Revolution, he focused primarily on the diplomatic causes and effects of that upheaval. His was not a science of the past, but an effort to recover the essence of each past time through meticulous analysis of available documents. It was through his faith in and love for diplomatic documents that Ranke developed his notion that the historian's duty was to reconstruct history "wie es eigentlich gewesen."[22]

20. Hegel [1830].
21. Ibid.
22. Leopold von Ranke, "The Great Powers" [1833], in Ranke, *The Secret of World History: Selected Writings on the Art and Science of History,* trans. and ed. Roger Wines (New York, 1981); Ranke, *Weltgeschichte,* 8 vols. (1883–1887). Peter Novick, in a thorough and enlightening discussion of the flawed translation of Ranke's tradition into American universities, argues that "Ranke's greatest contribution was to apply to modern history those documentary and philological methods which had been developed for the study of antiquity... and in his development of the seminar for the training of scholars." Novick, *That Noble Dream: The "Objectivity Question" and the American Historical Profession* (New York, 1987), 26.

With Ranke as a practitioner who ultimately became its paragon, the profession of academic history took form in the German universities of the nineteenth century. Recognizing, in sharp contrast to the philosophically focused historians of earlier generations, the need to document their arguments, these historians selected arenas in which to develop expertise. The result, with the assistance of editors and publishers, was a series of great compendia, assembling regional and civilizational summaries into multivolume collections of the elements of world history.[23] Thus Ranke's approach to history, if allied in principle with the thought of Hegel, in practice came to be aligned with more pragmatic views. His separation of political from economic and social factors and his analysis of processes by breaking them down with respect to discrete empirical documents led subsequently to his work being associated with the doctrine of "positivism" enunciated by French philosopher Auguste Comte.

Comte published his *Cours de philosophie positive* in six volumes from 1830 to 1842.[24] In it, he sought to review every area of scientific and intellectual endeavor according to common standards. He concluded that in each area of study, from mathematics to history, the analysis went through three stages or methods of philosophizing involving interpretation through theological, metaphysical, and positive reasoning. In the first, the essence of things was sought in supernatural influences; in the metaphysical stage, the essence and cause of things was sought in abstract forces; and in the positive stage, the search for absolute knowledge came to be replaced by scientific study of the laws of phenomena. Comte's positivism privileged quantitative over qualitative knowledge. He was a firm believer in progress—not only progress in human knowledge, but progress in the human social condition. He traced the development of deduction in mathematics, observation in astronomy, experiment in chemistry and physics, and comparison in biology.

Volumes four through six focused on "social physics." The fourth volume, on sociology, emphasized the distinction between studies of social statics and dynamics. The last two volumes centered on the three stages of history. Comte portrayed the second or metaphysical stage of historical study as an era in which Catholic interpretations were challenged by Protestant and deist views of the past, followed by a positive phase in which a positivist historical method would complement the rise of industrial society. He concluded that society needed to be

23. German compendia on world history began in the early nineteenth century, peaked late in that century, and continued into the twentieth century. See, for instance, Heinrich Leo, *Lehrbuch der Universalgeschichte*, 6 vols. (Halle, 1835–1844); Wilhelm Oncken, ed., *Allgemeine Geschichte in Einzeldarstellungen*, 32 vols. in 4 series (Berlin, 1879–1890); Ranke 1883–1887; Hans Delbruck, *Weltgeschichte*, 5 vols. (Berlin, 1931). For a single-author, multivolume work translated from German into English, see H. G. Helmolt, ed., *The History of the World*, 8 vols. ([1899] New York, 1901–1907).

24. Auguste Comte, *The Positive Philosophy*, trans. Harriet Martineau [1855], introduction by Abraham S. Blumburg (New York, 1974). This was originally published as *Cours de philosophie positive*, 6 vols. (Paris, 1830–1842). For a useful selection from all of Comte's works, see Gertrud Lenzer, ed., *Auguste Comte and Positivism: The Essential Writings* (Chicago, 1975).

reorganized, and that such reorganization should be carried out by an elite possessed of scientific knowledge.

For much of his career, Comte corresponded with the English philosopher John Stuart Mill. Mill expressed admiration for the order and comprehensiveness of Comte's analysis, and the two shared an approach that was pragmatic, analytical, and elitist. The division between them came after the later 1840s, when Comte's thought took another turn and he wrote a second multivolume study that reformulated positivism as a religion, though one without a God.[25]

While Comte largely sidestepped the historical philosophy of Hegel, the younger Karl Marx proposed to "stand him on his head." The work of Marx and its influence on twentieth-century conceptions of social change and world history have been documented and debated widely.[26] Marx's thinking was structured by the currents of positivism and Eurocentrism so influential in that time, but at the same time he was unusual in his emphasis on what would later be called interactions and systems. It is traditional to note the categories of Marx's training, and this list of categories does indeed help place his thinking in context. Marx, along with German intellectuals from Herder to Spengler, was trained in the classics and in philosophy. To this grounding, as he noted, he added English political economy and French socialism; the totality both reflected and generated the originality of his thought. As a nineteenth-century thinker, he developed a scheme of successive stages; he focused on the conflict of social classes and on the contradictions within a system for accumulating wealth. Marx's world historical vision concentrated on the modern period—the period of capitalism. Like other writers on world history, he speculated on the future and sought to influence it.

The *Manifesto of the Communist Party* of 1848 was a work of global sweep, showing the range of issues Marx and Engels wished to explore.[27] This brief and powerful statement was a distillation of their study and analysis of the previous years. Within four years after the *Manifesto*, Marx had written the closely reasoned *Eighteenth Brumaire of Louis Napoleon*, a pamphlet conveying a global and multi-thematic approach to a short period of time, focusing on the period from 1848 to 1851 in France. In this narrative, Marx showed the relationships of economic interests and political actions, along with the ways in which various social and

---

25. Comte's second major work was *Système de philosophie positive*, 4 vols. (Paris, 1851–1854). Mary Pickering notes that Comte anticipated in the *Cours* the vision of a religion of positivism that he was later to emphasize in the *Système*. On this point and on the correspondence and debates of Comte and Mill, see Pickering, *Auguste Comte: An Intellectual Biography*, vol. 1 (Cambridge, 1993), 505–38, 678; and John Stuart Mill, *Auguste Comte and Positivism* ([1865] Ann Arbor, 1965).

26. Isaiah Berlin, *Karl Marx: His Life and Environment* (New York, 1939); David McLellan, *Karl Marx* (New York, 1975); Paul M. Sweezy, *Modern Capitalism and Other Essays* (New York, 1972); Anthony Giddens, *A Contemporary Critique of Historical Materialism* (Berkeley, 1981).

27. Karl Marx and Friedrich Engels, *Manifesto of the Communist Party* [1848], in *Karl Marx and Frederick Engels, Selected Works* (New York, 1968), 35–63. For Marx's earlier notes, see Marx, *Economic-Philosophical Manuscripts of 1844*, ed. Dirk J. Struik, trans. Martin Milligan ([1844] New York, 1964).

political factions entered and left the stage of activism. He demonstrated an acute sense of historical process and of timing—arguing that a given conjuncture could sometimes set patterns that last for generations—and emphasized the contingency resulting from the conflicting pressures in which political economy is sheathed.[28] In the years to follow, Marx turned from the broad patterns of social interaction to the specifics of economic life. In 1857–1858 he summarized his views on the dynamics of political economy in a set of notebooks that laid out the full plan for his economic studies. These notebooks were published most of a century later as the *Grundrisse*.[29]

Marx's principal effort focused on *Capital*, his three-volume study of the workings of the economic and social system of the nineteenth-century world. There he developed in greatest detail the logic of capital accumulation and of the resultant transformation in the structures of labor, capital, output, and the state. The strength of his work was its emphasis on tracing long-term change and then anticipating its direction. His very effort to make predictions, of course, drew the attention of both supporters and critics and launched debate on his logic and his calculations.[30]

Only a little of Marx's work was explicitly historical. But in his short list of historical writings, in his more detailed statements on the philosophy of history, and in his critique of political economy, he addressed many of the major issues in the interpretation of modern world history and left a clear mark on the literature. In another influential pamphlet, Friedrich Engels sought to summarize and advance the current understanding of early social structures. *The Origin of the Family, Private Property, and the State* was an attempt to apply a materialist conception of history to times before the development of conflicting social classes. Based, as were studies in that time, more on speculation than research, the study proposed a unilinear development of patriarchy and monarchy out of an underlying primitive communism.[31]

As Marx and Engels worked to develop an internationalist perspective for an audience of workers and radical political figures, a series of historians worked to enunciate a national perspective for a developing and patriotic audience—an increasingly affluent reading public that saw the benefits in the new national states. Many of the founding contributors to national history worked outside the academy, submitting their celebration and critique of the nation directly to the reading public rather than to other historians. These included Jules Michelet in France, Thomas Babington Macaulay in Britain, and George Bancroft in the United States. Vasilii Kliuchevskii, writing in Russia at the turn of the twentieth

---

28. Marx, *The Eighteenth Brumaire of Louis Napoleon* [1852], in Marx and Engels, *Selected Works*, 95–180.
29. Marx, *Grundrisse: Introduction to the Critique of Political Economy*, trans. Martin Nicolaus ([1857] London, 1973). The manuscripts were published in a limited edition in Moscow (1939–1941), and then in East Berlin in 1953.
30. Marx [1867, 1885, 1894].
31. Engels relied heavily on the work of the U.S. anthropologist Lewis Henry Morgan who, based on field research among North American peoples and broad reading, categorized societies into the levels of savagery, barbarism, and civilization. Engels 1968; Morgan 1877.

century, also gained fame with a wide audience.[32] Each of these national historians developed the technique of the edifying narrative—a tale of the past that reaffirmed national identity and brought lessons to bear on the present. The various versions of this narrative centered on past events, processes, or personalities. Their success as analysts and interpreters did much to establish the primacy of national studies in history and paved the way for professional historians who sought to develop a historical discipline that was scientific as well as national.[33]

The English sociologist Herbert Spencer, like Comte, Marx, and most other nineteenth-century social scientists who followed attentively the exciting developments in physical and biological sciences, hoped thereby to gain insights for the study of human society. Spencer, however, was unusual for the rigor with which he attempted to apply biological models to society. His *Principles of Sociology* treated society as an organism, not only at a general and metaphorical level, but down to such details as the functioning and interaction of specific organs. Such detailed modeling led to Spencer's approach becoming known as Social Darwinism, and it was more of the same that brought Spencer to apply the term *survival of the fittest* to human society. Many adherents to this approach failed to notice its gross logical error: in assuming that an elite, which had gained power by any means necessary, would logically reproduce itself in the next generation, Spencer was assuming the inheritance of acquired characteristics. His theory of social evolution was thus Lamarckian rather than Darwinian.[34]

The German universities of the late nineteenth century brought forth many bright lights in history and especially in the social sciences, notably in what was evolving into the field of sociology.[35] The most wide-ranging and prominent among these was Max Weber. In his studies of religion in China and of religion in the rise of capitalism, he returned to questions already debated by Herder, Hegel,

32. Jules Michelet, *History of the French Revolution,* trans. Charles Cocks, ed. Gordon Wright ([1847–1853] Chicago, 1967); Thomas Babington Macaulay, *History of England from the Accession of James II,* 4 vols. ([1849–1861] London, 1953); George Bancroft, *History of the United States of America from the Discovery of the Continent,* 10 vols. (Boston, 1873–1874); Vasilii Kliuchevskii, *A History of Russia,* 5 vols., trans. C. J. Hogarth ([1904–1922] London, 1911–1931).

33. Perry Tapper analyzed the development of three approaches within national history, focusing on event, process, and personality: the taking of the Bastille as written by Michelet and Mignet, the British expulsion of Acadians from Nova Scotia as written by Parkman and Bancroft, and the personality of Peter the Great as portrayed by Kliuchevskii and Plataonov. Perry M. Tapper, "Who Are We? Tales of National Identity" (M.A. thesis, Northeastern University, 1991).

34. Herbert Spencer, *Principles of Sociology,* 3 vols. (London, 1876, 1882, 1896); Robert L. Carneiro, ed., *The Evolution of Society: Selections from Herbert Spencer's Principles of Sociology* (Chicago, 1967); William Peterson, *Malthus* (Cambridge, Mass., 1979), 226.

35. For the thoughts of a German historian who criticized the national framework but continued to use it, see Karl Lamprecht, *What Is History? Five Lectures on the Modern Science of History,* trans. E. A. Andrews ([1904] New York, 1905). In sociology see, for instance, Ferdinand Tönnies, *Community and Society,* trans. Charles P. Loomis ([1887] East Lansing, 1957).

and others. His critique of Marx on the turning points in the rise of capitalism set a debate that has been renewed every generation.

Weber's initial work was on ancient Roman and Germanic economic systems, but in the course of his career he showed himself able to undertake detailed study of a broad field of historical circumstances. His 1904 book on the Protestant ethic and the rise of capitalism brought him wide attention. In 1916 and 1917, he published books on religion in China and India, and thereafter he published a study of Judaism.[36] His interest in bureaucracy developed at this time; his theses on the rise of this entity reflect the viewpoint of one surrounded by the transformations of the twentieth century.[37] Throughout his career he criticized the methods and interpretations of Marx, seeking to provide a distinct interpretation of capitalism based on a wider range of factors.[38]

From about 1911, Weber began work on the project that was to become *Economy and Society*. This magnum opus was to be a theoretical and interpretive summary and thus more than an accumulation of his many studies. Weber died in 1920, well before completing the work. His wife published partial versions in 1921 and 1926, but it required three more decades and the patient work of Johann Winckelmann to bring out a comprehensive, coherent version of the study, known as the fourth German edition, in 1956. In the words of the English edition editor, it was "the first strictly empirical comparison of social structure and normative order in world historical depth," and the comprehensive presentation of the three volumes made clear that Weber's emphasis was on "a sociology of dominance."[39] The appearance of this major work in the 1950s drew much new attention to Weberian analysis—coincidentally at the same time as the publication of the *Grundrisse* drew new attention to Marxian analysis. In sum, Weber, along with Marx, remains a rich source of theses on world history.

These changes in historical philosophy accompanied more general changes in knowledge, in institutions for historical study, and in the organization of industrializing society. During the nineteenth century, the available knowledge about life on earth and the history of human society expanded dramatically, and scientific study became increasingly concentrated in universities, which themselves

36. For studies of Weber's life and work, see Reinhard Bendix, *Max Weber, an Intellectual Portrait* (New York, 1960); and Dirk Käsler, *Max Weber: An Introduction to his Life and Work,* trans. Philippa Hurd ([1979] Chicago 1988).

37. Max Weber, *The Protestant Ethic and the Spirit of Capitalism,* trans. Talcott Parsons ([1904] New York, 1958); Weber, *The Religion of China. Confucianism and Taoism,* trans. Hans H. Gerth ([1916] New York, 1968); Weber, *The Religion of India. The Sociology of Hinduism and Buddhism,* trans. Hans H. Gerth and Don Martindale ([1916–1917] Glencoe, 1958); Weber, *Ancient Judaism,* trans. Hans H. Gerth and Don Martindale ([1917–1919], Glencoe, 1952).

38. H. H. Gerth and C. Wright Mills, ed. and trans., *From Max Weber: Essays in Sociology* (New York, 1946), 46–50, 65–69.

39. Max Weber, *Economy and Society: An Outline of Interpretive Sociology,* ed. Guenther Ross and Claus Wittich, trans. E. Fischoff et al., 3 vols. ([1956] New York, 1968), I: xxvii; Weber, *Wirtschaft und Gesellschaft, Grundriss der verstehenden Soziologie,* 4th ed., ed. Johan Winckelmann (Tübingen, 1956). Bureaucracy is the focus of a substantial section of the third volume of *Economy and Society.*

grew rapidly. Ancient history, for instance, continued to be dominated by classical studies, but its geographical and disciplinary boundaries expanded with the advances in archaeology and linguistics. Geological studies, along with discoveries on the evolution of plants and animals, provided a comprehensive context in which to place early human history. In modern history, the systematic collection and cataloguing of documents (in European and non-European languages), as well as the initial developments in anthropology, provided a basis both for detailed narratives and interpretive summaries. The German universities became the strongest centers for historical studies of ancient and modern times, but strong traditions grew in many other Western countries and significant contributions to history developed in many non-Western countries.

In this same century, nationalist ideology, national states, and national education systems, along with electoral politics, grew in importance, and each of these developments pressed on the field of history. The study of history, while continuing as an accumulation of political narrative and an expression of philosophy, now came to open out at once as a field of knowledge and an arena for political mobilization. In the United States of the late nineteenth century, with its newly developing high schools, the patriotic impulse led to the expansion of survey courses in U.S. history. At the same time, the impulse to develop advanced students out of the high schools led to the creation of more broadly based history courses, commonly entitled "General History." These courses, from about 1870, included biblical history and the ancient Mediterranean, followed by European history, with brief references to other areas of the world. General History provided selected young people with evidence on societies beyond their own, but it reflected more the compilations of history by political historians than the interpretations of social change by social theorists.

As public high schools first expanded their role in the U.S. educational system, the history curriculum expanded as well. The reports of commissions of university professors in 1892 and 1899 affirmed the importance of a four-year history curriculum in high schools: Greek, Roman, early medieval, and some "Oriental" history in grade 9; Europe in the Middle Ages and modern times in grade 10; English history in grade 11; U.S. history and government in grade 12. To the degree that these courses addressed the world, they explained it through the British Empire. This approach, sustained by a set of lively and cosmopolitan textbooks written by David Muzzey, reached its peak on the eve of World War I. From that point, and as high school education became compulsory rather than exceptional, the history curriculum contracted: the rise in wartime national sentiment, the focus on contemporary issues, the growth of vocational education, and the challenge to history by other social sciences led to a substantial cutback in the history curriculum in high schools.[40] The expansion of historical knowledge reached high school and college students only in attenuated form.

40. Reports by leading academic historians in 1893 and 1899 reaffirmed the approach of General History in the high schools, though decrying the tendency to focus secondary teaching on rote memorization. The 1899 report proposed a four-year high school program of Greek and Roman history, Middle Ages and modern Europe, English history, and U.S. history and government. Nash, Crabtree, and Dunn 2000: 34–35, 46–48.

At the level of scholarly interpretation, a parallel set of restrictions on world historical thinking emerged along with the reaffirmation of national identity. The grand philosophical statements of the early nineteenth century, as they were implemented in the industrializing world at the turn of the twentieth century, showed themselves vulnerable to reductionism and truncation. Hegel's idealism was streamlined into a triumphalist mind-over-matter justification for corporate and imperial expansion. The historical work of Ranke was stripped of its breadth and idealism and re-labeled, in the United States, as a "scientific" history that served mainly to create monographic studies of diplomacy, only along the narrowest of interpretive lines, and only within the national framework. Positivists remained socially liberal and believed in progress, still relying on elite control to bring about progress. Yet they had largely lost contact with Comte, and positivism was reduced to mechanical models of microlevel social processes, abstracting from social influences, as in the price theory of Alfred Marshall.[41] Materialism remained in closer contact with the original works of Marx, yet it had been reduced to an economic determinism at the macrolevel. Materialists were socially radical; they believed in progress and thought it would come through social struggle, gave emphasis to interaction and change, and applied their ideas especially in macrolevel analyses. In the emerging hierarchy of empires, nations, and colonies, the term *civilization* became part of the vocabulary of every philosophical camp. The term served as a double-edged weapon for confirming the primacy of European (and later, North American) nations in the world order. For pre-modern times, *civilization* referred to the succession of leading empires and societies, in contrast to each other and to the timeless barbarians beyond their limits. For modern times, *civilization* meant *the civilized world*, including the leading nations and imperial homelands but not the colonies.

With this heritage, world historians of the twentieth century naturally drew on the civilizations paradigm in their analyses. In one sense, this choice meshed with earlier thought, preserving the distinctions between Christendom, the Classical world, and other civilizations. In a second sense, it was a concession to racial theorists: "civilizations" could be mapped neatly into what intellectuals of the early twentieth century thought of as "races"—Caucasians (Nordic and Mediterranean), Mongols, Malays, the peoples of the Americas, and that race thought to be without civilization, the Africans. In a third and important sense, however, the formalization of the civilizations paradigm reflected a reaction against the strengthening of the national paradigm in historical studies.

From the Renaissance to the twentieth century, there remained a substantial continuity in the preoccupations, priorities, reasoning, and even empirical concerns of writers in the Western tradition who sought out a global perspective. Each age had its own concerns and style, yet the continuity of concerns is clear enough that one should not be surprised to see similar issues and approaches recur in later generations.

---

41. For a useful summary of this sort of positivism, see Collingwood [1946], 126–33.

Historical Traditions beyond Modern Europe

Universities based on European models dominate the intellectual life of our planet. In the study of history as of medicine and astronomy, Western models of scholarship and practice are the most prestigious. Indeed, it is in the realm of intellectual life that the global conquest of the Western order seems to have become the most complete. The ideas and examples of Plato, Aristotle, Galileo, and Hegel are relied upon to resolve interpretive debates in India and China, in Senegal and Brazil, as well as in the Netherlands. Yet one would be ill-advised to believe that some integrated Western tradition of historical interpretation has carried all before it. In the previous sections of this chapter, I have emphasized the continuity of perspectives in the Western tradition, along with the variations in outlook. By the same token, one must expect to find interpretive continuity as well as contending perspectives in historical interpretation outside the Western academic tradition.

The societal specificity of historical interpretation comes in part because historians in any society rely on more than formal academic training for their interpretive ideas. History, rather than a neatly technical field, depends on the historian's priorities and judgments on complex relationships within society and on the interest of audiences. The interpretation of history arises as well out of the deeper levels of identity and self-interest and out of basic philosophical assumptions built deeply into social practice. This reliance of historical thinking on the implicit priorities and values that structure social life suggests that each social situation will continue to produce its own interpretation of the present and the past. Alternatives to Western notions of history will continue to survive and will reassert themselves at every opportunity.

What follows is a preliminary sketch of the development of global historical thinking in traditions other than that of modern Europe.[42] The literary traditions of the world are the chief repository of inherited ideas on history. In addition to Latin and the vernacular languages originating in Western Europe, these include writings in Chinese, Arabic, Hindi, Sanskrit, Persian, Japanese, and Greek. These are all literary, philosophical, and historical traditions of long standing, each with its own debates and internal conflicts, and this list can be extended significantly by adding other written languages. The historical priorities worked into each of these sets of text will be recovered by those who read them.[43] They may also be

42. I look forward to the day when a more thorough review of world traditions in historical thought will permit a chronological interlacing of writings from various regions. This endeavor would go beyond the present summary to give a sense of global developments in historical thinking.

43. Priorities within historical communities are revealed in answers to questions such as the following: Are rulers appointed by the gods or sustained by their abilities? Is equality or hierarchy to be most valued? Are family values best summarized by obedience to the father, devotion to the mother, or nurturing of the children? The answers, developed out of the experience of varying social situations, are built into the interpretations of history. In turn, the recounting of historical narratives restates these values for the next generation. Each community, therefore, has both a prevailing consensus and a characteristic set of debates about philosophical priorities.

updated and brought into current historical discussion in manners similar to the regular reintroduction of ideas written two millennia ago in Greek and Latin.

As I noted in the previous chapter, the ancient Greek historians Herodotus and Thucydides started a debate over priorities among historians that has been reinvented many times since. Herodotus wrote a broadly cast history incorporating what he had seen and heard during his Persian travels. On the other hand, Thucydides focused on careful documentation of a narrower, though central, set of events (the Peloponnesian War).

The Chinese tradition of historical writing is anchored by the great compiler and synthesizer of ancient documents, Sima Qian.[44] In his early years as the official historian of the Han dynasty, Sima Qian gathered and assessed documents from early times and from distant regions. In his later years, he focused on biographies of Han court figures. One of the most distinctive aspects of his report was his personal assessment of each figure, placed at the end of the documentary narrative. Thus, Sima Qian wrote history along the path of Herodotus in the early part of his life; in his later years, he wrote along the path of Thucydides. The texts themselves and the priorities underlying them have remained influential in Chinese historical thinking from that time forward. Later dynasties produced such leading historians as Liu Chih-chi (d. 721) of the Tang and Ssu-ma Kuang (d. 1096) of the Song, who were both known for their critical assessment in historical analysis. Recent work in China on world history connects clearly to the long tradition of Chinese historical studies, but also to Marxian analysis.[45]

The Abbasid dynasty, which came to power in 750 and built its capital at Baghdad, supported an active campaign of translation into Arabic, and laid the groundwork for the cosmopolitan, multivolume universal histories of al-Tabari (d. 923) and al-Mas'udi (d. 956). Systematic geographers developed volumes of maps and descriptions, for instance in the geographical dictionary of al-Yakut (d. 1229). In the Arab West, the brilliant historical analysis of Ibn Khaldun (d. 1406) gained wide attention in his time and thereafter. In the eighteenth and nineteenth centuries, historical analysis developed further in the context of Arab efforts to develop national structures.[46] In Iran of the Il-Khans, Izz-u-Din Ibn-ul-Athir wrote al-Kamil, a history of the world to 1230, relying in part on the history of the Mongols by Ala-u-Din Juwayni. As the Ottoman Empire rose to prominence, the monarchy encouraged the writing of histories, and Mustafa Ali (d. 1599) wrote Kunh ul-Ahbar, a universal history and geography in four volumes from the time of Adam forward; the later Book of Travels by Evliya Çelebi (d. 1682) was a widely read account of the author's travels and diplomacy

44. Sima Qian [ca. 100 B.C.E.].
45. Witold Rodzinski, A History of China (Oxford, 1979); Ralph Croizrer, "World History in the People's Republic of China," Journal of World History 1 (1990), 151–169.
46. Philip K. Hitti, A History of the Arabs (London, 1937); Abu Ja'far Muhammad b. Jarir al-Tabari, History of Prophets and Kings, 39 vols., numerous translators ([915], Albany, 1987–1998); Ibn Khaldun [1377]; Peter Gran, Islamic Roots of Capitalism: Egypt, 1760–1840 (Austin, 1979); Nikki Keddie, Sayyid Jamal ad-Din "al-Afghani": A Political Biography (Berkeley, 1972).

throughout the vast Ottoman realm.[47] For India, literary and philosophical writings have been more numerous and more prominent than histories, but exploration of these texts conveys a sense of two millennia of debates and changing social values.[48] In Japan, accounts written in the eighth century, in Chinese, gave emphasis to the legitimacy of the empire; in the seventeenth century, historians and Shinto scholars returned to these documents and wrote detailed history, again in classical Chinese, to emphasize the centrality of the imperial family. These latter became part of a wider range of Japanese historical debate in the eighteenth and nineteenth centuries.[49] With the 1552 publication of the *History of the Indies* by Bartolomé de las Casas, historical writing on Latin America was launched into a tradition based on realities in the region (the decimation and enslavement of native populations), and yet linked to debates in Europe. By the eighteenth century Mexico had an established community of intellectuals, yet its members still found themselves on the defensive vis-à-vis Europe.[50]

The African continent too has a substantial archive of historical writings. Writings in Egyptian, Phoenician, Greek, and Latin languages address early times in northern Africa; Ge'ez and Amharic writings document the history of the Horn of Africa, and Arabic has been written and spoken over a growing segment of the continent since the seventh century. In recent centuries, significant literatures have grown up in Hausa and Swahili languages, written first in Arabic script and then in Latin script. Writings in Portuguese, English, French, and other Indo-European languages, some by visitors and others by locals, add to the corpus of writings.[51]

To take the argument a step further, I maintain that even without written documents, the world views of localized populations may be propagated across the centuries. African interpretations of history have passed over the generations

---

47. Stanford Shaw has provided an exceptionally thorough description of historical writing in Turkish from the fifteenth century to the twentieth. Shaw, *History of the Ottoman Empire and Modern Turkey*, 2 vols. (Cambridge, 1976).

48. Ainslee T. Embree, ed., *Sources of Indian Tradition*, 2 vols., 2nd ed. (New York, 1988).

49. Edwin Reischauer, *Japan: The Story of a Nation* (New York, 1970); Tessa Morris-Suzuki, *Re-inventing Japan: Time, Space, Nation* (Armonk, N.Y., 1998).

50. Bartolomé de las Casas, *História de las Indias*, 3 vols. ([1566] Caracas, 1986); Jorge Cañizares-Esguerra, *How to Write the History of the New World: Histories, Epistemologies, and Identities in the Eighteenth-Century Atlantic World* (Stanford, 2001). When the German scholar Humboldt spent five years in Latin America in the early nineteenth century, he was able to read numerous historical and scientific treatises (mostly unpublished) of scholars he visited; these became an important part of Humboldt's publications.

51. Key Arabic-language documents on the West African savanna of the early modern era include *Tarikh al-Kittab* and *Tarikh al-Fettash*. Nehemiah Levtzion, ed., trans. J. F. P. Hopkins, *Corpus of Early Arabic Sources for West African History* (Cambridge, 1981). On more recent times in Africa, see Ali A. Mazrui and Alamin M. Mazrui, *The Power of Babel: Language and Governance in the African Experience* (London, 1998).

through oral tradition, and historians of Africa made concentrated efforts in the last half of the twentieth century to record and interpret oral traditions of families, districts, and states.[52] For example, the epic of Sundiata—the tale of the youth and conquests of the founder of the thirteenth-century Mali empire, replete with comparisons to Alexander—survives in numerous, overlapping versions in the Mande-speaking regions of the West African savanna. The Guinean historian D. T. Niane translated a composite version into French, and in this form it has become widely known.[53]

Conclusion: Old Ideas and New Ideas

The works surveyed in this chapter—written before the twentieth century and addressing the broad contours of history—developed patterns and priorities in the interpretation of history that would be reproduced in succeeding societies. Most clearly, the Western historical philosophers of the eighteenth and nineteenth centuries set the approaches, agenda, debates, and even the canonical examples of world history. Writers from Vico to Weber sharpened such key issues as the transformations of the Renaissance and the Reformation, the monogenesis of the human species, and the development of religious traditions. Out of their debates and the compilation and comparison of documents came the notions of civilization and of European dominance. These writers identified patterns of long-term change in society, began to treat that change as "progress," and initiated a search for the causes of progress. These are some of the great debates on global history in the Western world that were articulated and discussed before the end of the eighteenth century. The theme of dominance and the effort to find roots and causes of dominance have thus been granted a secure place in world history by our intellectual ancestors. Herder and Hegel reviewed the issue of whether Protestantism provided the essential change in modern history, an issue that Max Weber was to take up in later years and that continues to be talked about today. To these were added, in the nineteenth century, the

---

52. See Jan Vansina, *Oral Tradition: A Study in Historical Methodology*, trans. H. M. Wright ([1961] Madison, 1965); for a skeptical critique of oral tradition, see David P. Henige, *Oral Historiography* (London, 1982). In the United States at much the same time, a widespread interest in oral histories grew up as part of the expansion of social history. In this genre, the interviewees spoke mostly of their own experience rather than that of their ancestors. See Studs Terkel, *Hard Times: An Oral History of the Great Depression* (New York, 1986).

53. The human historical events in the tale are interlaced with stories of the supernatural. Yet the story also conveys historical lessons and personal values, of which fidelity and honor to one's mother are outstanding. Thus, whether this epic is seen as history or literature, it provides a way to retrieve the approaches to history in this social setting. Niane [1960].

debates over human origins and evolution, class conflict, industrialization, and bureaucracy.[54]

For authors anywhere on the planet before the twentieth century, writing about world history inevitably required speculation on the details of evidence and heavy reliance on logic and philosophy in order to complete any story. The task of the world historian today is to link speculation, logic, and evidence into a coherent analysis with the goal of developing broad, interpretive, and well-documented assessments of past transformations and connections.

Of all the elements of history, philosophy governs so much of historical writing that one would do well to address philosophical assumptions explicitly rather than bury them in recondite narrative. Every narrative, every movement from the general to the specific or from one time to another, includes assumptions on the nature of change and the causes and connections of events. The heritage of the Western and other intellectual traditions bequeaths historians of today with a set of assumptions, debates, and priorities that leads us back into discussions of old issues as much as it opens up new issues.

For this reason, world historians will regularly encounter the persistent relevance of old ideas. We may in addition develop a growing respect for the ability of our predecessors to locate and analyze evidence. While we may each seek to discover new evidence and previously unknown relationships, we should acknowledge that we will spend much of our time updating the ideas and interpretations bequeathed to us by our forebears.

54. I wish to acknowledge the structural similarity of this argument to one developed recently by Maghan Keita in his review of more than a century of African American scholarship with regard to Africa. Afrocentrism, he found, is not a recent intellectual fad but the continuation of a discourse about the place of Africa in the world that has continued, with many important twists and turns in the debate, from as early as the nineteenth-century writings of E. W. Blyden and George Washington Williams. Keita, *Race and the Writing of History: Riddling the Sphinx* (New York, 2000).

Chapter 3

Grand Synthesis, 1900–1965

The legacy of the nineteenth century opened new possibilities and new needs for the interpretation of history. The growth of universities led to expanded studies of the past. In industrializing countries, new presses and broader readership helped create a market for popular histories. The overlapping of multiple transformations—broader communications, waves of migration, imperial conquests, industrial and commercial transformations, and especially the disasters of the Great War from 1914 to 1918—called for some stock taking on the human condition. In this complex environment of the early twentieth century, three remarkably erudite and insightful writers sought to encompass all human history. They wrote works of synthesis portraying the links of historical developments to each other and to an overall pattern of the past, in contrast to their predecessors in the late nineteenth century, who had taken an encyclopedic approach to world history, touching on numerous issues in separate entries.[1] The synthetic works brought echoes for decades, finally inspiring a fourth author to write an interpretation that could enter formal discussion in the academy. In the same era, a range of other intellectual luminaries produced broad interpretations of history that did much to set the research agenda for world historians in their wake.

Spengler, Wells, and Toynbee

World War I, that great cataclysm ending what some Europeans myopically called their century of peace, provided a great impetus for the study, writing, and reading of world history. Oswald Spengler's monumental study appeared during the war. Spengler was an independent scholar with no institutional affiliation but

---

1. For a review of major synthetic works of world history in the twentieth century, see Paul Costello, *World Historians and their Goals: Twentieth-Century Answers to Modernism* (deKalb 1993). Costello devotes chapters to H. G. Wells, Oswald Spengler, Arnold J. Toynbee, Pitirim Sorokin, Christopher Dawson, Lewis Mumford, and William H. McNeill. He also compares the philosophical outlook of these authors.

with a strong background in the Roman and especially the Greek classics and in philosophy. His *Decline of the West* is an original, wide-ranging, and erudite consideration of many themes in world history. Spengler did not necessarily read primary sources exhaustively for the many civilizations he discusses. Rather, he had available and used several encyclopedic compendia on world history compiled by groups of German historians in the late nineteenth century. His thesis on the Magian civilization, asserting the existence of a distinct Arabian civilization existing during the first fifteen hundred years of the Christian era, provides a clear example of how he was able to take materials recovered by other scholars and reorganize them into a new framework.[2]

Spengler argued that the study of world history is an exercise in philosophy. He traced the classification of history into ancient, medieval, and modern periods and criticized it as imposing a simplified, rectilinear progression on the past. By a similar logic, he opposed the notion of "Europe" in history. He argued that this amounted to a "Ptolemaic system" of history in which all the past revolved around modern Europeans. That is, to the old and new (as in the Old and New Testament) were added the "modern" epoch, so that history was given the sense of a progression. Spengler underscored his disdain for professional historians by arguing that they treated the past as a tapeworm with succeeding segments, while he saw history as an endless set of transformations.

He proposed to replace the "Ptolemaic system" with a "Copernican system" that admitted to no sort of privileged position among civilizations.[3] In addition, Spengler's Copernican system was organic, not mechanical. A culture, in his view, was a living entity with the properties of a species. The analysis, therefore, was to be biographical, accounting for the growth, maturation, and death of each civilization.

The subject matter of Spengler's analysis was elite culture: he chronicled the stages in the life course of each civilization through the quality of its architecture, literature, and sculpture. This bias is perhaps not surprising for one steeped in the classics, but the primacy he gave to cultural issues is distinct from the focus of most subsequent analysis of world history, which has centered on political and military prowess and on trade links as the criteria for civilizational strength.

Spengler eschewed a positivistic approach to his material. Distinguishing the fields of nature and history, he argued that the logic of causation was appropriate for studies of the physical world; there, time is simply one-dimensional chronology.

2. Spengler [1918–1922]. The first edition, published in 1918, consisted of what is now vol. 2; the revised German edition in two volumes appeared in 1922. Spengler's full development of the Arabian civilization is in volume 2 of the 1926 English translation of his book, based on the revised and expanded German edition of 1922. The first volume of this version was, however, entirely new. The chapters on the Arabian civilization thus account for roughly half of the original 1918 volume. See also the abridged version: Oswald Spengler, *The Decline of the West*, abridged by Helmut Werner, English abridged edition prepared by Arthur Helps (Oxford, 1932).

3. Spengler did not use the term *Eurocentric*, but his critique of historiography clearly anticipates the critique of Eurocentrism that was to become prominent late in the twentieth century.

But in human affairs, he argued, the logic of causation must be replaced with that of destiny. Time is multidimensional: to the progressive clock of the calendar must be added the biographical clock of each civilization, the annual clock of underlying peasant populations, and more. In this, as at other key points in his argument, Spengler turned for support to Goethe's scientific writings and, ultimately, to *Faust*.[4]

Spengler chose his title well, but it has an ironic twist.[5] The horrors of World War I brought despair to many who had been shouting the triumphs of the West over the rest, and a massive study on the decline of the West played to their anxieties. Spengler, however, took a diffident posture toward this decline. He traced it back to 1800 and argued that it had nothing to do with the policies of leaders but resulted from the biography inherent in the civilizational life-course. His comparisons of civilizations were intended to demonstrate the regularity and inevitability of such a turning point. Rather than struggle against one's fate, he thought it preferable to play, to the best of one's ability, the role assigned by history.

In contrast to Spengler's tragic, fatalistic sense of history, H. G. Wells emerged from the carnage of war with a voluntaristic vision that offered hope for a better world. Wells, best known as a novelist, began in 1911 to read for a broad study of history. He found his efforts redoubled and redirected by the war. But where Spengler set himself forth as the fount of historical wisdom, Wells took on the role of a conduit of knowledge. *The Outline of History* synthesizes the contributions of leading historical scholars up to 1920. It was written, as Wells said, "through a single mind" and for a general audience. It is, as the subtitle indicates, a "plain history."

Wells sought to integrate nineteenth-century developments in physical and biological sciences with those of archaeology, linguistics, and ancient and modern history. The book's opening line is "The earth on which we live is a spinning globe." Once he got up to the last few millennia, Wells focused on political and military affairs. Following the approach of the historians on whom he drew, he analyzed the rise and fall of individual states more than of whole civilizations.

The originality and the accessibility of Wells's volume show up especially in the maps and diagrams and, in the coffee-table version, the illustrations. Elegant diagrams explain contemporaneous developments in various world regions, and sweeping arrows indicate their connections in time and space. Skillfully drawn maps amplify the messages of the text: one shows the frontiers of Tang China; superimposed on it is the outline of Rome at its greatest extent. The Roman Empire, even including the waters of the Mediterranean, is dwarfed by the territories of the Tang realm.[6]

4. Spengler [1918–1922] I: 95–96, 155–57; II: 28–32.
5. The German original is *Der Untergang des Abendlandes, Gestalt und Wirklichkeit.* *Untergang* translates as twilight but also as collapse; *Abendland* translates literally as evening-land, but more prosaically and directly as the Occident.
6. Wells 1920. Wells's book appeared in a one-volume, a two-volume, and a coffee-table version with color illustrations, published by George Newnes of London. In the latter, the map of Tang China and Rome appears on page 397.

Wells's summary of existing knowledge, while in many ways judicious, was tempered by the prejudices of his age. For instance, he included virtually no references to Africa: not in discussing migrations, or in the voyages of discovery, or in slave trade, or in European colonization. That this was a deliberate choice rather than an oversight is underscored by the fact that one of his most valued advisors was Sir H. H. Johnston, who spent years as an imperial official in Africa and who wrote a number of books on Africa and Africans.[7]

Despite such widely shared blind spots about the past, Wells was ready to take a visionary approach to the future. He believed that the solution to the hatreds unleashed by war lay in formally establishing a global order. Thus, in the concluding chapter of the illustrated edition of the *Outline*, the reader encounters a vegetation-relief map of the world. Written across the map is the label, "United States of the World." As with so many other writers on world history, Wells was drawn beyond the limits of narrow professionalism to offer hope and speculate on the future of human society.[8]

In 1933, the first three volumes of Arnold J. Toynbee's massive *A Study of History* appeared. Three more volumes came out in 1939, the final four in 1948, and supplements in 1961.[9] Toynbee's study advanced beyond earlier multivolume narratives of world history because of the coherence of his civilizations framework. Whereas Spengler simply asserted the relevance of the civilization as the unit of analysis, Toynbee defended this framework in detail. He presented "societies" (i.e., civilizations) as the "smallest intelligible fields of historical study."

7. H. H. Johnston, *The Discovery and Colonization of Africa by Alien Races* (Cambridge, 1900); Johnston, *The Negro in the New World* (London, 1910). W. E. B. Du Bois relied significantly on Johnston's work in his admirable study in global history, *The Negro* ([1915] New York, 1970).

8. In the George Newnes color-print edition, facing page 752, the caption reads "The Old World in the Future," and on a map of the eastern hemisphere appear names of the continents and, more boldly, "The United States of the World." I confess to having a particular attachment to this book, as I occasionally took it down from my parents' shelves as a child. I particularly remember his map of the complexities of German migrations. I found the text less than memorable, but I think the graphics made world history seem like an exciting topic.

9. Toynbee 1933–1961. The volumes and their main contents are as follows: Vol. 1, 1934 (I – Introduction; II – The Genesis of Civilizations); Vol. 2, 1934 (The Range of challenge-and-response); Vol. 3, 1934 (III – The Growth of Civilizations); Vol. 4, 1939 (IV – The Breakdowns of Civilizations); Vol. 5, 1939 (V – The Disintegrations of Civilizations); Vol. 6, 1939 (V contd. – The Disintegrations of Civilizations); Vol. 7, 1954 (VI – Universal States; VII – Universal Churches); Vol. 8, 1954 (VIII – Heroic Ages; IX – Contacts between Civilizations in Space [Encounters between Contemporaries]); Vol. 9, 1954 (X – Contacts Between Civilizations in Time [Renaissances]; XI – Law and Freedom; in History; XII – The Prospects of the Western Civilization); Vol. 10, 1954 (XIII – The Inspirations of Historians); Vol. 11, 1959 (Historical Atlas and Gazetteer); Vol. 12, 1961 (Reconsiderations).

For the abridged version of Toynbee's study, see Arnold J. Toynbee, *A Study of History: Abridgement of Volumes I–VI*, by D. C. Somervell (Oxford, 1946); and Toynbee, *A Study of History: Abridgement of Volumes VII–IX*, by D. C. Somervell (Oxford, 1957).

Just as firmly as Spengler but from a different vantage point, he rejected the relatively rigid national limits into which historical studies fell during the nineteenth century.[10]

Toynbee's study centered on explaining the birth, rise, and fall of civilizations. Like Spengler, Toynbee treated civilizations as organisms. But where Spengler focused on exploiting the limits of his one biographical metaphor of the civilizational life course, Toynbee alternated among a number of metaphors. Included among these was the parallel between civilizations and men on a steep hillside—some remaining immobile on a ledge, others straining to reach the peak.[11]

Toynbee concluded that the patterns of breakdown are similar from one civilization to another and that the uniqueness of civilizations is in their rise—that is, in their successful response to challenges they face. He thus gives pride of place to those urban civilizations that were the first to develop in any region: in the valleys of the Nile, the Tigris-Euphrates, the Yellow River, on Crete, or in Mexico, the Yucatan, and the Andes. A declining civilization might give birth to a successor through the intermediary of a world religion created in the period of decline. In Toynbee's view, the great world religions all emerged from the decay of a second-generation civilization.

Both Toynbee and Spengler claimed to have gained profound insights from Goethe. For Spengler, Goethe guided his methodology, developing his sense of intuitive generalization. Toynbee echoed Spengler in criticizing the professional historians for their narrow and simplistic approach to human development: "Race and environment were the two main rival keys that were offered by would-be scientific nineteenth-century Western historians for solving the problem of the cultural inequality of various extant human societies, and neither key proved, on trial, to unlock the fast-closed door." But Toynbee was unsatisfied with Spengler's resolution of the dilemma of rise and fall in civilizations: "According to him, civilizations arose, developed, declined, and foundered in unvarying conformity with a fixed time-table, and no explanation was offered for any of this. It was just a law of nature which Spengler had detected, and you must take it on trust from the master: *ipse dixit*."[12]

Toynbee turned to Goethe for a historical thesis and found it in the response of Faust to the challenge of Mephistopheles. Whereas Spengler applied the label of "Faustian" to a Western Christendom symbolized architecturally by soaring Gothic towers, Toynbee expanded the Faustian metaphor to the experience of every new civilization.[13]

Toynbee mimicked—even diluted—Spengler in many ways, but his philosophy and sense of history were quite different. Whereas Spengler rejected a positivistic

10. Toynbee 1933–1961, I: 193–95.
11. This metaphor, cross-sectional in time, prefigured the notion Toynbee was later to develop on the philosophical contemporaneity of societies: see note 16. For an intellectual biography of Toynbee, see William H. McNeill, *Arnold J. Toynbee: A Life* (New York, 1989).
12. Arnold J. Toynbee, *Civilization on Trial* (New York, 1948), 10.
13. Ibid., 11–12; Spengler [1918–1922], II: 188, 198–200.

analysis of history, Toynbee's clear framework and regular summaries underscored his effort to develop a sociology of world history. Toynbee's writing style is more simple and direct than Spengler's. It conveys sociological distinctions in logical form rather than communicating intuitive impressions laden with nuances. Toynbee focused on the political strength of societies rather than on their cultural achievements.

More so than Spengler, Toynbee considered the interactions of civilizations with influences outside their boundaries. In his early chapters, he considered the role of barbarians in initiating or deflecting the history of civilizations. In his later volumes, he addressed "encounters" among civilizations. His "encounters in time" started with the European Renaissance and continued with an innovative generalization of the notion of renaissance to several other historical situations. His "encounters in space" focused on the modern-era influences of the West on other civilizations. The treatment, while extensive, was quite elementary: Toynbee considered only the one-way impact of the putatively strong on the weak rather than more complex interactions.[14] Still, Toynbee went on to introduce the concept of the *oikoumene*, the occasionally extended region of contact linking several civilizations in communication. This concept, however, did not assume the importance in his analysis that it was later to have for McNeill.[15]

As he completed his study, Toynbee wrote a number of commentaries on world history, addressing at the same time the perplexities of life at the end of World War II. Three of the issues he discussed there are worthy of particular note. First, in autobiographical terms, he addressed the relationship between the ancient world and the modern world in understanding world history. He restated his notion of the philosophical contemporaneity of civilizations, illustrating it with Thucydides' description of the Peloponnesian Wars, which Toynbee found to anticipate the conditions of World War I.[16]

Second, Toynbee addressed the difficulties of applying his framework of civilizations to the modern world. For the period after 1500, he spoke of "our world-encompassing Western civilization" but also of the entire world under one roof. Do

---

14. In his study of renaissances, Toynbee treats the experience as a drama of reenactment, explores the "consequences of necromancy," and explores the impact of Renaissance on political ideas, law, philosophy, language and literature. Toynbee 1933–1961, vols. 8 and 9.

15. *Oikoumene*, translated to become "ecumene," changed slightly in meaning for McNeill.

16. In "My View of History," Toynbee notes that his mother was trained in modern British national history, and he in ancient Greece and Rome. The advantage of ancient history to him was that the field "is not encumbered and obscured by a surfeit of information, and so we can see the wood—thanks to a drastic thinning of the trees during the interregnum"; its outlook is Oecumenical rather than parochial. From Thucydides' anticipation of the situation in World War I, Toynbee deduced a "vision—new to me—of the philosophical contemporaneity of all civilizations.... Thus history, in the sense of the histories of the human societies called civilizations, revealed itself as a sheaf of parallel, contemporary, and recent essays in a new enterprise: a score of attempts, up to date, to transcend the level of primitive human life at which man, after having become himself, had apparently lain torpid for some hundreds of thousands of years ..." (Toynbee 1948: 8–9).

other civilizations maintain their identity once civilizational boundaries are superimposed? In addition, Toynbee spoke obliquely of a "universal civilization." But when does this civilization come into existence? How does "our" past become part of "their" past and "theirs" part of "ours"? This problem—the meaning of the term *civilization* in both ancient and modern worlds—remains unresolved in the historical literature.[17]

Third, Toynbee became more explicit than before in expressing his belief that hope for the future of mankind lay in creating a broader spirituality—some synthesis and conflation of the traditions of the world's great religions. He argued that war and class had ended civilizations and their upper strata in previous times. War and class now threatened to end humanity itself, so it was necessary to abolish them. Inequality, as he observed, had now become an injustice. As before and since, the world historian was drawn into expressing hopes and fears about the future: universal religion, for Toynbee, was the requirement for progress. Debate continues as to whether, in focusing on morality and spirituality, Toynbee departed from the bounds of historical analysis.

As Toynbee's final volumes appeared, the distinguished Spanish philosopher José Ortega y Gasset offered a course on the study and soon published his lectures. While many reviewers of Toynbee (as earlier of Spengler) concentrated on refuting points of detail, Ortega focused on the broad outlines of Toynbee's argument. He criticized the challenge-response thesis as superficial and devoted little attention to Toynbee's emphasis on universal religion. Ortega accepted the validity of the civilizations paradigm, though he noted the disadvantage, in this framework, of leaving out discussion of the colonial world.[18] In another extended commentary on Toynbee, British sociologist Christopher Dawson published a set of essays focusing on the cross-fertilizing contacts of civilizations, the dynamics of culture, and the religious experience of mankind.[19]

Thematic, Regional, and Temporal Insights

Outside the domain of academia, other major figures of the twentieth century made statements that have become important in the discourse on world history. Woodrow Wilson, who as a historian participated actively in the nationalistic style of writing about American history, became a theorist for a new world order once he became president and a leader of the Allied war effort. His Fourteen Points, and

17. Von Laue responds to this unresolved problem in Toynbee's framework with the image of distinct cellars under the one roof. Theodore von Laue, *The World Revolution of Westernization: The Twentieth Century in Global Perspective* (New York, 1987).

18. "I am ashamed not to have studied colonial man in the libraries of the world, a man who has existed in every civilization, in each with his own particular characteristics; he becomes a man totally different from metropolitan man, but this is something that those who talk too lightly and ingenuously of our American sons completely ignore." José Ortega y Gasset, *An Interpretation of Universal History*, trans. Mildred Adams ([1949] New York, 1973): 187.

19. Christopher Dawson, *The Dynamics of World History*, ed. John J. Mulloy (New York, 1956). This is an anthology of studies that had been published between 1921 and 1954.

in particular his affirmation of the rights of nations to self-determination, were implemented in part in the Treaty of Versailles. Wilson's vision of the League of Nations contributed, in the minds of some, to the notion of a world government.[20]

At the same time, V. I. Lenin, in preparing and leading the Bolshevik Revolution, posed an alternative conception of the self-determination of nations, which took practical form in the replacement of the Russian Empire with the Soviet Union as a constitutionally voluntary association of distinct republics. His analyses of capitalism and of imperialism (the latter based significantly on the inspiration of the British anti-imperialist writer J. A. Hobson) posed a theory of modern world history based more on the mobility of capital than on the sanctity of national units.[21] Lenin's writings drew attention away from the concentration on European (and now American) centers of power and pushed it toward other regions of the world as past and potential sources of historical change.

For a popular audience in the United States, Hendrik Willem van Loon offered a concise version of world history appearing immediately after the fuller outline by H. G. Wells. Dutch-born but writing in the United States, van Loon became a narrator of historical tales. His *Story of Mankind* attempted to convey a sense of the sweep of human history through an episodic, anecdotal collection of tales, emphasizing individual figures and their personal values. Van Loon began the book with a story from his childhood: an uncle took him up the many book-filled stories of the St. Lawrence church in Rotterdam to the bell tower, from which he gained a view of the town and the horizon that would serve as a metaphor for the author's perspective. Van Loon's numerous and charming sketches—portraits, maps and landscapes, including a five-page "animated chronology" from 500,000 B.C.E. to 1920—distill and underscore the highlights of his narrative. The book ended with an expression of hope that Americans, with their new world leadership, would have better success than Europeans in resolving the problems of the Machine Age.[22]

Jawaharlal Nehru's *Glimpses of World History*, though distinct in its form and in its audience, joins the works of Wells and van Loon in its general character. The author, an erudite but amateur student of history, sought to draw lessons from the past for guidance in the present. Nehru's commitment to political action emerges on every page: the chapters were each written as prison letters in the late 1920s. Within these letters, Nehru, incarcerated by the British for his pro-independence activities, reviewed his notes from years of reading and offered a program of

20. For the definitive biography of Wilson, see Arthur S. Link, *Wilson*, 5 vols. (Princeton, 1947).

21. V. I. Lenin, *The Development of Capitalism in Russia* ([1899] Moscow, 1964); Lenin, *Imperialism, the Highest Stage of Capitalism: A Popular Outline*, trans. Yuri Sbodnikov, ed. George Hanna, in V. I. Lenin, *Selected Works*, vol. 1 ([1917] New York, 1967), 673–777; J. A. Hobson, *Imperialism: A Study* (London, 1902).

22. Hendrik Willem van Loon, *The Story of Mankind* (New York, 1921). This volume was awarded the Newbery Medal for the most distinguished American literature for children in the first year the prize was offered. Subsequent editions added to the story, and a final update to the 1951 edition came from the hand of van Loon's son Gerard.

instruction to his daughter Indira, to whom they were addressed.[23] Nehru made clear his reliance on Wells's *Outline* for information and orientation, but the many evaluative passages were Nehru's own. The book is an episodic narrative, from a viewpoint at once European and Indian. Nehru implicitly gave a good lesson in historical relativism by developing an Indian angle on almost every major issue in world history. The cosmopolitan tone of the book makes clear the problems inherent in nationalism, but Nehru pursued the expectation that a distinct destiny for the Indian nation would not be contradictory to a broader sweep of coexistence.

Sigmund Freud gave his response to the disillusioning events of World War I in *Civilization and Its Discontents*.[24] This concise, speculative book interpreted the gradual maturation of mankind as a species, as reflected in the development of civilization. As such, it argued for the regularity of progress in the human past but allowed for the possibility of a disastrous failure to handle the present challenge: a potential inability of humans, now in control of powerful technology, to repress the base tendencies to violent confrontation with conscious efforts to resolve conflicts.

Archaeological research contributed increasingly to the understanding of long-term human development, especially through the writings of V. Gordon Childe. Australian-born but working in Scotland at the University of Edinburgh, Childe specialized in the material culture of Europe and excelled in the synthesis of archaeological findings. After publishing a detailed and regionally specific interpretation of the development of civilization in Europe, he turned to a wider scale. In his best-known books, *Man Makes Himself* and *What Happened in History*, Childe developed a framework that gained wide currency: two cultural revolutions in human society that preceded the Industrial Revolution. The first of these was the Neolithic Revolution, in which man developed food production rather than food gathering; second, an Urban Revolution, in which cities and states arose. Childe presumed that in each case a great increase in human population followed the technological breakthrough. He illustrated his simple set of stages with rich description of specific cultural artifacts and practices, so that his interpretations served both to synthesize archaeological investigations (focusing on Europe and the Middle East) and to popularize archaeological studies.[25]

---

23. Jawaharlal Nehru, *Glimpses of World History* (London, 1934). Nehru served as prime minister of India from 1947 to his death in 1964. His daughter, as Indira Gandhi, served as prime minister of India from 1966 to 1977, and from 1980 to her assassination in 1984.

24. Sigmund Freud, *Civilization and Its Discontents* (New York, 1930).

25. V. Gordon Childe, *The Dawn of European Civilization* (London, 1925); Childe, *New Light on the Most Ancient East: The Oriental Prelude to European Prehistory* (London, 1934); Childe, *Man Makes Himself* (London, 1936); Childe, *What Happened in History* (Harmondsworth, 1942). In this terminology, Childe had adopted the terminology of the U.S. anthropologist Lewis Henry Morgan: that is, the Neolithic Revolution created barbarism out of savagery, and the Urban Revolution created civilization out of barbarism. For an excellent introduction to Childe and to archaeology in Childe's era, see Bruce G. Trigger, *Gordon Childe: Revolutions in Archaeology* (New York, 1980).

Lewis Mumford, by training an architect but by avocation an erudite amateur historian of technology and culture, adopted some of Freud's insights in his trilogy on world history. Mumford's volumes appeared in almost precisely the same period as Toynbee's, and he also used the civilizations paradigm.[26] These volumes focus on pre-modern times, with a cultural orientation. Mumford, along with Childe, was one of few major writers after Spengler to focus substantially on elite culture in world history; more than Spengler, Mumford centered his study on Western Civilization and its putative ancestors in the Mediterranean. In later books, Mumford used his architectural insights to investigate the city in history. He also wrote a brief interpretation of the transformation of the human psyche through civilization.[27]

Other scholars, coming from various points of the academic compass, presented significant interventions into global historical analysis in the interwar years. Owen Lattimore developed an analysis of the interplay of China and Central Asia that emphasized the frontier as a zone of interaction giving shape to societies on both sides of it. Frederick Teggart developed an analysis of Rome and Han China emphasizing the parallels and the links between the two empires.[28]

U.S.-based sociologists began to develop global analyses in the interwar years. Robert Park and the Chicago school of sociologists, focusing on cities, developed a vision of social change that could be applied to global interactions. Talcott Parsons emerged in the 1930s as a leading sociological theorist who sought to propagate the heritage of Max Weber but also put his own stamp on the work. He translated some of Weber's work and later published his own volume entitled *Economy and Society*, the same title Weber had used. Parsons focused entirely on the modern, industrial era in contrast to Weber's effort to span the ages, and developed the paradigm that came to be known as "modernization," focusing on

26. Lewis Mumford, *Technics and Civilization* (New York, 1934); Mumford, *The Culture of Cities* (New York, 1938); Mumford, *The Condition of Man* (New York, 1944).

27. Mumford, *The City in History* (New York, 1961); Mumford, *The Transformations of Man* (Gloucester, Mass., 1978).

28. On Lattimore as a contributor to the study of world history, see David M. Kalivas, "A World History Worldview: Owen Lattimore, a Life Lived in Interesting Times, 1900–1950" (Ph.D. dissertation, Northeastern University, 2000). David Kalivas took as his topic the writing and thinking of Owen Lattimore, one of the foremost writers on history of China and on Chinese relations with Inner Asia from the 1920s until about 1950. In its execution, this was a standard biographical study: Kalivas sought out Lattimore's numerous publications, the archival holdings of his travel journals, letters, and papers, and interviewed associates and family members. In its conception, the focus of this study was on the evolution of Lattimore's thinking, particularly on the development of his understanding of the Great Wall and the many other walls of Inner Asia as links between communities on both sides rather than barriers between alien societies. See also Owen Lattimore, *Inner Asian Frontier of China* (Boston, 1940); Frederick J. Teggart, *Rome and China: A Study of Correlations in Historical Events* (Berkeley, 1939).

the process by which modern social institutions diffused from the centers of their formation to other areas.[29]

Professional historians, whose work in the interwar years focused primarily on political history, were not absent from discussions of world history. Their contributions to the debate, however, were very particular and—one might venture— narrow. Aside from occasional reviews lambasting the synthetic works by amateur historians for overgeneralization and factual errors, the work of professional historians consisted of elaborate chronologies. A leading such chronology was the *Encyclopedia of World History*, edited by William L. Langer and published in 1940.[30] This volume was the lineal descendant of Ploetz's manual, which was published seventy years earlier in Germany and translated into English in 1883, and which has since had numerous reprintings and new editions. This genre, the polar opposite of Spenglerian generalization, comes as close as anything in historical writing to "one-damned-thing-after-another" history. The evidence was presented simply as facts: in these manuals, as in the great German compendia of the nineteenth century, there was no discussion as to how the categories were selected or ordered.[31]

One further area of study in which scholars laid groundwork for world history was in imperial history. Particularly in British studies of the British Empire, historians wrote monographs and surveys that compared and connected the far-flung territories of the empire in analysis of considerable depth, though they focused rather uniformly on the standpoint of imperial rulers. W. Keith Hancock's *Survey of British Commonwealth Affairs*, a multivolume survey, was an outstanding example of a work of very wide scope. The tradition of imperial history continued from the interwar years to the end of the century.[32]

The teaching of world history in the United States—or of any history courses beyond national scope—expanded and declined with a rhythm distinct from the development of the major writings. At the collegiate level, Columbia University took

29. Robert Ezra Park, *Race and Culture* (New York, 1950); Pitirim Sorokin, *Social and Cultural Dynamics*, 4 vols. (New York, 1937); Talcott Parsons, *The Structure of Social Action: A Study in Social Theory with Special Reference to a Group of Recent European Writers* (New York, 1937); Talcott Parsons and Neil Smelser, *Economy and Society: A Study in the Integration of Economic and Social Theory* (Glencoe, Ill., 1956).

30. William L. Langer, ed., *An Encyclopedia of World History, Ancient, Medieval, and Modern, Chronologically Arranged* (Boston, 1940). Revised most recently as Peter Stearns, ed., William L. Langer, compiler, *The Encyclopedia of World History*, 6th edition (Boston, 2001).

31. Harry Elmer Barnes, ed., William H. Tillinghast, trans., *Ploetz's Manual of Universal History* (Boston, 1925). Previously Carl Ploetz, *Epitome of Universal History*, or *Auszug aus der alten, mittleren und neueren Geschichte* (Berlin, 1880). For a compendium that might be seen as a successor to these manuals, see John A. Garraty and Peter Gay, eds., *The Columbia History of the World* (New York, 1972).

32. W. Keith Hancock, *Survey of British Commonwealth Affairs*, 2 vols. (London, 1937–1942); Robert Rene Kuczynski, *The Population of the British Colonial Empire*, 2 vols. (London, 1948–1949); Vera Anstey, *The Economic Development of India* (London, 1929); Lilian C. A. Knowles, *The Economic Development of the British Overseas Empire* (London, 1924).

the lead as James Harvey Robinson developed a war issues course during World War I. This course became Contemporary Civilization in 1919, and thereafter served as prototype for numerous introductory college courses on Western Civilization. Such courses expanded first at elite universities, and then gradually became a fixture of general education curricula at the college level. The Columbia course focused on Europe in the modern era, but Western Civilization courses gradually expanded to incorporate earlier times and additional regions.[33] Such introductory courses were virtually the college student's only experience in history at the transnational level. Upper-division courses focused on individual nations and on shorter time periods.

In the high schools, these changes in history became entangled in a wider transformation of disciplinary and social agendas. The boom in nationalistic sentiment resulting from the early days of World War I, which brought emphasis to recent rather than early history, reinforced the rising interest of educators in "social studies," meaning the fields of political science, economics, and sociology. As high school education expanded to become compulsory rather than an elite experience, a wave of vocational education programs developed. In a 1916 report of a Commission on Social Studies of the National Education Association, educators ratified these trends: social studies should be added to the curriculum at the expense of the four-year high school history curriculum, and history should emphasize the modern period rather than earlier times. History beyond that of the United States was then reduced to one year of high school instruction, usually in grade 10. Charles Beard and James Harvey Robinson, leaders of the "new history" and supporters of an emphasis on modern history, backed this curricular recommendation.[34] But in practice, they were accommodating to the separation of university-based historians from the teaching of history in the high schools. The National Council for the Social Studies, formed in 1921 with the support of the American Historical Association and the National Education Association, came to be the principal influence in the teaching of history and social studies: it promoted cross-sectional, current-affairs-oriented curriculum at the expense of historical breadth or depth. The teaching of world history survived thereafter more because of the efforts of individual teachers than because of any coordinated efforts of professional organizations.

A later commission on social studies, this one supported by the AHA and led in practice by George F. Counts and Charles Beard, began work in the 1920s and reported in 1934. Yet history in the schools became steadily marginalized by other social sciences, and historians went on the defensive. Harold Rugg, a writer of popular U.S. history textbooks who modeled himself after the progressive Beard, found his work attacked as anti-patriotic from 1936 by allies of the National

33. James Harvey Robinson, trained as a historian of medieval Europe, became a champion of "new history," focusing on the modern period, during his time at Columbia. Gilbert Allardyce, "The Rise and Fall of the Western Civilization Course," *American Historical Review* 87 (1982), 695–725; Robinson, *The New History* (New York, 1912).

34. Beard, a political scientist by training, best known for his economic interpretation of the U.S. Constitution, served as president of the AHA in 1933. Nash, Crabtree, and Dunn 2000: 37–38, 48; Beard, *An Economic Interpretation of the Constitution of the United States* (New York, 1913).

Association of Manufacturers.[35] Professional historians spoke increasingly to each other, rather than to schoolteachers or general audiences. Meanwhile, the surveys by H. G. Wells and Hendrik van Loon continued to sell to general readers.

During its course and in its immediate wake, World War II brought a set of publications on world history. (These studies conveyed reflections brought by the shock of global conflict in a fashion parallel to those provoked by World War I; only after the experiences of Cold War and decolonization did a more fundamental reconsideration of world history emerge.) Karl Polanyi's *The Great Transformation*, published in 1944, posed a vision of modern society in which the essential change was a focus not on the technology and social institutions of capitalism, as posited by Marx, but on the nineteenth-century emergence of the economic institutions of the price-making market. These institutions, in Polanyi's view, fundamentally disordered human society, brought about war and fascism in the twentieth century, and would lead necessarily to the creation of new institutions that could harness the marketplace.[36]

Eric Williams's *Capitalism and Slavery*, published in the same year, challenged the notion of industrialization as an autonomous European process by arguing that West Indian slavery provided capital significant in the rise of British industry and that the emergent British industrialists then led in abolishing slavery to help expand their wage-labor force.[37] In 1946, the British economist Maurice Dobb published an interpretation of the rise of capitalism in Europe. The subsequent debate between Dobb and the American economist Paul Sweezy, both Marxists, served in part to reenact the differences between Karl Marx and Max Weber over the transition from feudalism to capitalism: Dobb played the role of Marx, arguing that the rise of wage labor represented the turning point in the establishment of capitalism; Sweezy played the role of Weber, arguing that the earlier expansion of profit-making commerce denoted the emergence of capitalism.[38] In another

---

35. Counts, of Teachers College at Columbia University, also served as president of the American Federation of Teachers at a crucial juncture in the 1920s. Nash, Crabtree, and Dunn 2000: 38, 48–51; Novick 1988: 190–92; Marjorie Murphy, *Blackboard Unions: The AFT and the NEA, 1900–1980* (Ithaca, 1991).

36. Karl Polanyi, *The Great Transformation: The Political and Economic Origins of Our Time* (New York, 1944). In later years, Polanyi drew a group of colleagues at Columbia into a heavily ideological but vain effort to demonstrate that "price-making markets" had existed in almost no situations outside of nineteenth-century Europe. His efforts led to a controversy known as the "formalist-substantivist" controversy in economic anthropology. Karl Polanyi, Conrad Arensberg, and Harry W. Pearson, *Trade and Market in the Early Empires* (New York, 1957); Polanyi, *Dahomey and the Slave Trade, an Analysis of an Archaic Economy*, ed. Abraham Rotstein (Seattle, 1966).

37. Eric Williams, *Capitalism and Slavery* ([1944] Chapel Hill, 1994). Williams's interpretation broke important ground in world history by arguing for transformative interactions across regions: social and economic developments in the British West Indies and in England are each seen as having transformed the other.

38. Maurice Dobb, *Studies in the Development of Capitalism* (Oxford, 1946). The Dobb-Sweezy debate took place in the pages of *Science and Society* from 1950 to 1953. It has been reprinted in Paul Sweezy et al., *The Transition from Feudalism to Capitalism*, introduced by Rodney Hilton (London, 1976).

decade Paul Baran, Hungarian-born and educated and working in the United States, argued for the twentieth century in *The Political Economy of Growth* that ex-colonial countries would have to develop a new route to industrialization precisely because the industrial leaders had already occupied the leading spots in the industrial world.[39] All of these studies were to have great importance in the later development of global studies in economic history.

Fernand Braudel, while in prison during World War II, wrote *The Mediterranean and the Mediterranean World in the Era of Philip II*, which would popularize the notion of interaction in historical change because of the great range of human experiences he addressed.[40] Braudel, in taking a maritime basin as a unit of analysis, chose to ignore the putative boundaries of nations, cultures, and civilizations. His approach emphasized on the one hand environmental structures underlying human societies and, on the other hand, considering human interactions on many levels. This approach caused him to consider time as a multi-dimensional factor, and he formalized this view with a tripartite division. Braudel was a professional, and the wide approval for his work signaled the increasing readiness of professional historians to address history on a global scale. In the same era Charles R. Boxer published several volumes that, while written at monographic rather then synthetic levels, conveyed a wealth of information about the early modern Portuguese and Dutch empires and their connections with people in most regions of the world.[41]

Jacques Pirenne, also a professional historian, published a two-volume interpretation of world history in the same year that Braudel's *Mediterranean* appeared.[42] The influences of Wells, Spengler, and Toynbee are clear in this work, which adopts a civilizations framework and a chronological organization, but which Pirenne labeled as universal history. Pirenne emphasized a maritime metaphor—the currents of world history—and contrasted the style of maritime contact among civilizations with that of contact by mainland. For the mainland, a series of maps of the Eurasian land mass reveal Pirenne's particular focus on times when uninterrupted communication flowed among major civilizations. In short, Pirenne's work prefigured that of William McNeill in many ways; its English translation appeared the year before *Rise of the West*.[43]

Contributions of philosophers were still to be important in world history. Karl Jaspers, a social philosopher, had earlier published studies of Max Weber,

39. Paul Baran, *The Political Economy of Growth* (New York, 1957). Baran and Sweezy were later to collaborate on *Monopoly Capital: An Essay on the American Economic and Social Order* (New York, 1966).
40. Fernand Braudel, *The Mediterranean and the Mediterranean World in the Era of Philip II*, trans. Sian Reynolds, 2 vols. ([1949] Berkeley, 1995).
41. Among Boxer's works are *The Christian Century in Japan* (London, 1951); *Salvador de Sá* (London, 1952); *The Dutch in Brazil* (London, 1957); *The Golden Age of Brazil* (London, 1962); *The Dutch Seaborne Empire* (London, 1965); and *The Portuguese Seaborne Empire* (London, 1969).
42. Jacques Pirenne, *The Tides of History*, 2 vols., trans. Lovett Edwards ([1948] New York, 1962); first published as *Les Grands Courants de l'Histoire*.
43. Ibid.

Nietzsche, and Descartes. He emerged from the experience of World War II in Germany with an optimistic outlook expressed in a dozen books published between 1946 and 1950, of which *The Origin and Goal of History* gained wide attention.[44] Jaspers took from Hegel the idea that Jesus was the axis of world history, around which all events turned, but moved the axis earlier by five hundred years. He defined "the Axial Period" (centering on 500 B.C.E. and including the three centuries on either side of that date) as a time of breakthrough in knowledge and faith, with insights that were never to be exceeded until the present day. This was the time of Confucius and Lao Tzu in China, of the Buddha and the authors of the Upanishads in India, of Elijah and Jeremiah in Israel, and of Homer, Plato, and Thucydides in Greece. Jaspers sought the cause of the axial moment that he observed: he rejected the assertion of Alfred Weber that it had resulted from the influences of Indo-European horsemen dispersing across Eurasia, and concluded instead that it resulted from "many small States and small towns: a politically divided age engaged in incessant conflicts; ... questioning the previously existing conditions."[45] In Jaspers's view, the later religions of Christianity and Islam and the dynastic tradition of China served simply to carry forth the insights of this axial moment. The future to be brought by the changes of the present era, meanwhile, was to consist of liberty, socialism, world unity, and faith. The notion of the Axial Period entered into many textbooks on world history, though Jaspers's visions of the present and the future did not accompany it.

The United Nations, mirroring but expanding the postwar efforts at conciliation by the League of Nations, formed UNESCO (the United Nations Educational, Scientific, and Cultural Organization) in 1945 to develop international cultural collaboration. The University of Chicago historian Louis Gottschalk took a leading role in the planned production by UNESCO of a collaborative history of the world. In this, he took on the thankless task of organizing professional historians with conflicting national perspectives into presenting a common interpretation of world history. Only three volumes, covering aspects of ancient history, would ever appear.[46] In a more modest but more lasting effort, Robert R. Palmer published the first edition of his perennial textbook, *A History of the Modern World*.[47] This book drew on the tradition (well established in the United States) of interpreting modern European history and expanded it to include references to the colonies (especially in the Americas) and to the increased globalization of warfare and politics in the twentieth century.

44. Karl Jaspers, *The Origin and Goal of History*, trans. Michael Bullock ([1949] New Haven, 1953).

45. Ibid., 18.

46. UNESCO, *Interrelations of Cultures: Their Contribution to International Understanding* (Paris, 1953). On Gottschalk's contribution to these volumes, see Gilbert Allardyce, "Toward World History: American Historians and the Coming of the World History Course," *Journal of World History* 1 (1990), 23–76.

47. Robert R. Palmer, *A History of the Modern World* (New York, 1950). Several subsequent editions appeared, co-authored by Joel Colton.

In a major effort at anthropological study of world history, George Peter Murdock published *Africa: Its Peoples and Its Culture History* in 1960.[48] Murdock led in creating a massive database out of ethnographic reports on most of Africa's ethnic groups, organizing it into what became the Human Relations Area Files, as the database was known when extended to the whole world. By collecting and correlating ethnographic data, Murdock was able to make observations on what he called culture history. Perhaps most important of his discoveries in this book was the demonstration that agriculture had been invented independently in at least two regions of Africa.

Most studies of world history published from 1920 to the 1960s focused on the creation of a world community rather than on its development once created. For pre-modern times, this has reinforced the civilizations paradigm. For the period beginning with the voyages of discovery, most authors in this period carried forth the civilizations paradigm. The exceptions were those focusing on the rise of capitalism (Lenin, Polanyi, Braudel, and Dobb). Of all the writers discussed here, only Teggart, Palmer, Braudel, Pirenne, and Gottschalk were professional historians. The others wrote on their own, or in disciplines outside history, and often for a general audience.

Still, if world historians working today ignore these writers, it will be at their own peril. The authors of the early twentieth century addressed all the main issues in world history. Much of their work can now be seen as inconclusive, biased, uninformed, and erroneous. But if it is a catalogue of errors, it is nonetheless a distinguished and thoughtful catalogue. It reflects the consensus and the debates of that time, and it reflects serious efforts at conceptualization by well-educated authors in an era of recurring social crisis. The residue of their work lies in our libraries and in our minds, and we would do best to identify it explicitly. Understanding their efforts, their insights, and their failures is a prerequisite to the work of future generations in developing original and appropriate concepts and generalizations for understanding world history.

McNeill

William McNeill's *The Rise of the West* served at once as the culmination of this era of grand synthesis in world history and as the opening of a new era in the study of world history.[49] Much of what it says was not new, yet its enthusiastic

---

48. In addition to Murdock, Alfred Kroeber, summarizing a life's work in anthropology, wrote interpretive summaries of the historical transformations of societies. At the same time in sociology, Norbert Elias developed a broad interpretation of the "civilizing process." George Peter Murdock, *Africa: Its Peoples and Its Culture History* (New York, 1959); A. L. Kroeber, *An Anthropologist Looks at History* (Berkeley, 1963); Norbert Elias, *The Civilizing Process*, 2 vols., trans. Edmund Jephcott ([1939] New York, 1978). Elias's two volumes are entitled *The History of Manners* and *Power and Civility.* See also Robert van Krieken, *Norbert Elias* (London, 1998).

49. W. McNeill 1963.

reception served as the opening step in establishing the formal recognition of world historical study in North America.[50]

The book is organized chronologically around a set of four succeeding paradigms: an urban paradigm for the creation and diffusion of civilizations, an ecumenical paradigm for the key periods of intercommunication among civilizations, an expansion-of-Europe paradigm for the creation of the modern world, and a Cold War paradigm for the twentieth century. The second and third of these excited the most discussion and the most imitation.

For the period before 1500, the distinction between McNeill's analysis and that of Toynbee is that while McNeill followed Toynbee in making civilization the unit of analysis, his paradigm was set at a wider level. Thus, the communication between Rome and Han China, through the Parthian and Sacian states, was not simply a set of encounters of civilizations but a century or so of transhemispheric contacts. McNeill's periodic closure of the ecumene represents the formal addition of a new dynamic to the history of civilizational growth: it was a phenomenon noted before but not presented so systematically. With the passage of time, the ecumenical paradigm has stimulated considerable discussion and new work.[51]

In the short run, however, it was the expansion-of-Europe paradigm that attracted the most attention. The book's title centered on this aspect of its story, and in the era of modernization theory and the Peace Corps, an explanation of how the West had come to dominate the world was of wide interest. For McNeill and his readers, the history of the world since 1500 was that of the creation of a new and permanently closed ecumene by the actions of Europeans.

McNeill assumed a cause-and-effect relationship between European history and world history and built it into his chronology. Thus, in discussing the world from 1500 to 1700, he included European developments only to 1650. In discussing the world from 1700 to 1850 ("the tottering world balance"), he included European events from 1650 to 1800.[52] This is one way to handle relations between the West and the rest. It assumes an asynchrony: a cause-and-effect relationship with a sizable lag. The approach is similar to the way in which histories of Europe include sixteenth- and

---

50. McNeill's debts to earlier writers were not made explicit but are easily picked out. He owed the title (but little else) to Spengler. The photos, maps, and early passages in the book reveal traces of H. G. Wells; the maps of the Eurasian ecumene recall those of Jacques Pirenne. The sketches suggest traces of van Loon. The emphasis on politics and trade in the rise and fall of civilizations, and the use of the term *ecumene*, reflect Toynbee's approach. These, of course, are matters of form; it is in the book's substance that its importance lies.

51. In retrospect, McNeill seemed particularly attached to this part of his argument. In his 1990 revisitation of the book, he described his undervaluation of the advances of late-medieval China (because essential research remained unpublished) as its most serious weakness. W. McNeill, "*The Rise of the West* after Twenty-Five Years," *Journal of World History* 1 (1990), 1–21.

52. McNeill, providing explicit reasoning in support of this organization, notes that "each chapter will go back in time to consider some of the transformations of European life which antedated and in large degree provoked the new phase of world development." W. McNeill 1963: 567–68.

seventeenth-century scientific developments in chapters centered on eighteenth-century political and social change, again assuming a certain cause-and-effect relationship and a sizable lag. Different organizations and different interpretations might result if one assumed neither a straightforward cause-and-effect relationship going from Europe outward nor one from intellectual to social and political life.

The expansion-of-Europe paradigm has not so much been validated as supplemented and in some sense superseded by subsequent work. Courses known as "Expansion of Europe" in the 1960s became "The Modern World" in the 1970s, as the result of new research and a more cosmopolitan perspective brought, in part, by application of the expansion-of-Europe paradigm.

McNeill's fourth paradigm, the Cold War approach to the twentieth century, reflected the ideological polarization of the age in which it was written. I have no wish to suggest that his focus on the contrast between dictatorship and democracy was anything other than a central issue for this century. But for him to map that distinction onto the difference between "liberal" and "Communist" societies was to downplay not only European fascism but also the experience of colonial rule, warlordism, and caudillismo on other continents. In his 1990 discussion of the book, McNeill gracefully notes that he would not now argue the same.[53] More than that, McNeill has lived and written for forty years since his magnum opus, so we can see not only the workings of his mind but their changes over time. My discussion of his work continues into later chapters.

Conclusion: Civilization and Transformation

The authors of grand syntheses raised the banner of world historical interpretation higher than ever before. In earlier centuries, a number of writers had developed frameworks for analyzing world history, identifying stages and turning points, and debating major issues in world history. Historians of the nineteenth century wrote compendia and chronologies that served as raw material for encompassing interpretations, and the first wave of textbooks summarized the high points of the compendia. But only in the twentieth century were authors able to propose detailed narratives and interpretations spanning large portions of world history. And then, once the genre of world historical synthesis finally gained a recognized place in the historical profession, the creation of such works lapsed. In the wake of this era came studies that, while clearly world historical in scope, focused more on depth and less on span.

In the same era of the grand syntheses, a small but significant collection of transnational studies came into print. Some of these were by historians, but most were from writers based in other fields, Sigmund Freud, V. I. Lenin, Frederick Teggart, Jacques Pirenne, Karl Polanyi, Eric Williams, Owen Lattimore, and Alfred Kroeber among them. Their studies, each of which seemed unique and isolated at the time, in retrospect can be seen as having created the connective tissue among issues in world history. They were the precursors to the monographic studies of world history that developed at the end of the century.

---

53. W. McNeill 1990a.

Chapter 4

Themes and Analyses, 1965–1990

McNeill's synthesis opened the door for professional historians to participate in global thinking. Especially in North America, *The Rise of the West* set the stage, enabling academic discussion of world history: the players assembled in the succeeding decades. Two key elements in McNeill's achievement were his use of a chronological rather than thematic framework (in contrast to Spengler and Toynbee) and his linking of his analysis to academic debates (in contrast to Wells). McNeill's rephrasing of a global synthesis in chronological and academic terms connected it to the optic of professional historians at a time when historical studies were expanding in thematic and regional terms.[1]

The same mid-century era brought innovations in interactive thinking.[2] The scope of these analyses was less than planetary, but their contribution was to identify patterns of interaction more complex than the diffusion of influence from the powerful to the weak. With the work of Andre Gunder Frank, Latin America entered the discourse on modern world history, and the argument was taken back to the initial stages of Iberian colonization. Frank, born in Germany, educated in the United States and having worked in South America, coined the phrase "development of underdevelopment" in arguing that Chile and other South American nations owed their poverty not to an inherited lack of resources, but because the Spanish colonial regime had created new institutions that drained wealth from its colonies.[3] Frank's

1. Following the publication of *Rise of the West*, McNeill's own work went in two directions—in synthesis, as he wrote a series of textbooks for students and a general audience, and in thematic analysis, where he completed studies of power and disease. See W. McNeill, *Plagues and Peoples* (New York, 1976), for an effective example of the development of a theme in world history.
2. Peter Novick, in his survey of a century of transformations in historical studies in the United States, highlights the rapid changes since the 1950s, though he gives little emphasis to the areas of world history and area-studies history that are the focus here. Novick 1988: 415–629.
3. Frank, "The Development of Underdevelopment," *Monthly Review* 18 (1966), 17–31; Frank, *Capitalism and Underdevelopment in Latin America* (New York, 1967).

was an explicit argument for a complex interaction in the history of continents in contact. He reaffirmed earlier arguments that Spain and Portugal and extracted wealth from Chile and Brazil, and he added that the metropoles had established social and fiscal systems that not only inhibited development but created underdevelopment in their colonies.[4]

Global thinking and interactive thinking were able to develop among historians in part because of changing social circumstances. The post–World War II era was a time of political, economic, and academic transformation. World politics and economics were dominated by two great confrontations: the Cold War conflict between two great blocs led by the United States and the Soviet Union, and the series of decolonization movements that led to recognition of new nations in Asia, Africa, and the islands. The expansion of university education in the postwar era of economic growth, including the development of area-studies programs addressing all regions of the earth, provided intellectual space for broader historical thinking.

More generally, the world historical analyses of the last third of the twentieth century turned away from grand synthesis to the identification and analysis of a series of themes, each of enormous scope, yet each clearly an aspect of world history rather than an overview of the whole. Development of world-historical themes became the principal emphasis of scholars during this formative period.[5] For much of this chapter I will interpret this scholarship through the work of three key contributors to the expanding literature, in an effort to identify their most lasting contributions. The three are Philip D. Curtin, Immanuel Wallerstein, and Alfred W. Crosby. The chapter then concludes with discussion of the thematic work of other scholars in this era and with a survey of the development of world history teaching and textbooks in the same period.

4. The lineage of Frank's argument extends in various directions. On one side it extends to Raúl Prebisch and the economists of the United Nations Economic Commission for Latin America, who sought from the 1950s to establish an intellectual basis for Latin American economic autonomy. On another side it extends to Paul Baran's argument that India could not now develop as England had earlier, because England was already developed. In addition, and through Baran's association with Paul Sweezy, Frank was linked to the debates between Sweezy and Dobb on the transition from feudalism to capitalism. Frank's work paralleled the interactive approach of Eric Williams to the history of slavery in the British Caribbean. E. Williams [1944]; Ian Roxborough, *Theories of Underdevelopment* (Atlantic Highlands, N.J., 1979).

5. By *theme* I mean a selected set of relationships or processes in the past. A theme may be distinguished from a *topic*, which refers more to historical subject matter rather than relationships within it. A theme may also be distinguished from the scholarly discipline through which it is studied. Thus forced migration is a theme, the Atlantic slave trade is a topic, and demography is a discipline for studying both. The term *theme* is also used in teaching of world history, where it helps distinguish among chronological, thematic, and regional organization of courses. For a detailed analysis of themes in world history, see Deborah Smith Johnston, "Rethinking World History: Conceptual Frameworks for the World History Survey" (Ph.D. dissertation, Northeastern University, 2003).

Global Monographs: Curtin

Philip D. Curtin has been a prolific and notably original historian, actively publishing from the 1950s through the 1990s on Africa since 1500 and its place in the
world. He is doubly a founding father of current historiography, both in African
history and in world history. The field of African history organized itself formally
in the 1950s, then grew with dramatic rapidity in the 1960s and 1970s, with
Curtin as one of its leading lights; from the 1980s he began contributing explicitly
to world history. For both African history and world history, Curtin has been
important as an author of trend-setting studies and as an architect of key institutions for graduate study.[6]

In this assessment of Curtin's impact, I emphasize his importance in broadening the historian's path to world history through research. His research and teaching focused, first, on making African history part of world history, emphasizing
the potential of research in African history and setting Africa in context and thus
in connection with other world areas. Second, he developed analyses in world history from an area-studies perspective, not only through his research on Africa but
through his teaching in Expansion of Europe and his leadership in a graduate program of comparative history. His studies focused on economic and social history
especially, but also intellectual history and medical history. Third, through his
consistent emphasis on research design and research methods, he developed a
clear pattern of monographic study in world history. At least five and perhaps as
many as eight of his books can be considered monographs in world history.[7]

Beginning in the 1950s, Curtin taught African history at undergraduate and
graduate levels, at the University of Wisconsin–Madison. Defining African history
presented major problems. The issue was to address established presumptions
that Africans had neither a history of significance nor documents with which to

6. This section is a condensed version of my "The Monograph in World History: Philip
Curtin's Comparative Approach," *World History Bulletin* 15 (Spring 1999), 12–17, which
itself was based on a presentation at the World History Association meeting in
Philadelphia, 1992. For a review of Curtin's work touching on a wider range of issues and
setting his work in more detailed context of the literature, see Craig A. Lockard, "The
Contribution of Philip Curtin and the 'Wisconsin School' to the Study and Promotion
of Comparative World History," *Journal of Third World Studies* 11 (1994), 180–82,
199–211, 219–23. For a discussion of Curtin's work in graduate study, see chapter 20.
7. The books Curtin authored, which I would label as monographs in world history,
include Curtin 1964; Curtin 1969; *Cross-Cultural Trade in World History* (Cambridge,
1984); *Death by Migration: Europe's Encounter with the Tropical World in the Nineteenth
Century* (Cambridge, 1989); and *Disease and Empire: The Health of European Troops in
the Conquest of Africa* (Cambridge, 1998). Others that might also be called monographs
in world history include *Economic Change in Precolonial Africa: Senegambia in the Era
of the Slave Trade*, 1 vol. and supplement (Madison, 1975); *The Rise and Fall of the
Plantation Complex: Essays in Atlantic History* (Cambridge, 1990); and *The World and
the West: The European Challenge and the Overseas Response in the Age of Empire*
(Cambridge, 2000). The latter two books brought much of Curtin's teaching from his
Expansion of Europe courses into print.

analyze its past. Curtin's two American Historical Association pamphlets on African history, published in 1964 and 1974, were influential in propagating and defining survey courses on African history.[8] In them, he emphasized African history as a segment of world history. Even the uniqueness of Africa, such as the "lag" of Africa behind other civilizations, appeared in Curtin's treatment as an issue in the dynamics of world history rather than as a parochial problem. His first book on Africa, *The Image of Africa*, traced the intellectual history of British involvement with Africa in the "precolonial century." After publishing an edited collection of narratives by Africans who had written or recounted stories of their enslavement, Curtin turned to a project of Africanist field research on the economic history of Senegambia from the seventeenth to nineteenth centuries. This project, involving archival research, field work, and oral tradition, appeared in 1975.[9] Curtin's studies, along with those of other historians of Africa, had by the end of the 1980s built a field with a strong set of monographs and recognition for its work throughout the historical profession.[10]

Curtin's work on the Atlantic slave trade began as one more exercise in articulating the history of Africa by linking the continent to other regions. His conceptual study, "Epidemiology and the Slave Trade," appeared in 1968.[11] The article begins by invoking the world historical context of the South Atlantic System, arguing that the immunities in African populations, resulting from their exposure to a difficult disease environment, gave African adults a demographic advantage over Amerindians and Europeans in the New World.[12] Curtin documented his argument with evidence on crew and slave mortality in the Middle Passage, and with the comparative military mortalities to which he would later return. In sum, epidemiology did much to explain demographic patterns of the Atlantic slave trade.

8. Curtin, *African History* (New York, 1964); Curtin, *Precolonial African History* (Washington, 1974). For a more detailed statement on African historiography up to about 1973, see Curtin, "Recent Trends in African Historiography and their Contribution to History in General," UNESCO, *General History of Africa* vol. 1 (London, 1981), 54–71.

9. Curtin 1964; Curtin, ed., *Africa Remembered: Narratives by West Africans from the Era of the Slave Trade* (Madison, 1967); Curtin 1975.

10. Some of the leading and best-recognized monographs in African history include: A. I. Asiwaju, *Western Yorubaland Under European Rule, 1889–1945* (London, 1976); Frederick Cooper, *From Slaves to Squatters: Plantation Labor and Agriculture in Zanzibar and Coastal Kenya, 1890–1925* (New Haven, 1980); Eugenia W. Herbert, *Red Gold of Africa: Copper in Precolonial History and Culture* (Madison, 1984); John Iliffe, *The African Poor: A History* (Cambridge, 1987); Charles van Onselen, *New Babylon, New Nineveh: Studies in the Economic and Social History of the Witwatersrand*, 2 vols. (London, 1982); Jan Vansina, *Paths in the Rainforests: Toward a History of Political Tradition in Equatorial Africa* (Madison, 1990); and Ivor Wilks, *Asante in the Nineteenth Century: The Structure and Evolution of a Political Order* (Cambridge, 1975).

11. Curtin, "Epidemiology and the Slave Trade," *Political Science Quarterly* 83 (June 1968), 190–216.

12. My notes on lectures from his survey classes suggest that the idea was clear in his mind by 1963; I was in Madison as a graduate student from 1963 to 1966.

The Atlantic Slave Trade, published in 1969, is Curtin's most influential book, partly because of its results but more fundamentally because of the design of his research.[13] He showed that it was possible for a single researcher to pull together the national and local literatures on the volume of the Atlantic slave trade, to ask similar questions of each body of data, and to assemble them into a comprehensive picture of the rise and fall and the regional specificity of slave trade. Aside from the skillfully expressed preface and first chapter, the book reads like a series of explications of tables. Yet it not only relaunched the debates on the volume of the slave trade, but served over the years to link the literatures on North America, the Caribbean, South America, Africa, and beyond. Curtin showed that the number of Africans transported across the Atlantic in chains was roughly 10 million persons, rather than the much larger numbers often cited, and that most of the captives were delivered to Brazil and the Caribbean. In his preface, Curtin sought to offer reasons for the "gap between monograph and synthesis," and presented his analysis as "an intermediate level of synthesis…written with an implicit set of rules that are neither those of monographic research, nor yet those of a survey."[14] For my present purposes, and without contesting his label, I propose to call the book a monograph in world history.

The Atlantic Slave Trade underwent the sort of debate that is usually reserved for major monographs: Curtin was rebutted on sheer numbers in the slave trade, with results modifying his estimates in several important particulars and yielding a scholarly consensus that increased his estimate of the total by perhaps a million persons. The design of the research and its impact in launching a wave of transatlantic studies gained widespread praise.[15]

Curtin was elected president of the American Historical Association for 1983 in recognition of his contributions as an Africanist. His presidential address, however, was that of a world historian. That address, "Depth, Span, and Relevance," can be seen as a plea for broad and cross-cultural (rather than narrowly national) monographic studies.[16] In seeking a balance of depth and span, he criticized the trend toward overspecialization in historical studies, noting that Africanists had become as overspecialized as Americanists. He noted the efforts to develop a new

---

13. Curtin 1969.
14. Ibid., xvii.
15. Modifications to Curtin's estimates have come particularly through work by Roger Anstey, Johannes Postma, David Richardson, and Jean Mettas; the sharpest controversy came with the debate initiated by Joseph Inikori's assertion that Curtin's figures were generally too low. All of these points are summarized in Paul E. Lovejoy, "The volume of the Atlantic slave trade: a synthesis," *Journal of African History* 23 (1982), 473–501. See also J. E. Inikori, "Measuring the Atlantic slave trade: an assessment of Curtin and Anstey," *Journal of African History* 17 (1976), 197–233; Curtin, "Measuring the Atlantic slave trade once again," ibid., 17 (1976), 595–605; and Inikori, "Measuring the Atlantic slave trade," ibid., 607–27.
16. Curtin 1984a. Curtin's presidency was celebrated with a series of African historical panels at the conference, at which I presented one on "The Relevance of Quantitative Methods in African History." Though his presidential address focused on global issues, the conference as a whole did not.

survey course but wagered that world history surveys would not work as well in the current day as the Western Civilization survey did in its day. He praised Arnold Toynbee's search for the appropriate unit of historical analysis but expressed disagreement with Toynbee's satisfaction that the "society" (or "civilization") provided the answer to his quest. Instead, Curtin suggested, historians should search for "relevant aggregates" of variables in setting up their projects. The term relevant, in this usage, indicated not relevance of an issue in the consciousness of the public, but relevance of historical factors to an analysis undertaken by a researcher.

The 1984 volume on *Cross-Cultural Trade* was Curtin's first book explicitly intervening in the discussion of world history; it was nearly complete at the time of his AHA presidency.[17] The method of presentation is that of a comparative set of case studies, from Armenian merchants in Tibet to Mande merchants in the Niger Valley. The systematic analysis of trade diasporas gives the book a mono-graphic dimension. Still, there is a synthetic element to the book, as Curtin argued that the phenomenon of the trade diaspora—using family and ethnic ties to sustain long-distance commerce—was appropriate within certain types of broader social organization and that it died out with industrialization. Applying a case-study approach to a problem in world history required a rethinking of all the basic categories in historical scholarship. What are primary sources and what are secondary authorities? What is a monograph and what is a synthesis? If these questions remained unanswered for a time, the book was successful in enabling many readers to visualize world history in terms both broad and specific.

Curtin's subsequent study of nineteenth-century military demography, published as *Death by Migration*, applied a more focused comparative research design and yielded brilliant results. The study traced the mortality rates of British and French military personnel serving throughout the world in the nineteenth century and verified the "relocation cost," the additional mortality brought by service away from home. Each pairing of birthplace and region of military service (as of a given period of time) served as a distinct case in the broad grid of Curtin's analysis. The striking result was the pattern of sharp and simultaneous decline in mortality throughout the world, in the 1860s and 1870s, before major advances in medical knowledge. A comparative study of localities thus revealed a global result. This study did not and could not explain the causes of that mortality decline, but it raised unmistakably the question of why and how health conditions for European military personnel improved so suddenly.[18]

---

17. Curtin 1984b. The book is neither sustained narrative nor a comprehensive analysis. Instead it is a series of illustrations of a single theme—the ethnically organized trade diaspora—and its reappearance in societies widely separated in time and space. In this and other work, Curtin has picked out key issues and has treated them comparatively, thereby giving emphasis to the fine structure of world history rather than its overall contours.

18. Curtin was able to speculate that greater attention to nutrition, cleanliness, and water quality were important in the change, but it would take another study to verify these points. Curtin 1989. For a later study pursuing related issues, see Curtin 1998.

The Rise and Fall of the Plantation Complex was a venture into synthesis, drawing on secondary sources and presenting a broad and general interpretation rather than focusing on some crucial but narrower issue.[19] It spans half a millennium, surveying the plantations of Brazil and the Caribbean and the networks of trade and migration linking plantations to Africa and Europe. It relies heavily on Curtin's own monographic work and his conception of how to integrate monographs into synthesis in world history. Its success suggests that we may look forward to the development of a style in world-historical synthesis, dense in interactive texture yet firm in outlining basic directions of change, reliant on the sort of world-historical monographic work of which Curtin has done much to pioneer.[20]

Philip Curtin's work shows us the nature of the world-historical monograph and demonstrates its power in influencing the thinking of historians. His work has made the field of world history into a practical and attainable arena for historical research. His emphasis on the basic skills of the professional historian, reinforced by his practical sense of how to reach far in analysis and when to stop reaching, has provided practical clues for his students and for other researchers. Yet the question of the balance between monographic and synthetic work in world history needs further debate, as is indicated in the range of Curtin's own works. In world history, we are still working out the appropriate contours of the monographic study. At the same time, there is a positive role to be played by historians willing to run the risk of making planetary generalizations. For this approach, we may turn to the work of Immanuel Wallerstein.

Political Economy: Wallerstein

Immanuel Wallerstein broadened the historians' path to world history through analysis. The opening volume of his *The Modern World-System* (1974) provided a new paradigm for world history that owed much to the Expansion of Europe paradigm, but which was inspired more by the underdevelopment theses of Andre Gunder Frank and Paul Baran than by the modernization approach adopted by McNeill.[21] His thesis was that the European economy led in creation of a world-system during the sixteenth century and that it underwent phases of expansion and stagnation thereafter. He contrasted this world-system—a single economy spanning many states—with the alternative of a world-empire, referring particularly to the effort of the Emperor Charles V to control all of Europe and its

19. Curtin 1990.
20. Curtin has been modest about this book, labeling it as the "pirating" of lectures he had given for years. Philip Curtin, in conversation with the author. For further discussion, see Manning 1999a.
21. Immanuel Wallerstein, *The Modern World-System: Capitalist Agriculture and the Origins of the European World-Economy in the Sixteenth Century* (New York, 1974). Subsequent volumes were published as *The Modern World-System, II: Mercantilism and the Consolidation of the European World-Economy, 1600–1750* (New York, 1980); and *The Modern World-System, III: The Second Era of Great Expansion of the Capitalist World-Economy, 1730–1840s* (New York, 1989).

colonies. His reasoning adopted a center–periphery terminology and encompassed a Marxian focus on evolution and transformation in the system, a Weberian focus on trade and bureaucracy, and a Braudelian emphasis on multidisciplinary analysis.

Wallerstein's unit of analysis was the world-system: the whole of a region linked by a single economic network, and divided into the sub-regions of the core, the semi-periphery and the periphery. Much of the argument, for instance, centered on relations between the Western European core and the Polish periphery. The unit of analysis was unusual, therefore, in that it was unique (only one world-system to a planet), its boundaries were not political frontiers but economic limits, and the system grew significantly with the passage of time.

This scheme elicited wide interest and sustained a lively dialogue on global political economy on at least three fronts: with Marxists, with modernization theorists, and with specialists on Third World studies. Among Europeanists, Robert Brenner provided the most detailed critique, in the form of a long article emphasizing the importance of class structures in social and economic transformations. Brenner argued that an interpretation based mainly on commercial developments failed to correspond to the timing of major social changes.[22] This was a critique from the standpoint of Marxian specialists on European capitalist development.

In a more general fashion, scholars working on Third World areas complained that the world-system paradigm tended to privilege the center and to treat the periphery as a passive recipient of external influence. Nonetheless, world-system analysis appeared as an alternative and response to the modernization paradigm that had grown in area studies and in political science and sociology. Modernization, a vision of social change inspired by the work of Talcott Parsons, tended to accept a national or case-study approach to the world, in which development took place once it overcame local resistance; Wallerstein's world-system assumed structural relationships among regions.[23]

Wallerstein himself, trained as a sociologist in the era when Weber's work had grown in popularity, did his first studies on Africa in the era of independence. Bogumil Jewsiewicki, a historian of Africa, has argued that there is a great continuity in Wallerstein's intellectual agenda despite the apparently large leap from twentieth-century Africa to sixteenth-century Europe. Another key underpinning to The Modern World-System was Wallerstein's involvement in the radical student

---

22. Robert Brenner, "Agrarian Class Structure and Economic Development in Pre-Industrian Europe," Past and Present 70 (1976), 30–74. Brenner implicitly underscored the sociologist Wallerstein's position as a parvenu in the historical literature by referring mainly to earlier debates on similar issues; Wallerstein, perhaps anticipating this response, had preemptively covered roughly 30 percent of each page of his book with footnotes. See also Robert S. DuPlessis, "The Partial Transition to World-Systems Analysis in Early Modern European History," Radical History Review 39 (1987), 11–27.

23. Parsons 1937; Gabriel A. Almond and James S. Coleman, The Politics of Developing Areas (Princeton, 1960).

protests of 1968 at Columbia University and his close ties to Paris and the left in France at the same time. Wallerstein left Columbia for McGill University in Montreal, and it was there that he wrote *The Modern World-System*.[24]

Terence Hopkins, a colleague and ally at Columbia, had moved to Binghamton, and with the success of *The Modern World-System* was able to bring Wallerstein to Binghamton and obtain university funding for establishment, in 1976, of the Fernand Braudel Center. The division of labor was that Wallerstein directed the center and edited its journal, *Review*, while Hopkins headed the doctoral program in the Department of Sociology. Hopkins and Wallerstein, along with others who joined them, developed a system of working groups focused on agreed-upon topics, some of which were eligible for outside funding. For over a quarter of a century, this center was the only research center for global historical studies in the United States and, arguably, the world.[25]

The impact of the world-system paradigm was clear in the first detailed narrative of the emergence of the Third World, Leften Stavrianos's *Global Rift*.[26] For Stavrianos, the Third World was the periphery of Wallerstein's interpretation: as a result, it shifted and expanded with time. In the period from 1400 to 1770, the Third World included Eastern Europe and Latin America; after 1770 the Middle East and India were incorporated into the Third World; and after 1870 Africa, China and Russia were added. After 1914 the character of the dynamic changed, and Third World areas struggled to gain independence from the center. This version of the framework includes some evident weaknesses: it traces the incorporation of areas into the Third World, but not their graduation into the semi-periphery or the center; the analysis centers on European-dominated commerce and downplays domestic events in areas peripheral or external to the system.

Historical studies by sociologists under Wallerstein's tutelage began appearing in the 1970s, focusing especially in applying world-system thinking to colonial areas. Responses in reviews by area-studies scholars were critical of the scarce room left in these studies for agency by local actors. One response of world-system analysts to this line of criticism was to embark on studies linking households to the world-system.[27]

24. Bogumil Jewsiewicki, "The African Prism of Immanuel Wallerstein," *Radical History Review* 38 (1987), 50–68; Immanuel Wallerstein and Paul Starr, *The University Crisis Reader* (New York, 1971); Wallerstein, *Africa, the Politics of Independence: An Interpretation of Modern African History* (New York, 1961); Wallerstein, *Africa: The Politics of Unity; an Analysis of a Contemporary Social Movement* (New York, 1967).

25. Fernand Braudel Center for the Study of Economies, Historical Systems, and Civilizations (www.fbc.binghamton.edu). In a visit to Binghamton in the winter of 1997, I had the opportunity to speak with both Wallerstein and Hopkins about their program and its development. The university, founded as the State University of New York at Binghamton, has become known as Binghamton University.

26. L. S. Stavrianos, *Global Rift: The Third World Comes of Age* (New York, 1981).

27. Rhoda E. Howard, *Colonialism and Underdevelopment in Ghana* (New York, 1978); Joan Smith, Immanuel Wallerstein, and Hans-Dieter Evers, eds., *Households and the World-Economy* (Beverly Hills, 1984).

When the two remaining volumes of *The Modern World-System* appeared, in 1980 and 1989, they gained rather less attention than the first volume had. The overall logic of the paradigm was by now widely recognized, so that the details of the later volumes were of interest mainly to the scholars working within the world-system paradigm. The second volume focused on the era of Dutch hegemony and on the core-region struggle for hegemony between France and Britain. The third volume focused first on the struggle for hegemony between France and Britain, then concentrated on the expansion of the areas under European control and on "settler decolonization" in the Americas.[28]

The Fernand Braudel Center managed to sustain doctoral and postdoctoral research within the world-system paradigm, and its journal *Review* succeeded in engaging critics as well as practitioners of the approach in an ongoing dialogue. Among the long-term projects of the center were analysis of the economic history of the Ottoman Empire, studies of Kondratieff waves or long waves in economic activity, and quantitative studies in world-system analysis. William McNeill and Andre Gunder Frank contributed occasional articles, along with regulars Giovanni Arrighi and Samir Amin. A 1989 debate between Sanjay Subrahmanyam and Ravi Arvind Palat documented the controversy about whether world-system analysis could be applied effectively to South Asia.[29] Thus the center and the journal have not only affirmed and developed the paradigm, but have allowed for rethinking and change.

Wallerstein's work, though centered in the adjoining discipline of sociology, brought major advances to the study of world history. It centered on the creation of a modern system of global dominance by European-led economies, a topic of debate for at least two centuries, but developed a new perspective on this old issue. The notion of the modern world-system opened a major front in the struggle to surmount the national framework for historical interpretation. It brought a major shift in perspective in debates overlapping world history and modern European history. Wallerstein caused historians to investigate the logic of world-systems analysis. He and his allies brought historians to read Marx, Weber and more recent theoreticians, to focus on long cycles and hegemonic shifts, and to consider the interplay among social science theories, with all of this analysis focused on the single great empirical issue of the unfolding and transformation of what he defined as the European-centered capitalist world-system.[30] The world-system

28. Wallerstein 1980; Wallerstein 1989.
29. Wallerstein's many other publications include a set of essays and a concise survey: Wallerstein, *The Capitalist World-Economy* (Cambridge, 1979); Wallerstein, *Historical Capitalism* (London, 1983). The Fernand Braudel Center has sponsored publication of volumes based on a number of its conferences. *Review* includes both thematic issues and individual articles. In its first dozen years, it included studies in the economic history of the Ottoman Empire (in vols. 1, 8, and 11) and studies of long-term cycles in economic activity (in vols. 1, 7, and 11).
30. Wallerstein, *The End of the World as We Know it: Social Science for the Twenty-first Century* (Minneapolis, 2001); Wallerstein et al., *Open the Social Sciences: Report of the Gulbenkian Commission on the Restructuring of the Social Sciences* (Stanford, 1996). Wallerstein served as president of the International Sociological Association from 1994 to 1998.

framework did indeed address many historical issues. At the same time, however, new arenas for world historical study continued to open up.

Ecological History: Crosby

Alfred Crosby took a great step toward clearing the scientific-cultural path to world history with the publication of his *Columbian Exchange* in 1972. This book gained a wide audience, conveying a global historical approach to an audience beyond historical specialists. His geographical framework reached all the shores of the Atlantic, and his work illustrated the relative ease with which biological and ecological phenomena escape the limits of nations and civilizations. His volume traced the movements of diseases, plants, and animals in both directions across the Atlantic after 1492 and explored their consequences.[31] The opening chapter provided a multidisciplinary history of the initial peopling of the Americas and the concurrent elimination of large mammals in the very regions where horses and camels had first developed. He showed how to convey environmental issues in historical terms and to do so in transregional scale. With this work, Crosby became a founding father of ecological history and global history.

Crosby's information was not all new, but the work of assembling it and presenting it as a central aspect of history in the Age of Discovery was new. He commented modestly on his crossing of disciplinary frontiers in his preface: "I am the first to appreciate that historians, geologists, anthropologists, zoologists, botanists, and demographers will see me as an amateur in their particular fields." He did not label his study as world history, yet argued that "although the Renaissance is long past, there is great need for Renaissance-style attempts at pulling together the discoveries of the specialists to learn what we know, in general, about life on this planet."[32]

The book established the term *Columbian exchange* as part of the literature and the teaching lexicon. It provided a look at disease, crops, weeds, and animals as they moved in both directions once transatlantic contacts opened. While the analysis addressed questions of intentions, it reached beyond human agency to show the importance of unconscious, biological forces in human history. On the other hand, Crosby's subtitle emphasized cultural as well as biological consequences of 1492: in practice, his analysis of culture focused on material culture and on lifestyles conditioned by available plants and animals. Crosby's analysis concentrated not on complex interactions, but on simple linkage of an unexpectedly wide range of factors, which added up to a distinctive narrative of history.

His next book, an analysis of the 1918 influenza epidemic in the United States, gained less attention. Only with the emergence of later influenza epidemics and the AIDS pandemic did interest in the topic arise, and Crosby's book was reissued

31. Crosby 1972. The book was an instant success once published, but Crosby had had to submit it to numerous publishers in order to get it into print. Crosby's first book was *America, Russia, Hemp, and Napoleon: American Trade with Russia and the Baltic, 1783–1812* (Columbus, OH, 1965).
32. Crosby 1972: xiv–xv. Curtin 1968 covered some of the same ground, but Crosby found a way to convey the multidisciplinary framework.

in 1989. Even then, since the volume refers primarily to the United States, world historians took little notice of it.[33]

Crosby then turned to a study in ecological history that was in some senses more fully global: it appeared in 1986 under the title *Ecological Imperialism*.[34] It focused on the growing connections among the temperate regions of the world that Crosby labeled "neo-Europes," because of their ecological similarity to and ultimate domination by Europe. Because Europe was the most densely populated and most connected of the temperate regions, its diseases, plants, and animals were more likely to spread overseas than the reverse. Tracing these phenomena over a millennium, Crosby finished with a flourish by using New Zealand as a case study and a metaphor for the whole set of neo-Europes, showing the full range of phenomena for a small region in a compressed time. The insights in this study exceeded the thousand years of its formal analysis. The story could not be told without a review of continental drift and the evolution of plants and mammals in the period since the supercontinent of Pangaea began to break up 180 million years ago.[35]

Consciousness about ecological issues expanded in the 1970s and 1980s, in the United States and most everywhere in the world, for reasons having to do with contemporary atmospheric changes, water pollution, and loss of forest cover. In one sense, this expanding concern for ecology created wide interest in Crosby's analyses, and Crosby strengthened the connection with the engaging quality of his writing and his effectiveness in conveying the long time-frame and broad geographic sweep of ecological changes. On the other hand, the subject matter and the style of analysis in Crosby's works were dramatically different from the standard emphases of political or social history, and the number of scholars working in ecological history, especially at the global level, was very small. As a result, Crosby's interpretations did not gain substantial representation in textbooks, and he did not receive recognition by professional organizations similar to that for Curtin, McNeill, or Wallerstein.[36]

Nevertheless, the very distinctiveness of the historical patterns that Crosby portrayed served to reveal the potency of the scientific-cultural path to world history. New topics, new dynamics, documented through work in disciplines normally far from history, gradually caused historians to rethink their periods,

---

33. Crosby, *Epidemic and Peace, 1918* (Westport, 1976); reissued as *America's Forgotten Pandemic: The Influenza of 1918* (Cambridge, 1989). Crosby underscores the remarkable forgettability of influenza by noting that the 1918 convention of the American Historical Association was canceled because of the epidemic, but that two decades later the AHR blamed the cancellation on postwar congestion of the railroads. Crosby, "The Past and Present of Environmental History," *American Historical Review* 100 (1995), 1177.

34. Crosby, *Ecological Imperialism: The Biological Expansion of Europe, 900–1900* (Cambridge, 1986).

35. In addition, Crosby hinted at a conclusion, suggested by some other scholars, that the sweet potato had been transported by Polynesian mariners of the first millennium from South America to the Pacific, where it permitted New Zealand to develop its dense Maori population. Ibid., 222.

36. Crosby did, however, become co-editor of a major book series in ecological history at Cambridge University Press, along with Donald Worster, an environmental historian specializing on the United States.

regional frameworks, and analytical priorities. For Crosby, it was conveying an effective narrative of global environmental processes that stood out and suggested that new ways of looking at historical documents could bring substantially different perspectives on the past. In this sense, Crosby's achievement was parallel to that of William McNeill, who broadened the historians' path to world history by developing an overall narrative of world history attractive both to general readers and academic historians. Crosby did not attempt a comprehensive narrative of the ecological history of the world, though he produced studies of sufficient breadth to suggest that such a narrative could be developed. Perhaps more importantly, he provided convincing examples of how global studies of ecological issues could be carried out.

The Later McNeill and other Writers

William McNeill had launched thematic and global study by professional historians with his 1963 *Rise of the West*. In one sense he remained on the pedestal he had created, offering periodic restatements of his synthesis in textbook form: his work in developing the form of the textbook in world history is discussed in the final section of this chapter.

More productively, he joined with the scholars he had inspired and wrote a number of thematic investigations in world history, in which he explored new issues and developed new ideas. In *Plagues and Peoples* (1978), he explored the interaction between disease and civilization—that is, the concentration of diseases among densely settled peoples, and the relative advantage they had when encountering less densely populated and less disease-ridden peoples.[37] *The Pursuit of Power* (1982) explored further the issue of dominance in the second millennium C.E., tracing changes in military technology first in China and then in Europe.[38] In a third volume, McNeill focused on a shorter period of time and a wider geographic era, as he surveyed *The Age of Gunpowder Empires*.[39]

Meanwhile, McNeill became president of the American Historical Association in 1985, and his presidential address, along with nine other essays, appeared in print in 1986.[40] In his presidential address, McNeill sought to reconcile the inevitability and social function of myth in history with a gradual success of the search for historical truth. He attempted to persuade historians to take the risk of writing broadly defined studies. Two further chapters make the case for world history in research and teaching. The final five chapters address four historians: Lord Acton, Toynbee, Carl Becker, and Fernand Braudel. These essays, written between 1961 and 1985, reveal at once McNeill's commitment to global analysis and his sense of the continuity in historical writing.

As McNeill turned from grand synthesis to more focused yet wide-ranging studies, other scholars stepped forward to join him in creating a generation's work

---

37. W. McNeill 1976.
38. W. McNeill, *The Pursuit of Power: Technology, Armed Force, and Society since A.D. 1000* (Chicago, 1982).
39. W. McNeill, *The Age of Gunpowder Empires* (Chicago, 1990).
40. W. McNeill, *Mythistory and other Essays* (Chicago, 1986).

in transnational scholarship. Barrington Moore's 1966 study, *The Social Origins of Dictatorship and Democracy*, owed more to the social analysis common to the Western Civilization paradigm than to the economic focus of the Expansion of Europe framework. Moore compared the relations of lord and peasant in several cases from seventeenth-century Europe to twentieth-century China to seek out the essential components of a democratic political order. Theda Skocpol used a similarly comparative framework to consider the most prominent social revolutions of the modern world.[41]

Andre Gunder Frank's studies in this era focused on underdevelopment and on capitalist accumulation in the early modern era.[42] Among the other scholars who made a Third World concentration into an approach to world history, the economist Paul Bairoch published a series of interpretations of the weakening position of African, Asian, and Latin American economies in the nineteenth and twentieth centuries, the economist Samir Amin theorized the changing modes of production in Third World regions, and the historian Walter Rodney applied the underdevelopment thesis to African history.[43]

Fernand Braudel broadened his now-famed interpretation of the sixteenth-century Mediterranean with a 1967 volume entitled *Capitalism and Material Life*. This volume sketched an argument that he was later—in 1979—to present in three volumes.[44] The general theme was the rise of capitalism from the sixteenth through the eighteenth centuries, and Braudel pursued it through a wide range of specific issues, from population and food crops to styles of dress. The framework emphasized interactions among all the regions of the world, though in practice most examples focused on European data and on European impetus to change.

From a more traditionally Marxist viewpoint and for an earlier period, Perry Anderson published two imposing volumes in 1974, one tracing the European transition from the ancient world to the feudal order and the other tracing the political dimension of the transition from feudalism to capitalism.[45] These two

41. Barrington Moore, *The Social Origins of Dictatorship and Democracy* (New York, 1966); Theda Skocpol, *States and Social Revolutions* (New York, 1979).

42. Andre Gunder Frank, *Latin America: Underdevelopment or Revolution* (New York, 1969); Frank, *World Accumulation 1492–1789* (London, 1978); Frank, *Crisis: In the World Economy* (New York, 1980).

43. Amin began by identifying limits on economic growth in postcolonial West Africa, and expanded to global studies in modes-of-production and political economy analysis, with a focus on tributary modes of production. Rodney, though a strong critic of the slave trade, started his analysis of underdevelopment of Africa with the late-nineteenth-century European colonization of Africa. Paul Bairoch, *Le Tiers-monde dans l'impasse* (Paris, 1971); Samir Amin, *Accumulation on a World Scale: A Critique of the Theory of Underdevelopment*, trans. Brian Pearce ([1970] New York, 1974); Samir Amin, *Unequal Development: An Essay on the Social Formations of Peripheral Capitalism*, trans. Brian Pearce ([1973] New York, 1976); Walter Rodney, *How Europe Underdeveloped Africa* (Kingston, Jamaica, 1972).

44. Fernand Braudel, *Capitalism and Material Life*, trans. Miriam Kochan ([1967] New York, 1973); Braudel, *Civilization and Capitalism, 15th–18th century*, trans. Sian Reynolds, 3 vols. ([1979] New York, 1981).

45. Perry Anderson, *Passages from Antiquity to Feudalism* (London, 1974); Anderson, *Lineages of the Absolutist State* (London, 1974).

studies, while centered in Europe, show a clear effort to link political and economic changes into a global pattern.

Marshall Hodgson, a colleague of William McNeill at the University of Chicago, prepared a three-volume study on the civilization of the Islamic world; Hodgson died in 1968, but the energies of his colleague Reuben W. Smith brought the study into print in 1974.[46] Hodgson's approach was to treat the Islamic world not as a single civilization but as a world community, both because of the cultural variety within it and because of the wider range of civilizations linked to each other through its central position. Hodgson's study covered the full span of time from the birth of Islam to the twentieth century, but his arguments for the pre-modern era attracted the most attention. As such, his was one of few major interventions of the 1970s in world history before 1500. His line of argument has since been taken up by other scholars.

In 1982 anthropologist Eric Wolf, well known for his studies of peasantries, published a historical and theoretical interpretation of the modern world. *Europe and the People Without History* provided an interpretation of the integration of Africa, Asia, and Latin America into the European economic orbit up to 1900.[47] The book gained wide attention and was initially well received. Over the longer run, it seems to have been most successful in demonstrating the difficulty of carrying out a coherent analysis of the topic. Wolf provided a thorough critique of social science approaches to the world community and announced an intention to show interactions throughout the world. In practice, however, he focused on linking a narrative of the world market and a theory of capitalist development to processes of local development. The emphasis, in short, is the impact of Europe on the world outside Europe. Wolf's units of analysis are social classes and ethnic groups; these, however, serve more as recipients of change than as interacting groupings.

The book is made hard to follow by the alternations in themes: a general introduction, a survey of the world in 1400 (ranging as much as several hundred years earlier), an analysis of modes of production in general, a survey of Europe's prelude to expansion, and chapters of European impact elsewhere. While the work focused on the creation of the world community as a capitalist order, it provided no clear chronology on the creation of that community. Wolf did include a few compelling examples of economic interactions, including one on the market for cotton fiber and cotton textiles in the nineteenth century. A strong concluding section underscored the need to examine the history of culture in the modern world. But the text itself looked mainly at economics and the influence of Europe beyond the seas.

Several studies of the Atlantic basin appeared, addressing that immense region through various perspectives, ranging from the politics of the pan-African movement to the history of the African diaspora generally to the politics of eighteenth-century imperial expansion. In a study that was narrower both topically and regionally, Elizabeth Fox-Genovese and Eugene Genovese investigated merchant capital in eighteenth- and nineteenth-century France and the

---

46. Marshall G. S. Hodgson, *The Venture of Islam: Conscience and History in a World Civilization,* 3 vols. (New York, 1974); Hodgson, *Rethinking World History: Essays on Europe, Islam, and World History,* ed. Edmund Burke III (Cambridge, 1993).

47. Eric Wolf, *Europe and the People Without History* (Berkeley, 1982).

United States.[48] In this collection of their articles, the authors addressed the property relations in American slavery, French agriculture and Franco-American trade, as well as the political and ideological positions of property owners. In a wide-ranging introduction they sought to lay the groundwork for an economic and juridical theory of bourgeois property in the rise of capitalism.

Peter Worsley, whose 1968 *The Third World* did much to launch the term among English-speakers (it originated in the French language), returned in 1984 with *The Three Worlds* to address contemporary problems of culture and world development.[49] The book addressed a range of world-historical issues for the past two centuries, but through the cross-sectional organization of the sociologist. Worsley sought to highlight cultural issues by delivering a critique of base-superstructure analysis in Marxism while seeking to maintain and apply other elements of Marxist analysis. To this end he took as his units of analysis classes (peasants and workers) and ethnic groups. In practice, his analysis of nationalism did not get far into the cultural issues he targeted.

Charles Tilly, employed as both a sociologist and a historian, published in 1984 an engaging review of the methods of large-scale historical analysis, focusing on their application to European history. Tilly began with a critique of nineteenth-century social theory, showing it to have concentrated on differentiation as the putative key to progress. He then identified four levels of interpretation—world-historical, world-system, macrohistorical, and microhistorical—and announced his focus on the macrohistorical level, meaning generalizations at the level of the European continent. He went on to identify four types of comparison, each defined by the objective of the analyst—individualizing, universalizing, variation-seeking, and encompassing—and elected to look at variation-seeking comparisons. Tilly argued that world-historical and encompassing generalizations were beyond the current abilities of historians, and he documented his case by showing that various analysts turned back from the big questions they started to address, focusing in practice on smaller ones. He gave examples to show the tendency of very broad analysis to get caught up in endless variables and relationships.[50]

48. Joseph E. Harris, ed., *Global Dimensions of the African Diaspora* (Washington, D.C., 1982); Imanuel Geiss, *The Pan-African Movement: A History of Pan-Africanism in America, Europe, and Africa,* trans. Ann Keep ([1968] New York, 1974); Peggy K. Liss, *Atlantic Empires: The Network of Trade and Revolution, 1713–1826* (Baltimore, 1983); Elizabeth Fox-Genovese and Eugene D. Genovese, *Fruits of Merchant Capital: Slavery and Bourgeois Property in the Rise and Expansion of Capitalism* (New York, 1983).
49. Peter Worsley, *The Three Worlds: Culture and World Development* (Chicago, 1984); Worsley, *The Third World* (London, 1968).
50. Tilly's point is well taken: the research strategy for a very broad study requires a method for deciding which of the many possible variables and interactions are worth analyzing in detail. I would argue, however, that the present situation requires a more earnest effort to understand world-historical processes, since they are becoming steadily more determinative of our lives; further, I would argue that systematic, critical (and preferably collective) efforts at analysis can lead to substantial progress in historical interpretation, even on a broad scale. Charles Tilly, *Big Structures, Large Processes, Huge Comparisons* (New York, 1984).

Eric Hobsbawm produced several studies based in European history with world-historical implications. His *Age of Revolution*, for instance, centered on the period from the French Revolution through the European risings of 1848 and suggested broader implications of these events. His edited collection on *The Invention of Tradition*, prepared in collaboration with the African historian Terence Ranger, compared European and non-European experiences to develop a broader perspective on nationalism. Hobsbawm's *Nations and Nationalism* carried this approach forth into a synthetic approach to nationalism at a nearly planetary level. This interpretation of nationalism owed part of its breadth to the work of Benedict Anderson, whose studies of Southeast Asia permitted him to visualize nations as "imagined communities": as Anderson suggested and Hobsbawm confirmed, this approach made it easier to see European and non-European nationalisms as varieties of a single beast rather than as distinct species.[51]

Other studies focused on Western hegemony, in economic, social, and cultural arenas. Nathan Rosenberg and L. E. Birdzell illustrated the continuing strength of the Western Civilization paradigm in economic history in a widely read book that focused on innovation as the key to European growth in the early modern period and since then and that assumed that European interactions with other regions may safely be neglected in their interpretation.[52] Theodore von Laue developed an

---

51. Eric J. Hobsbawm, *The Age of Revolution: Europe 1789–1848* (New York, 1962); Hobsbawm, *Industry and Empire: From 1750 to the Present Day* (Harmondsworth, 1969); Hobsbawm, *The Age of Capital, 1848–1875* (New York, 1975); Hobsbawm, *Nations and Nationalism since 1780: Programme, Myth, Reality* (Cambridge, 1990); Hobsbawm, *The Age of Extremes: A History of the World, 1914–1991* (New York, 1994); Hobsbawm and Terence Ranger, eds., *The Invention of Tradition* (Cambridge, 1983); Benedict Anderson, *Imagined Communities* (London, 1983).

52. Nathan Rosenberg and L. E. Birdzell, *How the West Grew Rich* (New York, 1986). John Hicks was influential in launching the literature of which this volume is a part: it is theoretically sophisticated and largely neoclassical in orientation, and it holds rather tightly to assumptions of the Western Civilization model. An earlier English tradition in economic history, led by Eileen Power and M. M. Postan and focusing on medieval and early times, remains an important reference point for recent and more theoretically trained economic historians. W. W. Rostow set forth a monocentric, diffusionist view of historical evolution that was popular for a time but came under increasing criticism. Douglass North and Robert Paul Thomas's subsequent interpretation made a much stronger effort to fit institutional change into this neoclassical framework. Marxian and institutionalist critique of this approach pushed North to dig deeper in an attempt to get beyond an interpretation of world economic history based simply on markets, and in 1981 he published a theory of institutional change in economic history. William N. Parker's view of the twentieth-century economy centers firmly on the North Atlantic. John D. Hicks, *A Theory of Economic History* (Cambridge, 1969); W. W. Rostow, *The Stages of Economic Growth: A Non-Communist Manifesto* (Cambridge, 1960); Douglass C. North and Robert Paul Thomas, *The Rise of the Western World: A New Economic History* (New York, 1973); North, *Structure and Change in Economic History* (New York, 1981); William N. Parker, *Europe, America, and the Wider World: Essays on the Economic History of Western Capitalism*, 2 vols. (Cambridge, 1984–1991); Rondo Cameron, *A Concise Economic History of the World from Paleolithic Times to the Present* (New York, 1989).

interpretation of the twentieth-century world that centered on Europe and on the diffusion of its influence. In von Laue's view, European impact came suddenly and with devastating effect to the lands of Africa, Asia, and Latin America: the result, while revolutionary, is above all tragic.[53]

Michael Geyer and Charles Bright proposed a sharply contradictory interpretation of the twentieth century, based on a different paradigm. Where von Laue applied the expansion of Europe paradigm (and assumed, further, that the most significant European expansion took place in the twentieth century rather than previously), Geyer and Bright utilized a world community paradigm. That is, they assumed that the incorporation of the various regions into a global order had taken place before 1900 and that twentieth-century changes consisted of renegotiating the control and the benefits of that system. The central dynamic, in their view, was the contradiction between the trend toward centralization and homogenization in the economic order, on the one hand, and the affirmation of new cultural differences, on the other hand, by peoples and regions seeking to assert their autonomy.[54]

The most widely read book on world history in this period was Paul Kennedy's *The Rise and Fall of the Great Powers*.[55] Kennedy's paradigm centered on great powers (that is, on nations), on military power, and on the economic underpinnings of battlefield strength. While most of the book is European history writ large, Kennedy's framework led him easily into a twentieth-century world in which Japan, China, and the Soviet Union play major roles. Kennedy noted the uniqueness of the period of a bipolar Cold War world and suggested that a polycentric world is to be expected for the future. Partly because of the simplicity of his tale, but also because of the care and skill with which he told it, Kennedy succeeded in conveying a global narrative to his readers, whereas other authors have either lost readers in details or left them with only hints of the story.

At the end of the 1980s, a generation's work on thematic studies in world historical studies had expanded the application of the preceding period's emphasis on grand synthesis. The pattern in these thematic studies is clear: emphasis on the

---

53. Von Laue 1987.
54. Michael Geyer and Charles Bright, "For a Unified History of the World in the Twentieth Century," *Radical History Review* 39 (1987), 69–91. Some neoclassical economic historians, notably A. G. Hopkins and A. J. H. Latham, have developed alternative views centered in the Third World, and the Marxist Samir Amin has written widely on the global economic order. E. L. Jones, in a comparison of Europe and Asia, focused on what he called "very long-term growth" and argued that environmental change figured in the establishment of European economic dominance; he later argued that growth was not unique to the modern West. A. G. Hopkins, *An Economic History of West Africa* (London, 1973); A. J. H. Latham, *The International Economy and the Underdeveloped World, 1850–1914* (London, 1979); Patrick Manning, *Slavery, Colonialism, and Economic Growth in Dahomey, 1640–1960* (Cambridge, 1982); Amin 1974; Amin 1976; E. L. Jones, *The European Miracle: Environments, economies, and geopolitics in the history of Europe and Asia* (Cambridge, 1981).
55. Paul Kennedy, *The Rise and Fall of the Great Powers: Economic Change and Military Conflict from 1500 to 2000* (New York, 1987).

"expansion of Europe," meaning a focus on early modern times, with the development of capitalist economic systems and the spread of European empire, but also on nineteenth- and twentieth-century empire and colonialism. McNeill contributed to it with *Pursuit of Power*, but he also pursued other avenues with *Plagues and Peoples*. Curtin's work on enslaved Africans fit in at the margins of this debate, but his work on trade diasporas and on mortality rates went beyond this discussion. Wallerstein's modern world-system was at the core of the discussion. Crosby's work, while explicitly linked to the political economy of an expanding Europe, sought out new issues in ecological history through new techniques and disciplines. Groundwork was being laid for a broader vision of world history, but the principal discussion remained where James Harvey Robinson and Charles Beard seventy years earlier thought it ought to be focused: on the modern political and economic evolution of a world centered on the North Atlantic.

Janet Abu-Lughod proposed that the Islamic heartland of the thirteenth century, along with the regions linked to it, could be seen as a world-system in terms quite similar to those proposed by Wallerstein for the later European world-system.[56] Partly because of the direct and articulate expression of her interpretation, *Before European Hegemony* was influential in enabling historians to sustain a global vision as broad as that of Wallerstein and yet free themselves of a focus on the modern period and on European powers.

Teaching and Textbooks on World History

There was no bright Monday morning on which thousands of students and teachers showed up in school to begin study of world history, no preceding Friday on which the newly printed textbooks were delivered to the classrooms. But if there were such a day, it would have been in the mid-1970s, and it could have taken place in almost any part of the United States.

This wave of world-historical study brought unprecedented levels of teaching, discussion, research, and writing on the past in global terms. The dramatic expansion in study of world history centered to a remarkable degree in the United States. The social basis of world history lay in the "middle class" (as it is known in the United States), and not in the elite or among workers. This broad stratum of people with solid education and regular employment, often in the professions— and including a small but growing number of immigrants, African Americans, and Latinos—came to include a significant number who found world history to be of interest. World history gained its institutional base through courses at state universities and liberal-arts colleges. It then expanded to public schools and community colleges. Elite schools at secondary and college levels were slow to explore

56. Janet Abu-Lughod, *Before European Hegemony: The World System A.D. 1250–1350* (New York, 1989). Her earlier work had centered on cities of the Muslim world. Andre Gunder Frank responded to Abu-Lughod's interpretation by suggesting that the world-system concept could reasonably be projected even further back in time. Frank, "A Plea for World System History," *Journal of World History* 2 (1991), 1–28.

world history. The formation of the World History Association in 1982, along with its bulletin and later its journal, provided new institutions and allowed for connections among college and high school teachers of world history.

Nevertheless, there were to be constraints on world history. Despite the opportunities and the rapid development of world history as a field of study, the new field was restrained by a range of old patterns. The intellectual interplay of this era set a powerful impetus for global conceptualization of the past against an almost equally powerful mixture of inertia and backlash that repeatedly stymied any coordination in study of world history. Much of this contest was played out in the classroom.

At the university level, there was no easy path for developing studies of world history. The history profession in U.S. universities is organized into national and (to a lesser degree) regional or thematic categories. Professional history is dominated by research: historians base hiring and promotion primarily on research monographs. World history fit uneasily into this structure. Few departments were ready to give up national or regional slots to define world history positions, so that virtually no promotions were based on research in world history. Instead, departments added world history teaching assignments to the jobs of national historians (though infrequently to those specializing in the United States). Most college teachers of world history to this day have never taken a course in world history, and certainly not a graduate course. For reasons such as these, it is difficult for national historians to accept world history as an intellectual enterprise equivalent to theirs.

In community colleges, which expanded dramatically in this era, appointments were less structured, and it was easier for administrators to open classes in world history when there was a demand. The teacher, of course, was self-trained, commonly a part-time employee teaching numerous courses at several institutions at very low pay. In those high schools where curricula were open to revision, courses in world history opened up in this era, and enthusiastic teachers trained themselves. The relative employment security of high school teachers enabled them to develop improved materials and course formats over the years.

Books surveying world history for general audiences appeared well before H. G. Wells published the *Outline of History*, and they continued long after. As popular interest in world history recurred, the British writer Hugh Thomas published a history of the world in 1979, and J. M. Roberts followed promptly with a history of the modern world.[57] Each of them followed a slightly broadened history of the West, with Roberts focusing principally on politics. But now, in addition, there began to appear textbooks in world history for college and high school students.

William McNeill's *A World History* (1967) was an early entry into this field.[58] It may be seen as a condensation of *Rise of the West*, with chapters on successive

57. Hugh Thomas, *A History of the World* (London, 1979); J. M. Roberts, *History of the World* (New York, 1976).
58. W. McNeill, *A World History* (New York, 1967; later editions 1971, 1979, 1999); W. McNeill, *History of the Human Community* (Englewood Cliffs, N.J., 1987); previously *Ecumene: Story of Humanity*.

regions of the world, divided into four sections at 500 C.E., 1500, and 1800. Leften Stavrianos followed in 1971 with *Man's Past and Present: A Global History*. Three chapters on man before civilization were followed by fourteen chapters on Eurasian civilizations to 500 C.E. Africa, the Americas, and Australia were introduced just before 1500. For Stavrianos, as for McNeill, the issue of Western dominance was the focus of world history after 1500; the narrative was regional to 1900 and global in the twentieth century. Stavrianos's organization and terminology changed substantially with time, so that by the time he published *Lifelines from our Past* in 1989, his narrative was organized around kinship societies, tributary societies, capitalist societies, and human prospects.[59]

Kevin Reilly, another early leader in the creation of world history textbooks, published a thematically organized text entitled *The West and the World*. Aimed especially at community college audiences, it was unusual in drawing materials together from widely different times and places to illustrate several main themes in world history.[60] In addition to publishing his own texts and collections of readings, Reilly coordinated an effort to collect and publish syllabi from a wide range of college courses and some high school courses.[61] These syllabi became well known, and were one of the principal means for conveying methods for teaching world history to those entering the field.

The textbooks of world historians evolved gradually. From an early focus on political history and great powers, they added information on trade and ethnicity, and they introduced cultural issues through surveys of major religious traditions. They highlighted the general notions of the ecumene and of expansion of Europe, and they began to identify more specific connections among regions. All of these texts and their authors worked, explicitly if incrementally, toward encouraging a global understanding of history.[62]

Gilbert Allardyce, in his elegant survey of the rise of the world history course, emphasized the priority given to teaching in this period: "The work of McNeill and Stavrianos in particular inspired the rise of the World History Association (WHA), formed by young historians in 1982 to take over the cause of world history from these men of the older generation. This organization wants historians to turn the leading scholarship of the men studied here into effective world history courses through the art of classroom teaching. *Their message, in short, is this: the way to make world history possible is to teach it.*"[63]

The judgment may be correct, but implementation of this approach gave priority to building teaching programs and put almost no systematic effort into

59. Leften Stavrianos, *Man's Past and Present: A Global History* (Englewood Cliffs, N.J., 1971); Stavrianos, *Lifelines from our Past: A New World History* (New York, 1989).
60. Kevin Reilly, *The West and the World: A Topical History of Civilization* (New York, 1980).
61. Kevin Reilly, ed., *World History: Selected Reading Lists and Course Outlines from American Colleges and Universities* (New York, 1985), with later editions.
62. These early leaders in world history, in addition to textbook writing, visited many schools, addressed teacher organizations, and wrote pedagogical articles.
63. Allardyce 1990: 25–26; emphasis added.

supporting new research. The very success of world history teaching made the field vulnerable to new pressures.

For now the publishers of Western Civilization texts, noting the growing sales of texts by McNeill, Stavrianos, and Reilly, entered the market for world history. The Western Civilization course, which in many cases spanned the time from classical antiquity to the twentieth century, seemed to provide a model that could apply to the teaching of world history. From the early 1980s, publishers began taking their established Western Civilization texts, adding sections on areas of the world beyond their traditional Mediterranean and Western European focus, and publishing them as histories of the world.[64]

In the Western Civ approach to world history, courses followed an inherited framework. Nations had histories, which were presented one after another. Textbooks generally clung to established patterns. While early textbooks in world history showed considerable originality, the move of major textbook publishers into world history stifled that trend.[65] Relying on the formulas and the actual language of their texts for Western Civilization, publishers imposed a view of the world that relied heavily on national and civilizational history, with a few comparisons. They treated European societies as "our heritage" and the rest of the world as "other cultures." The textbooks began each time period with a chapter on Europe and ended with a chapter combining Africa and Latin America; within each chapter, the text began with a political narrative and ended with some comments on culture. The message was clear in each case: Africa and Latin America absorbed the results of world history but made no impact themselves; cultural patterns were the results of politics and trade and possessed little significance. From the table of contents of such textbooks, one sees that the overall story still came down to the primacy of the city and the primacy of Europe.

"Teach first, study later" might just as well have been the slogan for this approach to world history. The rush to meet the immediate demand for courses in world history led implicitly to the assumption that world history required no new thinking, just assembly of the facts already known about the nations of the world. "World history" meant textbooks, introductory surveys, and very little more.[66]

World history, as a field dominated by teaching, remained thinly researched and weakly conceptualized. As part of this same wave of expansion in world history, a number of research monographs and original interpretations appeared

64. For a Western Civ-based approach to world history, see John P. McKay, Bennett D. Hill, and John Buckler, *A History of World Societies*, 3rd ed. (Boston, 1992).

65. Early textbooks by W. McNeill 1967; Stavrianos 1971; Reilly 1980; Kevin Reilly, ed., *Readings in World Civilizations*, 2 vols. (New York, 1988).

66. Textbook after textbook was proclaimed as "the first truly global" interpretation. My favorite was the text labeled as "the first truly global" text, for which the graphics on the same page showed five well-separated continents, each written up by a separate author. The authors are distinguished scholars in their fields; my point is that the publicists assumed that the world could be "truly" represented as an accumulation of discrete regions. Mark Kishlansky, Patrick Geary, Patricia O'Brien, and R. Bin Wong, with Roy Mottahedeh, Leroy Vail, Ann Waltner, Mark Wasserman, and James Gelvin, *Societies and Cultures in World History* (New York, 1995).

in print. Virtually all of these, however, were written by senior scholars who themselves had been trained in the history of one locality or another and not as world historians.[67] Universities offered introductory surveys but almost no upper-division or graduate courses in world history.[68]

There were of course critical evaluations of teaching in these years, and research in world history crept along through individual efforts if not through organized programs.[69] Yet for original approaches, it was sometimes better to cross the boundaries of the United States. Victor Julius Ngoh, trained in the United States but teaching and writing in his homeland of Cameroon, published a world history of the twentieth century that included a substantial emphasis on international organizations—surely one of the most important themes for world history in the late twentieth century, and one that was almost uniformly neglected in other textbooks.[70]

Conclusion: Impetus and Constraint

My assessment of world history from the 1960s through the 1980s, if positive on balance, is tinged unmistakably with ambivalence. Research and teaching in world history expanded substantially and changed significantly in this period, yet some fundamental weaknesses in both research and teaching remained to be addressed.

In research, the efforts of individual scholars brought a range of thematic monographs and analytical studies that were broader than national histories but more specific than planetary syntheses. Curtin, Wallerstein, and others explored politics and economics, trading systems, empire, and colonization, with growing attention to areas of the world beyond the North Atlantic. This was the historians' path to world history, and it focused on the creation and transformation of the modern world. It addressed migration but otherwise did not partake deeply of the current fascination with social history. A much smaller number of studies broke open the scientific-cultural path to world history, focusing on ecological and then on cultural history. With few exceptions, the research along both paths was conducted by senior scholars, not by junior scholars or graduate students.

The teaching of world history expanded dramatically from the 1960s through the 1980s, but it did not advance much. The content and analysis of world history

---

67. Curtin 1969; Wallerstein 1974; Crosby 1986.
68. Similar patterns can be observed for the early expansion of African American history and women's history in the United States, as these were also fields in which the teaching imperative sometimes obscured the need for research.
69. Ross E. Dunn, "The Challenge of Hemispheric History," *The History Teacher* 18 (1985), 329–338; Donald Johnson, "The American Educational Tradition: Hostile to a Humanistic World History?" *The History Teacher* 20 (1987), 519–544. The latter is reprinted in Ross E. Dunn, ed., *The New World History: A Teacher's Companion* (Boston, 2000), 329–49.
70. Victor Julius Ngoh, *The World Since 1919: A Short History* (Yaounde, 1989). For a more standard and less praiseworthy history of the twentieth century, see J. M. Roberts, *Twentieth Century: The History of the World, 1901 to 2000* (New York, 1999).

as taught and written in this period mainly recapitulated—for a broader audience—the lessons of world historians of the previous century. At best, the breadth of Toynbee and Wells, the comparative energies of Weber, the metaphorical skills of Spengler, the analytical insights of Marx, and the chronological synthesis of McNeill were now popularized and disseminated to students and general readers. But the inevitable need to simplify the vast terrain of world history led to courses and texts that focused primarily on having a few words to say about every major area of the world. For students previously trapped within national boundaries, encounters with previously mysterious areas of the world provided excitement. But when the thrill of discovery fades, will the experience have imparted any substantial learning about processes in the past? Can world history get beyond travelogue to convey logic and facilitate debate?

The promise and the dilemmas of world history were thus each clarified as the twentieth century neared its end, and as the global convulsion of democratization movements began in 1989 to bring millions to new levels of global awareness. World history, for better or worse, was now to become an academically and socially significant subject.

Chapter 5

Organizing a Field since 1990

In recent years the literature in world history has continued to grow and thrive. There is no shortage of exciting new contributions. World historians, in addition to publishing increasingly confident studies of global historical themes and patterns, have come to identify themselves as a group, and to establish institutions providing more solid support for teaching and research in their field. A more compelling indication of the changes came with decisions by academic and educational leaders in the United States—if not yet in other countries—to give formal recognition to the field of world history by choosing to require its instruction in many institutions at secondary and college levels.

In this survey of recent studies in world history, I begin by reviewing the changing institutions of world history: I balance the rapid creation of structures for research and teaching by world-history enthusiasts against the inertia of the academic system and the difficulty of locating resources to invest in world history. I turn then to reviewing a selection of the numerous publications in world history since 1990 in four categories: area-studies approaches to world history, thematic studies in world history, conceptual studies at the global level, and, in a concluding section, the linkages and comparisons of world history with the large literatures on U.S. and European history.[1]

Institutions

Teachers, professors, and other enthusiasts of world history moved rapidly to establish new structures for the field. Institutions established from 1990 included programs of graduate study, academic journals, electronic discussion lists and websites, conferences, stronger professional associations, legislation of statewide public education standards requiring world history, series of books and pamphlets,

---

1. Some of these same works, along with other recent works, are discussed in later chapters from two perspectives: their place in the methods of world history in part II, and their place in the interpretation of world history in part III.

teaching workshops, and the awarding of private and government grants for research and teaching materials. The new institutions were mostly precarious and sometimes fleeting, but they appeared in sum to suggest that world history would become established as an academic field of research and teaching.

Formal graduate study addressing issues in world history began as early as the 1960s and 1970s at the University of Wisconsin–Madison, under the leadership of Philip Curtin. Beginning in 1977, the doctoral program in sociology at Binghamton University trained historical sociologists in association with the Fernand Braudel Center. In the late 1980s, world history became an examination field or a teaching field for students in doctoral programs at the University of Hawaii, Ohio State University, Rutgers University, and, in history of the Early Modern World, at the University of Minnesota. In 1994, Northeastern University began to offer a Ph.D. in world history, and additional graduate programs in world history began at Georgia State University and at the University of California campuses at Riverside, Irvine, and Santa Cruz. Doctoral programs in Atlantic history opened at Florida International University and University of Texas–Arlington.[2]

Following the launching of graduate training, the founding of a scholarly journal was the next step. The *Journal of World History* first appeared in 1990, edited by Jerry Bentley and published by the University of Hawaii Press. A biennial journal, it won an award from the American Library Association as the best new academic journal for 1990.[3]

The World History Association held the first ten years of its annual meetings in association with the American Historical Association.[4] In June 1992 the WHA held its first annual conference, at Drexel University in Philadelphia. Two related developments were the beginning of a pattern of alternating WHA meetings in the United States and outside—beginning with the 1995 meeting in Florence and the 1997 meeting in Pamplona—and the growth in strength of regional associations of world historians affiliated with the WHA. Most of these regional associations were in the U.S., but some were beyond.[5]

---

2. See chapter 19 for a more detailed discussion of graduate programs in world history.
3. The *World History Bulletin*, a periodical publication of the World History Association, was created along with the formation of the association, and has included news, short articles, and teaching materials. Other historical journals with a transregional emphasis are *Comparative Studies in Society and History* and *Itinerario*.
4. Presidents of the World History Association, elected for two-year terms from 1984, were Kevin Reilly (1982–1984), Ross Dunn (1984–1986), Kevin Reilly (1986–1988), Arnold Schreier (1988–1990), Marilynn Jo Hitchens (1990–1992), Raymond Lorantas (1992–1994), John A. Mears (1994–1996), Judith P. Zinsser (1996–1998), Heidi Roupp (1998–2000), Carter Findley (2000–2002), and Ralph Croizrer (2002–2004).
5. WHA conferences were held as follows: Philadelphia (1992), Honolulu (1993), Aspen (October, 1994), Florence (1995), Pomona (1996), Pamplona (1997), Ft. Collins (1998), Victoria (1999), Boston (2000), Salt Lake City (2001), Seoul (2002). For further information on the WHA in the days of its formation and through the 1980s, see the *World History Bulletin*, which provides information on the Wingspread meetings sponsored by the Air Force Academy, the Cameroon trip of world historians, and the 1980s meetings at AHA conferences. Regional affiliates of the WHA are active for Australasia,

Meanwhile, pioneering work in the professional development of teachers opened up. In Princeton, the Woodrow Wilson Foundation provided support for summer institutes from 1991 through 1993, in which nationally selected groups of about fifty teachers read widely and exchanged ideas on interpreting and teaching world history. Many of the participants became prominent in world history affairs thereafter; one of them, Michele Forman of Vermont, became the U.S. national teacher of the year for 2001.[6] In the same period, Howard Spodek of Temple University led workshops during the school year for secondary and college teachers of world history in the Philadelphia area, and Tara Sethia of California Polytechnic State University in Pomona conducted an institute on South Asia in world history; both of these were supported by the National Endowment for the Humanities.

A debate on world history burst onto the U.S. national scene in the fall of 1994, though the discussion had been in preparation for years. The national administration of President George Bush in the late 1980s had called for the establishment of national standards (meaning guidelines for teaching but not strict requirements) in the major fields of public instruction: mathematics, science, English language arts, and history. For history, commissions were established to prepare standards in U.S. and world history, and the National Center for History in the Schools at UCLA, under the direction of Gary B. Nash, was established to coordinate the preparation of the standards under a grant from the National Endowment for the Humanities. In both cases, comprehensive discussions by widely representative groups of teachers and scholars had developed consensus documents. The national standards studies in history, meanwhile, were both preceded and supplemented by other reports: the 1988 Bradley Commission report on the teaching of history, the 1994 National Council of Social Studies report on the social studies curriculum, and a 1989 report by Charlotte Crabtree.[7]

With the September 1994 announcement of the reports on U.S. history and world history, Lynne V. Cheney, former director of the National Endowment for

Continued

    California, Canada and Northwestern United States, Europe, Mid-Atlantic, New England, Ohio, Rocky Mountains, Southeast United States, and Texas. Further details on the regional affiliates are available through the WHA website, www.thewha.org.

6. Ross Dunn directed the 1991 Wilson workshop, focusing on the world in the sixteenth century; Murdo McLeod directed the 1992 workshop, focusing on the seventeenth and eighteenth centuries; Dunn directed the 1993 workshop, focusing on the world in the nineteenth century. Four teachers were selected out of each group to give weekend institutes during the subsequent school year. Among those selected were Michele Forman in 1992, Jean Johnson, and Heidi Roupp.

7. National Council for History Education, *Building a World History Curriculum: Guides for Implementing the History Curriculum Recommended by the Bradley Commission on History in the Schools* (1988); National Council for Social Studies, *Expectations of Excellence: Curriculum Standards for Social Studies* (1994); Charlotte Crabtree, *Lessons from History* (Los Angeles, 1989). The principal author of the Bradley Commission report was Paul Gagnon.

the Humanities, led an active public campaign to discredit both reports, arguing that they were politically biased, undermining American patriotism and the Western heritage.[8] The debate raged for over six months: the U.S. Congress formally declined to approve the reports, and ultimately a revised set of national standards was prepared, including a number of cuts and concessions to the critics of world history. In the short run it appeared that world history had received a public rejection. Nevertheless, the National Center for History in the Schools continued its activities, turning to preparation of teaching units in world history.[9] And within another three years, quite a different picture had appeared about the place of world history in U.S. schools.

The 1990s were also the decade of the internet, and world historians began rapidly to rely on electronic communication. The first discussion list was the unmoderated World-L, based at a server at the University of Kansas and directed by Haines Brown of Central Connecticut State University. This list was supplemented in 1994 by the moderated H-WORLD list, edited initially by Patrick Manning and Daniel Segal. H-WORLD grew rapidly to 600 subscribers, to over 1,100 subscribers by the end of 1998, and to 1,500 subscribers by the end of 2000. Much of the 1994–1995 debate over the national standards took place or was reflected on H-WORLD. Websites in world history grew in the wake of the discussion lists: Haines Brown set up a Gateway to World History, and it was followed by websites for H-WORLD, for the World History Center at Northeastern, for various of the regional WHA affiliates, and by individuals.[10]

As readers' interest in world history grew, publishers began to create series of monographs and edited collections in world history, in addition to the textbooks and document readers they had begun to publish earlier. Westview, Markus Wiener, M. E. Sharpe, and Cambridge University Press had each established series in world history by 1995, and a larger number of publishers had launched series on globalization in other social sciences. On the other hand, many publishers and bookstores continued to use the rubric "World History" as a grab bag into which to deposit books addressing regions beyond the frontiers of European and U.S. history. Meanwhile, the first substantial book exhibit at a WHA conference took place in Boston in 2000.

The American Historical Association began to involve itself in world history from 1995 through publication of several series of pamphlets, through announcing a project for a CD-ROM on world history, and through an effort to coordinate exchange among area-studies historians. The CD-ROM project did not materialize,

8. National Center for History in the Schools, *National Standards for World History: Exploring Paths to the Present,* Grades 5–12, expanded edition (Los Angeles, 1994); Nash, Crabtree, and Dunn 1997.

9. National Center for History in the Schools (www.sscnet.ucla.edu/nchs). Teaching units are discussed in chapter 21.

10. World-L was discontinued in 1997; see H-WORLD (www.h-net.msu.edu/~world), the World History Center (www.worldhistorycenter.org), and the World History Association (www.thewha.org).

but the series of pamphlets became widely used, and was followed by an AHA-supported project of publishing volumes on the history of women in Africa, Latin America, the Middle East, and Asia.[11]

An effort to connect historians and professional organizations in the study of world history was launched by the American Historical Association. Initial efforts to arrange cooperation among the AHA and area-studies associations did not bear fruit, but the initiative, known for a time as Globalizing Regional Histories, resulted in two successful workshops for community college teachers and a conference convening twenty-seven scholars to present research on interactions in history.[12]

Meanwhile, from the time of the national standards debate, legislatures and departments of education in several major states had been at work on preparing new curricula mandating world history in the public schools. Despite the public controversy about the national standards, world history benefited from the broader trend toward standards-based education in the public schools, which called for establishment of formal, statewide outlines, sometimes accompanied by state testing. State legislatures and departments of education drew eclectically on the reports of the national standards, the Bradley Report, and the NCSS report to develop their requirements. Thus, rather quietly, by the late 1990s formal requirements for one to four years of world history had been established in states including the great majority of the U.S. population. While the term *world history* had been used in U.S. public schools for a century, the meaning now came to be different: these were to be courses of study addressing every continent, thousands of years of time, and a wide range of issues. The states of particular prominence in this transformation were California, Texas, Virginia, New York, and

11. The AHA pamphlet series includes "Essays in Global and Comparative History" (ed. Michael Adas), "Women's and Gender History in Global Perspective" (ed. Bonnie Smith), "Essays on the Columbian Encounter" (eds. Carla Rahn Phillips and David J. Weber), and "Historical Perspectives on Technology, Society and Culture" (eds. Robert C. Post and Pamela O. Long).

12. Renate Bridenthal of Brooklyn College and Patrick Manning of Northeastern University, chair and co-chair of the American Historical Association program committee for the 1996 annual meeting, met in 1994 with James Gardner, interim executive director of the AHA, and thereafter with executive director Sandria Freitag, to propose a campaign of discussions between AHA and the area-studies associations in an attempt to organize joint sponsorship of a series of panels. While discussion with area-studies associations did not lead directly to action, a tentative call for papers led to a significant response. Thereafter, AHA successfully applied to Ford Foundation for grants to conduct summer workshops on world history for community college teachers, under the direction of Jerry Bentley, and conferences on "Interactions" in March 2001 and on "Seascapes" in February 2003, co-chaired by Renate Bridenthal and Jerry Bentley. In the course of these developments the AHA and WHA had to work out their relations: WHA as a fledgling group, and AHA moving to express interest in world history. S. Tune to P. Manning, 22 February 1994; J. Gardner to area-studies association executive directors, 3 May 1994; R. Bridenthal to P. Manning, 19 May 1994; P. Manning to J. Gardner, 28 May 1994. Letters held in files of the World History Center.

Massachusetts.[13] There was to be new curriculum for students, new topics to learn and lessons to plan for teachers, and new textbooks for publishers to prepare and sell.

At a somewhat slower pace, the College Board long considered and finally acted on a plan to create an AP World History course. This advanced placement course, for which students would take a national exam in May, was to join the established AP courses on U.S. and European history. The College Board itself postponed action for several years and then approved the course design in April 1999, with the first exam given in May 2002. The course design adopted by the Development Committee included a five-week Foundations section for background, and thirty-one weeks to cover the period from 1000 C.E. to the present.[14]

As course materials and exam questions were prepared, a controversy sprung up from 1999 to 2001, in which critics of the AP World History course argued that it failed to lead students through "proper historical analysis" before the year 1000 and was therefore a course in modern history that left out the essential grounding in earlier times. The designers and supporters of the course argued that restricting the time period made it easier to present this daunting course to high school students, and that a global approach to history could be conveyed for any time period. After a series of meetings of representatives of the contending groups in late 2000 and early 2001, it was agreed that the course would be unchanged for its first two years (with exams in 2002 and 2003), and that thereafter the time frame of the course would lengthen, and an additional week would be added to the early period.[15] With this, the first round of debates over the AP World History course subsided.

13. For a survey of the standards in each state, including an excellent overview of their development, see Susan Douglass, *Teaching About Religion in National and State Social Studies Standards* (Fountain Valley, CA, 2000). It is published online by the Council on Islamic Education at www.cie.org/pdffiles/CIE_Report.pdf. The author has prepared an update, as yet unpublished, of changes in state standards to 2003.

14. Lawrence Beaber and Despina Danos of the Educational Testing Service, an affiliate of the College Board, oversaw development of the plan along with a committee of teachers and professors chaired by Peter Stearns, editor of the *Journal of Social History.* The College Board had considered creating such a course for years but delayed repeatedly, then gave its approval in 1999; when the course was first given, over 20,000 students took the exam in May 2002. I served as a member of the AP World History Development Committee from 1999 to 2002. For a cogent statement of a practical educational philosophy that was influential in structuring the AP World History course, see Peter N. Stearns, *Meaning over Memory: Recasting the Teaching of Culture and History* (Chapel Hill, 1993).

15. Most vocal among this group was Jerry Bentley, who led in obtaining passage of a resolution by the World History Association Executive Council condemning the course in its current form. Underlying this dispute lay a wider range of issues. First, in the complex structure of the College Board (based in New York) and Educational Testing Service (its affiliate, based in Princeton), world historians Bentley, Michele Forman, and Maghan Keita had served for years on the College Board Advisory Committee on History, whose repeated calls for approval of the AP World History by the College Board had been ignored and reversed by the College Board leadership. Once the course

Many other institutional changes took place in the same period. The World History Association instituted a World History Book Prize in 1999, awarded each year for the best study in world history appearing in the previous year.[16] In addition a review section on "Global and Comparative" books appeared in the *American Historical Review*; teaching and other positions in world history began to be listed in the AHA *Perspectives* and online at the H-Net job register; and the New England Regional World History Association held a series of electronic conferences.[17] Projects with a longer time frame included two nationally funded curriculum projects led by Ross Dunn; nationally funded projects led by WHA President Heidi Roupp for teacher preparation and for summer teaching institutes.[18]

In sum, the institutions created during the 1990s provided a dramatic advance in the professional organization of and support for world historians. At the same time, the new institutions were limited primarily to coordinating the efforts of individual scholars and teachers of world history. Absent was any substantial effort by government, foundations, or universities to provide support for research in world history. The WHA, with the potential of expanding to a membership similar to that of the AHA or the National Council of Social Studies, and with the potential of becoming an international organization, remained stalled at a membership of 1,400, and its move after 2000 to establish a headquarters was slowed by tight finances. Heidi Roupp, in her term as president of WHA, sought energetically to contact major foundations and gained significant support from the National Endowment for the Humanities. But world historians had no regular connections to such policy-setting bodies as the American Council of Learned Societies or the Social Science Research Council.

Continued

    was approved in 1999 and the Development Committee (working with the Educational Testing Service) formalized, the Advisory Board was never put in direct contact with the Development Committee. Second, the strong expressions of interest by teachers, parents, and students in the new course meant that it would be the largest new AP course ever launched, and its importance raised tensions over its precise organization. Third, the course raised the question of the links between secondary and college-level teaching of world history survey courses—whether there was to be a common standard for the two.

16. WHA Book Prize awardees: for 1999, Andre Gunder Frank, *ReOrient: Global Economy in the Asian Age* (Berkeley, 1998); for 2000, James E. McClellan III and Harold Dorn, *Science and Technology in World History: An Introduction* (Baltimore, 1999); for 2001, awards both to J. McNeill, 2000 and Kenneth Pomeranz, *The Great Divergence: China, Europe, and the Making of the Modern World Economy* (Princeton, 2000).

17. Transcripts online at www.worldhistorycenter.org/nerwha.

18. The National Endowment for the Humanities supported all of these projects except the Woodrow Wilson Foundation teaching institutes of 1991–1993. NEH provided, among others, awards in world history to Temple University for teacher institutes, to Northeastern University for a collaborative college-level project on teaching, to the World History Association for teaching institutes, to Northeastern for a comprehensive world history website, to George Mason University for a website on documents in world history, and to San Diego State University for a web-based world history curriculum.

Area-Studies Approaches

Scholars trained in area-studies history—the history of regions outside Western Europe and North America—became increasingly active contributors to the literature on world history after 1990. Most of this work followed the historians' path to world history, exploring issues in political, commercial, and civilizational studies and aspects of social history. Some of the area-studies work, however, took the scientific-cultural path, bringing innovations in linguistics, anthropology, and ecological studies to world history.[19]

Area-studies historians have used several tactics for addressing world history. Lynda Shaffer has taken a case-study approach, having written successive studies on China, Southeast Asia, and the Mississippi Valley and, with work in progress on West Africa, emphasizing world-historical connections for each region. Taking a critical approach, a mix of Africanists and Latin Americanists contributed to a volume entitled *Confronting Historical Paradigms*, in which they explored linkages and commonalities among their regions in the nineteenth and twentieth centuries, to challenge the restrictions of area-studies paradigms. John Wills, in the synthetic approach of his global snapshot of the world in 1688, leads readers on a visit to region after region of the world, in each case noting the connections among them and the way those connections highlight the similarities and contrasts among regions.[20]

More commonly, the work of area-studies scholars has focused on a single region, but in so doing has elucidated global dynamics. S. A. M. Adshead wrote volumes showing global connections both for China and for Central Asia. Stanley Burstein relied on his strength as a historian of the Eastern Mediterranean in classical times to show connections with the Red Sea, the Horn of Africa, and India. Richard Hovannisian placed Armenian communities in world-historical perspective, and Ben Finney proposed an outline for the history of the Pacific.[21]

The strongest connection between area studies and global approaches to history in the 1990s came in the study of East Asia. Scholars trained in Chinese

19. Studies of the U.S. and Western Europe, having developed earlier, are not organized into cross-disciplinary programs. The historical literature on the United States and Europe is addressed in a later section of this chapter.

20. Lynda N. Shaffer, *Mao and the Workers: The Hunan Labor Movement, 1920–1923* (New York, 1982); Shaffer, *Native Americans before 1492: The Moundbuilding Centers of the Eastern Woodlands* (Armonk, N.Y., 1992); Shaffer, *Maritime Southeast Asia to 1500* (Armonk, N.Y., 1996); Cooper et al. 1993; John E. Wills, Jr., *1688: A Global History* (New York, 2001).

21. S. A. M. Adshead, *Central Asia in World History* (New York, 1993); Adshead, *Material Culture in Europe and China, 1400–1800* (New York, 1997); Stanley Burstein, *Graeco-Africana: Studies in the History of Greek Relations with Egypt and Nubia* (New Rochelle, 1995); Burstein, ed., *Ancient African Civilizations: Kush and Axum* (Princeton, 1998); Ben Finney with Marlene Among, *Voyage of Rediscovery: A Cultural Odyssey through Polynesia* (Berkeley, 1994); Finney, *Hokulea: The Way to Tahiti* (New York, 1979); Richard G. Hovannisian, *The Armenian People from Ancient to Modern Times*, 2 vols. (New York, 1997).

history, partly in response to the intervention of globalists, began to reinterpret Chinese dynasties as rulers of empires comparable to other empires.[22] Demographic and economic studies of the Chinese economy in early modern times showed it to be dynamic, innovative, growing, and well connected to other regions. R. Bin Wong and Kenneth Pomeranz published comparative studies focusing on China and Europe, each gaining wide discussion. James Lee's work on Chinese demography, while not comparative beyond China, provided a basis for interregional comparison; Richard Von Glahn's long-term analysis of money similarly provided a basis for global monetary comparisons. Sucheta Mazumdar's study of sugar production in south China gave emphasis to the strength and flexibility of smallholder agriculture, while Robert Marks's study of the environment of the same region focused instead on the strength of market relations. The 1990s saw continuing publication of works on Chinese science and technology initiated by Joseph Needham, and scholars in many fields relied on G. William Skinner. The global impact of historical research on Japan during the 1990s was less than that for China, but Akira Iriye balanced the two nations nicely in a study of international politics.[23] Alongside these advances in research, developments in outreach and teaching resources disseminated the new scholarship widely.[24]

Global studies of South Asia focused in part on elaborating interpretations of the Indian Ocean region and in part on the growing ties of South Asia to Britain. In the focus on the regional economy, outstanding studies were Sanjay Subrahmanyam's analyses of Indian Ocean politics and trade, R. J. Barendse's detailed analysis of trade in the western Indian Ocean, and Patricia Risso's study of merchants and religion. Parallel to these studies, but emphasizing links on the mainland, Richard Foltz published two analyses of links between India and

22. Peter Perdue, *Exhausting the Earth: State and Peasant in Hunan, 1500–1850* (Cambridge, Mass., 1987).

23. R. Bin Wong, *China Transformed: Historical Change and the Limits of European Experience* (Ithaca, 1997); Pomeranz 2000; James Z. Lee and Wang Feng, *One Quarter of Humanity: Malthusian Mythology and Chinese Realities, 1700–2000* (Cambridge, Mass., 1999); Flynn and Giráldez 1995; Richard Von Glahn, *Fountain of Fortune: Money and Monetary Policy in China 1000–1700* (Berkeley, 1996); Frank 1998; Sucheta Mazumdar, *Sugar and Society in China: Peasants, Technology and the World Market* (Cambridge, Mass., 1998); Robert Marks, *Tigers, Rice, Silk, and Silt: Environment and Economy in Late Imperial South China* (Cambridge, 1998); Joseph Needham, Robin D. S. Yates, Krzysztof Gawlikowsky, Edward McEwen, and Wang Ling, *Science and Civilisation in China. Volume 5, part 6, section 30. Military Technology: Missiles and Sieges* (Cambridge, 1994); G. William Skinner, *Marketing and Social Structure in China* (Tucson, 1964–65); Akira Iriye, *China and Japan in the Global Setting* (Cambridge, Mass., 1992); Tessa Morris-Suzuki, *Reinventing Japan: Time, Space Nation* (Armonk, N.Y., 1998).

24. Ainslie T. Embree and Carol Gluck, eds., *Asia in Western and World History: A Guide for Teaching* (Armonk, N.Y., 1997). In addition, H-Asia (www.h-net.msu.edu/~asia) is a very active discussion list, and *Education About Asia* provides regular information on new teaching materials. The Freeman Foundation, based in Vermont, provides substantial funding for teaching and research on Asia, especially East Asia.

Central Asia. Among those emphasizing comparison and linkage to Britain, Prasannan Parthasarathi published studies arguing that Indian textile manufacturers of the eighteenth century not only produced cotton textiles at lower prices than did British mills, but they also equaled Britain in productivity and in the level of wages paid to workers. Christopher Bayly encompassed these and other perspectives in an analysis of British empire in South Asia. A very different sort of global analysis arose with subaltern studies, a critical analysis of the colonial situation focusing especially on Bengal. Ranajit Guha's formative labors in assembling subaltern studies laid the groundwork for Dipesh Chakrabarty's recent restatement of Europe's place in history and intellectually life.[25]

In the same era but with less public visibility, Central Asia was the topic of the most outstanding work during the 1990s in linking a region to world history. The independence of Soviet Central Asia in 1992 and the long series of wars in Afghanistan drew political attention to the region and may have stimulated some of this work. Central Asia, which appeared as a backwater for much of the twentieth century, had nonetheless played a key role in linking Eurasian regions over the centuries, and a remarkable series of books appearing in the 1990s made this point in several fashions. The work of multidisciplinary teams headed by Victor Mair traced the movements and activities of Indo-European speakers in the region several thousand years ago. Xinru Liu documented connections of trade and religion along the silk road, Jonathan Lipman analyzed the place of Muslims under Chinese rule, and Andre Gunder Frank underscored the "centrality" of Central Asia.[26]

25. K. N. Chaudhuri, *Asia Before Europe: Economy and Civilisation of the Indian Ocean from the Rise of Islam to 1750* (Cambridge, 1990); Sanjay Subrahmanyam, *The Political Economy of Commerce. Southern India 1500–1650* (Cambridge, 1990); R. J. Barendse, *The Arabian Seas: The Indian Ocean World of the Seventeenth Century* (Armonk, N.Y., 2002); Barendse, "Trade and State in the Arabian Seas: A Survey from the Fifteenth to the Eighteenth Century," *Journal of World History* 11 (2000), 173–226; Patricia Risso, *Merchants and Faith: Muslim Commerce and Culture in the Indian Ocean* (Boulder, 1995); Richard Foltz, *Mughal India and Central Asia* (Karachi, 1998); Prasannan Parthasarathi, *The Transition to a Colonial Economy: Weavers, Merchants and Kings in South India, 1720–1800* (New York, 2001); Parthasarathi 1988; Mrinalini Sinha, *Colonial Masculinity: the "Manly" Englishman and "Effeminate" Bengali in Nineteenth Century India* (New York, 1995); Christopher A. Bayly, *Imperial Meridian: The British Empire and the World, 1780–1830* (London, 1989); Bayly, *Indian Society and the Making of the British Empire* (Cambridge, 1988); Bayly, *Empire and Information: Intelligence Gathering and Social Communication in India, 1780–1870* (Cambridge, 1996); Ranajit Guha and Gayatri Chakravorty Spivak, eds., *Selected Subaltern Studies* (New York, 1988); Ranajit Guha, *Dominance without Hegemony: History and Power in Colonial India* (Cambridge, Mass., 1998); Dipesh Chakrabarty, *Provincializing Europe* (Princeton, 2000).
26. For times before the Common Era, see Victor Mair, ed., *The Bronze Age and Early Iron Age Peoples of Eastern Central Asia*, 2 vols. (Philadelphia, 1998); and Julian Baldick, *Animal and Shaman: Ancient Religions of Central Asia* (New York, 2000). For surveys into the second millennium C.E., see David Christian, *Inner Eurasia from Prehistory to*

For Southeast Asia a sizable literature on global connections developed as well. Of these contributions, Anthony Reid's substantial volumes on trade and politics confirm ties of that region to East Asia, the Indian Ocean, and the Atlantic in the early modern period.[27] Studies on the Middle East in global perspective, while they continued to appear, were less prominent in this decade than in previous times.[28]

The Pacific region, far more vast in extent than the Indian Ocean and the Atlantic basin (not to mention the Mediterranean), is just as obviously a possible unit for historical analysis. Recent studies in Pacific perspective have included analyses of early migrations, the voyages of Captain Cook and their aftermath,

Continued

the Mongol Empire, vol. 1 of *A History of Russia, Central Asia and Mongolia* (Oxford, 1998); Richard N. Frye, *The Heritage of Central Asia: From Antiquity to the Turkish Expansion* (Princeton,1996); and M. S. Asimov and C. E. Bosworth, eds., *History of Civilizations of Central Asia*, Volume IV, *The Age of Achievement: A.D. 750 to the End of the Fifteenth Century. Part One: The Historical, Social and Economic Setting* (Paris, 1998). On trade, religion, and migration before 1200 C.E., see Xinru Liu, *Ancient India and Ancient China: Trade and Religious Exchanges, AD 1–600* (Oxford, 1995); Liu, "Silks and Religions in Eurasia, c. A.D. 600–1200," *Journal of World History* 6 (1995), 25–48; Liu, "Migration and Settlement of the Yuezhi-Kushan: Interaction and Interdependence of Nomadic and Sedentary Societies," *Journal of World History* 12 (2001), 261–292; and Richard C. Foltz, *Religions of the Silk Road: Overland Trade and Cultural Exchange from Antiquity to the Fifteenth Century* (New York, 1999). On Tibetan empire, see Christopher I. Beckwith, *The Tibetan Empire in Central Asia: A History of the Struggle for Great Power among Tibetans, Turks, Arabs and Chinese during the Early Middle Ages* (Princeton, 1987). For research on more recent times, see Steven G. Marks, *Road to Power: The Trans-Siberian Railroad and the Colonization of Asian Russia, 1850–1917* (Ithaca, 1991); Jonathan Lipman, *Familiar Strangers: A History of Muslims in Northwest China* (Seattle, 1997); Gorm Pederson and Ida Nicolaisen. *Afghan Nomads in Transition: A Century of Change Among the Zala Khan Khel* (New York, 1995); Korkut A. Erturk, ed., *Rethinking Central Asia: Non-Eurocentric Studies in History, Social Structure and Identity* (Ithaca, 1999); Feride Acar and Ayse Gunes-Ayata, eds., *Gender and Identity Construction: Women of Central Asia, the Caucasus and Turkey* (Leiden, 2000). For the best general survey, see Adshead 1993. The widely respected works of Joseph Fletcher have now been published together in Joseph Fletcher (ed. Beatrice Forbes Manz), *Studies on Chinese and Islamic Inner Asia* (Aldershot, U.K., 1995). See also Andre Gunder Frank, *The Centrality of Central Asia* (Amsterdam, 1992).

27. Anthony Reid, *Southeast Asia in the Age of Commerce, 1450–1680*, 2 vols. (New Haven, 1988–1993); Anthony Reid, ed., *Southeast Asia in the Early Modern Era: Trade, Power, and Belief* (New York, 1993); Shaffer 1996.

28. John Voll, "Islam as a Special World-system," *Journal of World History* 5 (1994), 213–226; Richard Eaton, *Essays on Islam and Indian History* (New Delhi, 2000); Eaton, *The Rise of Islam and the Bengal Frontier, 1204–1760* (Berkeley, 1993); Eaton, "Islamic History as Global History" (Washington, 1990). For a regional study of religion with global implications, see Mary Boyce, *Zoroastrianism: Its Antiquity and Constant Vigour* (Costa Mesa, CA, 1992).

cultural and environmental change in recent centuries, and the invention of the Pacific Rim.[29]

Historians of Africa added several distinctive works within area-studies perspective that brought implications for global studies. Christopher Ehret's interpretation of East Africa, relying predominantly on linguistic data, linked the societies of this region to global patterns in technological change and to the expanding commercial networks of the "Classical" age, from 500 B.C.E. to 500 C.E. George Brooks developed a broad overview of West African environmental history in the early second millennium C.E.[30] In an explicit linkage of local and global, Donald Wright traced the transformation of a Gambian locality over three centuries of global connection. Paul Lovejoy and Jan Hogendorn documented the immense system of slavery inherited by British Northern Nigeria from its conquest of the Sokoto Caliphate. Iris Berger and E. Frances White reviewed the history of women in Africa.[31]

The study of Latin America, while curiously neglected in most treatments of world history, nonetheless produced a series of significant transregional studies during the 1990s. In social history of the early Spanish colonial era, Ida Altman traced a thread of migrants from a Spanish to a Mexican town and observed how social structure changed in the process. Jorge Cañizares-Esguerra demonstrated the significance of intellectual history for the Spanish colonies, tracing the development of what he called a "patriotic epistemology" among intellectuals in Mexico over the course of complex debates on the interpretation of historical sources written by Amerindians. Luiz Felipe de Alencastro interpreted the formation of Brazilian society in terms of its interaction with Angola. In addressing the early national period, Jeremy Adelman traced the creation of the Argentine nation in Atlantic context, focusing especially on changing understandings of law.

---

29. John R. McNeill, "Of Rats and Men: A Synoptic Environmental History of the Island Pacific," *Journal of World History* 5 (1994), 299–350; Gananath Obeyesekere, *The Apotheosis of Captain Cook: European Mythmaking in the Pacific* (Princeton, 1992); David Chappell, *Double Ghosts: Oceanian Voyagers on Euroamerican Ships* (Armonk, N.Y., 1997); Eric Jones, Lionel Frost, and Colin White, *Coming Full Circle: An Economic History of the Pacific Rim* (Boulder, 1993); Arif Dirlik, ed., *What is a Rim? Critical Perspectives on the Pacific Region Idea* (Boulder, 1993).

30. Christopher Ehret, *An African Classical Age: Eastern and Southern Africa in World History, 1000 B.C. to 400 A.D.* (Charlottesville, 1998); see also Bernd Heine and Derek Nurse, *African Languages: An Introduction* (Cambridge, 2000). George Brooks, *Landlords and Strangers: Ecology, Society, and Trade in Western Africa, 1000–1630* (Boulder, 1993); James McCann, *Green Land, Brown Land, Black Land: An Environmental History of Africa, 1800–1990* (Portsmouth, N.H., 1999).

31. Donald R. Wright, *The World and a Very Small Place in Africa* (Armonk, N.Y., 1997); Paul E. Lovejoy and Jan S. Hogendorn, *Slow Death for Slavery: The Course of Abolition in Northern Nigeria, 1897–1936* (Cambridge, 1993); Frederick Cooper, *Decolonization and African Society: The Labor Question in French and British Africa* (Cambridge, 1996); Iris Berger and E. Frances White, *Women in Africa: Restoring Women to History* (Bloomington, 1999). On the African diaspora, see Judith Carney, *Black Rice: The African Origins of Rice Cultivation in the Americas* (Cambridge, 2001).

Patricia Seed compared British and Spanish America, emphasizing the contrasting understandings of the importance of labor and land in the two imperial cultures, and traced the implications of those differences. John Russell-Wood surveyed the Portuguese empire worldwide, working from his specialization in Brazil.[32]

Russia and Eastern Europe, having undergone immense political turmoil at the turn of the 1990s, remained largely mired in exceptionalist scholarly interpretations. As with the Middle East, a legacy of long-term political struggle, for all the debate it creates, need not lead to creative and wide-ranging historical studies. Robert Strayer, however, sought to present a global view of the collapse of the Soviet Union aimed at a general audience.[33]

Thematic Approaches

Perhaps the strongest statement of the arrival of thematic approaches in world history was the massive effort of Variorum Publishers in producing thirty-one edited volumes on the period from 1500 to 1800 under the general title, "An Expanding World." These volumes, centered thematically on politics, economy, society, religion, technology, and health in early modern world history, drew monographic articles into a global framework. Produced through the energetic work of A. J. R. Russell-Wood, they demonstrated the range of scholarly studies relevant to world-historical issues in the early modern period.[34]

Thematic studies in world history, which in previous decades had contributed powerfully to making world history a practical subject of study, developed further in the 1990s. The themes of global political economy and ecology retained their earlier strength, for instance in the work of Janet Abu-Lughod and K. N. Chaudhuri. Abu-Lughod applied, to Eurasia and the Indian Ocean in the

32. Ida Altman, *Transatlantic Ties in the Spanish Empire: Brihuega, Spain and Puebla, Mexico, 1560–1620* (Stanford, 2000); Cañizares-Esguerra 2001; Cañizares-Esguerra, "New World, New Stars: Patriotic Astrology and the Invention of Indian and Creole Bodies in Colonial Spanish America, 1600–1650," *American Historical Review* 104 (1999), 33–68; Luiz Felipe de Alencastro, *O Trato dos viventes: formação do Brasil no Atlântico sul* (São Paulo, 2000); Jeremy Adelman, *Republic of Capital: Buenos Aires and the Legal Transformation of the Atlantic World* (Stanford, 1999); Patricia Seed, *American Pentimento: The Invention of Indians and the Pursuit of Riches* (Minneapolis, 2001); A. J. R. Russell-Wood, *A World on the Move: The Portuguese in Africa, Asia, and America, 1415–1808* (New York, 1993); Cooper et al. 1993.

33. Robert Strayer, *Why did the Soviet Union Collapse? Understanding Historical Change* (Armonk, N.Y., 1998); Andre Znamenski, *Shamanism and Christianity—Native Encounters with Russian Orthodox Missions, 1820–1917* (Westport, Conn., 1999); Daniel R. Brower and Edward J. Lazzerini, eds., *Russia's Orient: Imperial Borderlands and Peoples, 1700–1917* (Bloomington, Ind., 1997).

34. A. J. R. Russell-Wood, general editor, *An Expanding World: The European Impact on World History, 1450–1800,* 31 vols. (Aldershot, U.K., 1995–2000). The introductions to these volumes, with a few exceptions, provide excellent historiographical reviews.

fourteenth century, the logic of the world-system developed for the Atlantic world after 1500, and thereby launched widespread interest in world-historical analysis over the longer term. Chaudhuri, whose earlier work centered on the economic history of the Indian Ocean, now made explicit a paradigm drawn from the Annaliste tradition of Fernand Braudel. In *Asia before Europe*, Chaudhuri set forth a global interpretation, modeled in terms of set theory, in which the structures of everyday life are seen to link regions from East Africa to Japan, and in which the slow transformation of these structures sets far into the background the more short-term narratives of monarchies and empires.[35]

Studies in many other thematic areas appeared during the 1990s, addressing such issues as gender, technology, cycles and civilizational growth. Jerry Bentley's *Old World Encounters* offered an interpretation of cross-cultural connections from the first millennium B.C.E. to the middle of the second millennium C.E. In this analysis Bentley characterized types of exchanges among societies, and traced the transformations in cross-cultural connections over time.[36] This accessible work provided a survey of world history before 1500 that acknowledged the familiar list of civilizations but set them into a new and broader narrative by emphasizing their connections.

Andre Gunder Frank's interest in long-term cycles led him, with Barry Gills, to assemble an edited collection pushing back the temporal frontiers of the world systems framework: the editors proposed that a single world system has expanded cyclically for five thousand years. Discussion of this volume had the effect of adding additional emphasis to the importance of early times in world history. In a complementary volume edited by Stephen Sanderson, authors argued the relative advantages of world system and civilizational approaches to ancient history.[37]

For modern times, the theme of empire gained attention in several studies. Bin Wong led in comparing the Qing empire of China with European empires, and David Abernethy compared European empires to each other, while Patricia Seed traced the long-term implications of differing Spanish and British approaches to empire in the Americas. Lauren Benton brought the field of legal studies formally into world history through a set of case studies exploring the complexities of law in the overlapping cultures bridged by early modern empires. In intellectual history, David Armitage and Anthony Pagden conducted studies of imperial ideologies. Frederick Cooper and Ann Laura Stoler produced a collection of studies on social tensions within empires of the nineteenth and twentieth centuries. For the late twentieth century, analyses of international and

---

35. Abu-Lughod 1989; K. Chaudhuri 1990.
36. Jerry H. Bentley, *Old World Encounters: Cross-Cultural Contacts and Exchanges in Pre-Modern Times* (New York, 1993). For a later statement of the resultant periodization of world history, see Bentley, "Cross-Cultural Interaction and Periodization in World History," *American Historical Review* 101 (1996), 749–70.
37. Andre Gunder Frank and Barry K. Gills, eds., *The World-system: Five Hundred Years or Five Thousand?* (London, 1993); Sanderson 1995.

inter-imperial relations came from Noam Chomsky, Barry Buzan, and Akira Iriye.[38]

While historians grew in the significance of their contribution to the world history literature, scholars based in other disciplines continued to make important contributions. The continuing work in civilizational studies, coordinated by the International Society for the Comparative Study of Civilizations, brought together political scientists, sociologists, and historians. In sociology, in addition to Wallerstein, Christopher Chase-Dunn, and Thomas Hall led in developing long-term and comparative analysis, focusing on developments, pulsations, and transformations in civilizational systems.[39]

James Tracy's two edited volumes on early modern merchants pulled together materials from most regions of the world. In modern times, Giovanni Arrighi's long twentieth century provided an interpretation of political economy from the fourteenth century forward, and Peter Stearns provided a two-stage interpretation of industrialization at the global level. Wallerstein and his associates worked on commodity chains, while S. A. M. Adshead and others conducted studies of individual commodities in world history.[40]

Linkage of materials from East Asia, South Asia, Europe, and the Americas led to major debate on the early modern world economy. Andre Gunder Frank, in *ReOrient*, offered a summary and generalization of this work and thereby presented a forceful challenge to the previous orthodoxy, according to which Europe had been the economic center of global power from the sixteenth century. The result of this line of argument was to draw increased attention to the turn of the nineteenth century as the time at which the European economy would, in this

---

38. Jack A. Goldstone, *Revolution and Rebellion in the Early Modern World* (Berkeley, 1991); Wong 1997); Abernethy 2000; Seed 2001; Lauren Benton, *Law and Colonial Cultures: Legal Regimes in World History, 1400–1900* (New York, 2002); Anthony Pagden, *Lords of All the World: Ideologies of Empire in Spain, Britain and France, c. 1500–c. 1800* (New Haven, 1995); David Armitage, *The Ideological Origins of the British Empire* (Cambridge, 2000); Frederick Cooper and Ann Laura Stoler, eds., *Tensions of Empire: Colonial Cultures in a Bourgeois World* (Berkeley, 1997); Noam Chomsky, *World Orders Old and New* (New York, 1994); Barry Buzan, *People, States, and Fear* (New York, 1991); Akira Iriye, *Cultural Internationalism and World Order* (Baltimore, 1997).

39. Alan K. Smith returned to the issue most scrutinized in recent studies of world history, the creation of the modern world community, with a revised paradigm and new detail. In *Creating a World Economy*, he utilized a paradigm modified slightly from that of Wallerstein: he chose to distinguish peripheral regions of the world economy from dependencies of the European powers. Perhaps more important than the paradigmatic distinction was an empirical distinction: in the balance of European and overseas dimensions of the tale, Smith went further than Wallerstein in emphasizing events in overseas areas, and the particular characteristics of European colonies. Alan K. Smith, *Creating a World Economy: Merchant Capital, Colonialism, and World Trade, 1400–1825* (Boulder, 1991).

40. James D. Tracy, ed., *The Rise of Merchant Empires: Long-distance Trade in the Early Modern World, 1350–1750* (Cambridge, 1990); Tracy, ed., *The Political Economy of Merchant Empires: State Power and World Trade, 1350–1750* (Cambridge, 1991);

view, have finally surpassed that of East Asia in productivity. David Landes published, also in 1998, a bestselling reaffirmation of the vision of European global hegemony from the time of the voyages of discovery. Proponents of the two viewpoints debated over the internet and in person during the course of the year.[41] Overall, this discussion probably engaged a wider range of regions and interactions than any previous historical debate.

Migration was the aspect of global social history that got the most attention in the 1990s. Robin Cohen, a sociologist by training, edited an eclectic set of studies in the *Cambridge Survey of World Migration* and then wrote an imaginative and original survey of diasporas. Another sociologist, Thomas Sowell, wrote a broad survey of migration which, while drawing on a wide and representative literature, focused heavily on the history of migration to the United States. Wang Gungwu, working from a base in Southeast Asia, edited a collection of studies of migration that contributed more effectively to a global vision. David Northrup added a solid volume to the growing literature on the migration of indentured workers, mostly Chinese and Indian, in the late nineteenth century. Adam McKeown's analysis of the Chinese diaspora to the Pacific and the United States developed the concept of network, providing a critique of migration studies focused on destinations or on origins. My own contributions to thematic studies in migration included an analysis of the demography of slavery and slave trade combining Africa, the Americas, and the basins of the Mediterranean and Indian Ocean; and participation in an interpretation of modern world history through the issue of migration, in the form of a CD-ROM.[42]

Continued

Giovanni Arrighi, *The Long Twentieth Century: Money, Power, and the Origins of Our Times* (London, 1994); S. A. M. Adshead, *Salt and Civilization* (New York, 1992); Mark Kurlansky, *Salt: A World History* (New York, 2002); Sidney W. Mintz, *Sweetness and Power: The Place of Sugar in Modern History* (New York, 1985); Peter Stearns, *The Industrial Revolution in World History* (Boulder, 1993); Michael N. Pearson, *Port Cities and Intruders: The Swahili Coast, India, and Portugal in the Early Modern Era* (Baltimore, 1998); Gary Geriffi and Miguel Korzeniewicz, eds., *Commodity Chains and Global Capitalism* (Westport, Conn., 1994).

41. In June 1998 a debate on the two theses dominated four electronic discussion lists which briefly combined efforts for this purpose, and in December Frank and Landes met for a televised debate at Northeastern University. For details, see note 41, page 196.

42. Adam McKeown, *Chinese Migrant Networks and Cultural Change: Peru, Chicago, Hawaii, 1900–1936* (Chicago, 2001); Walter Nugent, *Crossings: The Great Transatlantic Migrations, 1870–1914* (Bloomington, Ind., 1992); Timothy J. Hatton and Jeffrey G. Williamson, eds., *Migration and the International Labor Market 1850–1939* (London, 1994); Patrick Manning et al., *Migration in Modern World History, 1500–2000* (Belmont, CA, 2000); Manning 1990; Robin Cohen, *Global Diasporas: An Introduction* (Seattle, 1997); Wang Gungwu, ed., *Global History and Migrations* (Boulder, 1997); David Northrup, *Indentured Labor in the Age of Imperialism, 1831–1922* (New York, 1995); David Eltis, Stephen D. Behrendt, David Richardson, and Herbert S. Klein, *The Trans-Atlantic Slave Trade: A Database on CD-ROM* (New York, 1999).

Other aspects of social history explored in broad regional or temporal context included Wally Seccombe's long-term studies of families in England, Aidan Southall's world-encompassing analysis of cities, and Frances Karttunen's biographical studies of translators and intermediaries in the early modern era of encounters.[43] Peter Partner reviewed religious controversy in terms of holy wars from the Assyrians to the present day.[44]

Studies of gender in world history began appearing in significant numbers, though an overall perspective on world history from a gendered point of view was slow to develop. Linda De Pauw's eclectic narrative of women in war showed the recurring significance of women as combatants and victims in armed conflict, but in it the author explicitly abstained from developing a gendered analysis. Regional studies of women in history, like those by Julia Clancy-Smith and Margaret Strobel, developed information relevant to the study of women in world history. A nexus of detailed studies formed around the issue of gender and colonialism, led by authors such as Strobel, Nupur Chaudhuri, and Ann Laura Stoler. It may be that the colonial situation provided a particular highlight for gender relations. That is, with a foreign power coming to govern an area, gendered issues in social relations, which were normally addressed in the private sphere, were brought into the public sphere by the conflict in values and the very need of the imperial power to impose its hegemony in every area of society. The practice of *sati*, or immolation of widows upon the death of their husbands, in India was paradigmatic, but scholars have located a far wider range of social and gender relations that were debated and documented in the course of colonization.[45]

Among the key contributions along the scientific-cultural path to world history have been histories of the various disciplines. Scholars in fields from anthropology and geography to literary studies, as they became more attentive to change over time, also came to focus a critical eye on changes in the methods, interpretations, and institutions in each field. Among the most significant and readable of these studies are two volumes by Adam Kuper on the development of anthropology.

43. Wally Seccombe, *A Millennium of Family Change: Feudalism to Capitalism in Northwestern Europe* (London, 1992); Seccombe, *Weathering the Storm: Working-Class Families from the Industrial Revolution to the Fertility Decline* (London, 1993); Frances Karttunen, *Between Worlds: Interpreters, Guides, and Survivors* (New Brunswick, N.J., 1994); R. Stephen Warner and Judith G. Wittner, *Gatherings in Diaspora: Religious Communities and the New Immigration* (Philadelphia, 1998); Karen Louise Jolly, ed., *Tradition and Diversity: Christianity in a World Context to 1500* (Armonk, N.Y., 1997).

44. Peter Partner, *God of Battles: Holy Wars of Christianity and Islam* (Princeton, 1998).

45. Linda Grant De Pauw, *Battle Cries and Lullabies: Women in War from Prehistory to the Present* (Norman, 1998); Nupur Chaudhuri and Margaret Strobel, eds., *Western Women and Imperialism: Complicity and Resistance* (Bloomington, Ind., 1992); Julia Clancy-Smith and Frances Gouda, eds., *Domesticating the Empire: Race, Gender, and Family Life in French and Dutch Colonialism* (Charlottesville, 1998); Lata Mani, *Contentious Traditions: The Debate on Sati in Colonial India* (Berkeley, 1998); Cynthia Enloe, *Bananas, Beaches, and Bases: Making Feminist Sense of International Politics* (Berkeley, 1990).

The results are full of implications for history.[46] Other cultural theorists presented critiques of the interpretation of history, including anthropologists Jean and John Comaroff, literary theorist Edward Said, and world-system theorist Anthony King.[47]

Studies of language, often in association with archaeological investigations, have been important in identifying the scale and direction of early migrations. Tracing the early stages of the Indo-European language divergence and migration remains a topic of debate; the current leaders of opposing camps are Colin Renfrew and J. P. Mallory. At a level still broader, Joseph Greenberg and his colleagues assembled interpretations of the connections of many of the major language groups, and of the migrations that brought about their present distribution.[48]

Studies of technology in world history developed impressively during the 1990s, though they remained separated into temporal and disciplinary subfields. Daniel Headrick continued his series of skillful analyses of the technology of imperialism with studies on telegraphy and radio (1850–1940) and on information at the turn of the nineteenth century; Joel Mokyr produced studies synthesizing work on technology in modern Europe, but also on the long-term contributions of technology to economic growth.[49] Major studies on agricultural technology included Zohary's study of plant domestication, Vasey's history of agriculture, Kenneth Kiple's encyclopedia of food, plus studies of individual crops

46. Adam Kuper, *The Invention of Primitive Society: Transformations of an Illusion* (London, 1988); Kuper, *The Chosen Primate: Human Nature and Cultural Diversity* (Cambridge, Mass., 1994); Robert H. Bates, V. Y. Mudimbe, and Jean O'Barr, eds., *Africa and the Disciplines: The Contributions of Research in Africa to the Social Sciences and Humanities* (Chicago, 1993).

47. Anthony King, ed., *Culture, Globalization and the World-System: Contemporary Conditions for the Representation of Identity* (Minneapolis, 1991); John Mackenzie, ed., *Imperialism and Popular Culture* (Manchester, 1992); Arjun Appadurai, *Modernity at Large: Cultural Dimensions of Globalization* (Minneapolis, 1996); Edward Said, *Culture and Imperialism* (New York, 1993); Richard Drayton, *Nature's Government: Science, Imperial Britain, and the "Improvement" of the World* (New Haven, 2000); Jean and John Comaroff, eds., *Modernity and Its Malcontents: Ritual and Power in Postcolonial Africa* (Chicago, 1993); John and Jean Comaroff, *Ethnography and the Historical Imagination* (Boulder, 1992); William H. McNeill, *Keeping Together in Time: Dance and Drill in Human History* (Cambridge, Mass., 1995). Cultural studies at the global level have focused particularly on intellectual culture, though studies have begun appearing on cultural encounters: see chapter 13 for discussion of this work.

48. J. P. Mallory and Victor Mair, *The Tarim Mummies: Ancient China and the Mystery of the Earliest Peoples from the West* (London, 2000); Mallory 1989; Renfrew 1988; Joseph H. Greenberg, *Language in the Americas* (Stanford, 1987).

49. Headrick, *The Invisible Weapon: Telecommunications and International Politics, 1851–1945* (New York, 1991); Headrick, *When Information Came of Age: Technologies of Knowledge in the Age of Reason and Revolution, 1700–1850* (New York, 2000); Joel Mokyr, ed., *The Economics of the Industrial Revolution* (Totowa, N.J., 1985); Mokyr, *The Lever of Riches, Technological Creativity and Economic Progress* (New York, 1990).

such as that of Zuckerman on the potato.[50] Despite the innovative summary by Arnold Pacey, there were no real moves to assemble an updated overview of technology in world history.[51]

Studies of environment and health continued to appear, notably through the work of such senior scholars as Philip D. Curtin and Alfred W. Crosby.[52] Brian Fagan's analysis of the El Niño phenomenon in the Pacific began as a study of the 1990s, but traced issues back several centuries before, and emphasized the links of the Pacific, Atlantic, and Indian Oceans in the complex mix of currents that brought El Niño as one type of result. Tracing these same issues in another direction, he produced a study of the Little Ice Age in Europe. In broader surveys of ecological history, Clive Ponting published his popular *Green History of the Earth*, Sheldon Watts surveyed the known history of epidemics, and John McNeill produced a survey of global environmental history for the twentieth century.[53]

Studies of world history in early times continued to be relatively low in number, at least those produced by historians. Nonetheless Liu Xinru, working from documentary records on India and China, and Christopher Ehret, working from linguistic evidence on East Africa, each provided wide-ranging yet detailed syntheses of these two regions in hemispheric context.[54]

In addition to strictly thematic approaches to the past, a growing number of studies began to emphasize connections crossing themes and regions at the same time. Notable among these were Mrinalini Sinha's work on connections of gendered images in Britain and India, and of Tessa Morris-Suzuki's study of contradiction and change in Japan from the eighteenth to twentieth centuries. Sinha was

50. Daniel E. Vasey, *An Ecological History of Agriculture, 10,000 B.C.–A.D. 10,000* (Ames, 1992); Daniel Zohary and Maria Hopf, *Domestication of Plants in the Old World* (Oxford, 1993); Kenneth Kiple and Krlemhild Conee Ornelas, eds., *The Cambridge World History of Food* (Cambridge, 2000); Larry Zuckerman, *The Potato: How the Humble Spud Rescued the Western World* (Boston, 1998).

51. Arnold Pacey, *Technology in World Civilization: A Thousand-Year History* (Cambridge, Mass., 1990); James Burke and Robert Ornstein, *The Axemaker's Gift: A Double-Edged History of Human Culture* (New York, 1995); Joel Mokyr, *Twenty-five Centuries of Technological Change: An Historical Survey* (New York, 1990).

52. Philip D. Curtin, "The Environment beyond Europe and the European Theory of Empire," *Journal of World History* 1 (1990), 131–150; Alfred W. Crosby, "Infectuous Disease and the Demography of the Atlantic Peoples," *Journal of World History* 2 (1991), 119–134.

53. Clive Ponting, *A Green History of the World: The Environment and the Collapse of Great Civilizations* (New York, 1991); Brian M. Fagan, *Floods, Famines, and Emperors: El Niño and the Fate of Civilizations* (New York, 2000); Aan G. Simmons and Ian G. Simmons, *Changing the Face of the Earth: Culture, Environment, History* (Oxford, 1996); Sheldon Watts, *Epidemics and History: Disease, Power and Imperialism* (New Haven, 1999); James O'Connor, *Natural Causes: Essays in Ecological Marxism* (New York, 1998); J. R. McNeill 2000; John R. McNeill, ed., *Environmental History in the Pacific* (Aldershot, U.K., 2001); John R. McNeill, *The Mountains of the Mediterranean World: An Environmental History* (Cambridge, 1992); John R. McNeill, *Atlantic Empires of France and Spain: Louisbourg and Havana, 1700–1763* (Chapel Hill, 1985).

54. Ehret 1998; Liu 1995a.

especially successful in documenting in documenting a process by which the English rulers of India transformed gender identities in Bengal, yet in which the same process also changed gendered identities in England. Morris-Suzuki showed the interplay of multiple themes in Japanese life, and in so doing confirmed the participation of Japan in global processes both before and after the Meiji Restoration.[55]

Studies of Broad and Conceptual Scope

As the literature in world history expanded, the nature and position of synthetic studies changed. Rather than summary statements on world history as a whole, synthetic studies became reviews of broad themes in world history, in interaction with monographic studies. In this new form, synthetic studies became important foci for world-historical debate. The *American Historical Review* affirmed in 1996 its recognition of the developments in world history by publishing a two-article forum on periodization. In the main article Jerry Bentley proposed a periodization dividing the history of the world from early civilizations into five main periods, focusing on changing patterns of cross-cultural interaction; in the response I sought to problematize, at a world-historical level, the concepts of culture and interaction.[56] Bruce Mazlish sought to distinguish the approaches of global history and world history, and *History and Theory* produced an issue focusing on approaches to world history.[57]

General statements on the structure of world politics, including the economic linkages of politics, came from Giovanni Arrighi, from Torbjorn Knutsen, and from David Gress. The world-historical vision of Marshall Hodgson, as presented by Edmund Burke III, showed the prescience of Hodgson's global analysis and the importance of the Islamic world as a source of world-historical insights and

55. Sinha 1995; Roxann Prazniak, *Dialogues across Civilizations: Sketches in World History* (Boulder, 1996); Sally Hovey Wriggins, *Xuanzang: A Buddhist Pilgrim on the Silk Road* (Boulder, 1996); Dauril Alden, *The Making of an Enterprise—The Jesuits in Portugal, its Empire and Beyond, 1540–1750* (Stanford, 1996); O. R. Dathorne, *Asian Voyages: Two Thousand Years of Constructing the Other* (Westport, Conn., 1996); Morris-Suzuki 1998.

56. Bentley 1996; Patrick Manning, "The Problem of Interactions in World History," *American Historical Review* 101 (1996), 771–82; William A. Green, "Periodization in European and World History," *Journal of World History* 3 (1992), 13–54; Segal 2000; Gale Stokes, "The Fates of Human Societies: A Review of Recent Macrohistories," *American Historical Review* 106 (2001), 508–25; Michael Adas, "From Settler Colony to Global Hegemon: Integrating the Exceptionalist Narrative of the American Experience into World History," ibid., 1692–1720. World history, as presented in the *American Historical Review*, reflects the gradual transition from a focus on synthesis and teaching to an expanded attention to research.

57. Martin W. Lewis and Karen E. Wigen, *The Myth of Continents: A Critique of Metageography* (Berkeley, 1997); Philip Pomper, Richard H. Elphick, and Richard T. Vann, eds., "World Historians and Their Critics," theme issue 34, *History & Theory* (1995); Bruce Mazlish, "Comparing Global to World History," *Journal of Interdisciplinary History* 28 (1998), 385–95.

patterns. Roland Robertson, a sociologist of globalization, sought to demonstrate the uniqueness of the current wave of globalization, but he also linked his analysis to the work of Norbert Elias on the civilizing process.[58]

Several broad and significant statements in intellectual history at the global level appeared during the 1990s. James Blaut, a geographer, carried on a fierce polemic against diffusionism, arguing in his last work that eight major historians had focused their work on Eurocentric assumptions. Paul Costello reviewed the work of the great synthesizers of world history, focusing on their philosophical assumptions rather than their historical reconstruction. Maghan Keita carried out a project that is in some ways parallel, reviewing the work of major African American and African writers of the nineteenth and twentieth centuries, and the role of Egypt and the interpretation of race in their views of the world.[59] Martin Lewis and Karen Wigen produced a critique of metageography, focusing especially on changing conventions for the notion of "continents" and calling for more geographic research at the global level.[60]

Stimulating as these studies were, historians were able to participate in them with the standard historical bag of tricks. But several interventions in world history at a broad and synthetic level required attention to knowledge and theories beyond the established topics of history. Most dramatically, the "big history" initiated by David Christian and pursued by Fred Spier drew on the full range of the natural sciences and the information they provided on the earth's past. Christian demonstrated that effective undergraduate courses could be taught on history from the Big Bang to the present, and Spier's concise volume on world history, centering on the concept of "regimes," led to wide exploration of the potential of this approach.[61]

In *Civilizations*, a work aimed at a broad audience, Felipe Fernández-Armesto wrote an extensive narrative of world history that is at once conventional and innovative. In an approach anticipated by Jacques Maquet and others working in African studies, he defined "civilization" as a style of life, rather than as a state or political-religious tradition. Fernández-Armesto went further to emphasize that each civilization developed out of a distinctive human transformation of nature. Thus he portrayed civilizations of ice, of alluvial soils, and of the edges of the sea. This effort to link the traditionally state-centered histories of civilization to the

58. Arrighi 1994; Torbjorn Knutsen, *The Rise and Fall of World Orders* (Manchester, 1999); Hodgson 1993; Roland Robertson, *Globalization: Social Theory and Global Culture* (London, 1992).

59. David Gress, *From Plato to NATO: The Idea of the West and its Opponents* (New York, 1998); Blaut 1993; Costello 1993; Maghan Keita, *Race and the Writing of History: Riddling the Sphinx* (New York, 2000).

60. M. Lewis and Wigen 1997.

61. David Christian," The Case for 'Big History'," *Journal of World History* 2 (1991), 223–38; Fred Spier, *The Structure of Big History: From the Big Bang until Today* (Amsterdam, 1996); Johan Goudsblom, Eric Jones, and Stephen Mennell, eds., *The Course of Human History: Economic Growth, Social Process, and Civilization* (Armonk, N.Y., 1996); Robert P. Clark, *The Global Imperative: An Interpretive History of the Spread of Humankind* (Boulder, 1997); King 1991.

discoveries of ecological history opened new possibilities for inclusive approaches to world history.[62]

William Durham published in 1991 a general statement on the notion of coevolution. It was intended as an evolutionary theory of cultural change, focusing on "the relationship between genetic and cultural dynamics." While he did not expand it to the level of a narrative, he explored community-level dynamics through such examples as the cultural evolution of incest taboos and inheritance systems.[63]

Jared Diamond, in *Guns, Germs, and Steel*, gained a wide popular audience for an approach to world history that included a good deal of technical information.[64] Diamond, a physiologist who had made numerous field trips to Papua New Guinea, combined his biological and ethnographic experience to develop a long-term interpretation of world history. Diamond focused first on the ecological conditions that created plants and animals suitable for human domestication, and then on the fate of those peoples who carried out the domestication. His point was that the extensive east–west zone of temperate climate in Eurasia permitted the development of a range of plants and animals that, once domesticated, provided the societies in those regions with a substantial edge over other societies in power to reproduce themselves and expand. Wheat, cattle, and horses figure centrally in his arguments. For instance, the very expansion of early populations in close contact meant that they developed diseases, but the eventual development of immunity meant that these communities, on meeting other and sparser communities, would expand preferentially. In one particularly skillful chapter, he described the various patterns by which viruses and bacteria mutate in order to locate hosts but not kill them off entirely.

Diamond's argument, while it has been contested by other scholars, is an elegant simplification of a major issue in world history and an effective illustration of long-term trends in history. Yet when he attempts to use the same reasoning to explain the comparatively short-term changes of imperialism and racism in recent centuries, his results are far less satisfactory.

In *The Dynamic Society*, Graeme Snooks developed an interpretation of global history that was equally broad in its conception. In what he was later to characterize

---

62. Fernández-Armesto acknowledged the contributions of Kenneth Clark and Norbert Elias to his conception of civilization. Meanwhile Africanists, faced with an earlier paradigm separating the civilized from barbarians and placing Africa at the bottom of a civilizational hierarchy, had proposed a vision of ecologically and technologically differentiated civilizations of the hoe and the bow. Felipe Fernández-Armesto, *Civilizations: Culture, Ambition, and the Transformation of Nature* (New York, 2001): 26; Jacques Maquet, *Civilizations of Black Africa*, trans. Joan Rayfield ([1962] New York, 1972). For efforts to set the lens at an even wider scope, David Fromkin, *The Way of the World: From the Dawn of Civilizations to the Eve of the Twenty-first Century* (New York, 1998), and J. Burke and Ornstein (1995) each offer overall views of human history.

63. William Durham, *Coevolution: Genes, Culture, and Human Diversity* (Stanford, 1991). Others with broad interpretation of great swaths of human history include Simmons 1996; and Peter Bogucki, *The Origins of Human Society* (Oxford, 1999).

64. J. Diamond 1997.

as a "global strategic-transition model," Snooks traced five great phases in the advance of technology to the limits of its possibility. In each case he then traced the development of a technological breakthrough in certain strategic areas of the world, followed by a generalization of the new technology and then a repetition of the cycle.[65]

Presumably not every world historian will be expected to be knowledgable about each of the topics explored in these wide-ranging works. Yet within the developing community of world historians, it will no longer be acceptable for one to develop a comfortable specialization and work on it in isolation from all other topics. The connections among issues, regions, and even time frames are now being argued with sufficient force to make it appear that historians will do better in their own specialization if they learn about its connection to others.

World, United States, and European History

I have postponed detailed discussion of U.S. and European history until this point, in order to develop a vision of world history as a meeting ground of academic specializations. For historical studies overall, the largest sections of the literature (especially in English) are in U.S. history and European history. These fields have strong traditions and substantial momentum and are not likely to be fundamentally reorganized through an encounter with world history. At the same time, the U.S. and European fields are also undergoing rethinking and development of new emphasis on historical connections and transnational interactions. The establishment of a dialogue among the literatures on U.S., European and world history represents, for world historians, one of the main remaining steps in creation of a genuinely comprehensive field of world history. So far, few studies have been very successful in linking interpretations of Europe and North America in global context. Alfred Crosby's *Ecological Imperialism* is one example of an adept integration of Europe and North America into the flow of world history.[66]

One way to link European and American history to world history is to project the former onto the latter: in a word, Eurocentrism. This was the idea that history outside "the West" was the story of Westerners away from home, or the history of Western impact on other areas of the world. The reaction against this sort of thinking has now succeeded in discrediting such parochialism and triumphalism among historians, though the approach continues to be well represented in

65. Graeme Snooks, *The Dynamic Society: Exploring the Sources of Global Change* (London, 1996); Andre Gunder Frank, "Materialistically Yours: The Dynamic Society of Graeme Snooks," *Journal of World History* 9 (1998), 107–16. Snooks appears to draw on the interpretive approach of V. Gordon Childe. In a flood of publication focusing on the application of these long-run ideas to economic theory and policy for the present day, Snooks published *The Ephemeral Civilization* (London, 1998); *The Laws of History* (London, 1998); *Longrun Dynamics: A General Economic and Political Theory* (London, 1998); and *Global Transition: A General Theory of Economic Development* (London, 1999). See also Childe 1942.
66. Crosby 1986.

textbooks. The result has been that world historians have tended to distance themselves from integrating Europe and North America into their perspective, and historians trained in the study of Europe and North America have tended to do a sort of global study that is restricted mostly to comparative studies of the North Atlantic.[67]

Beyond this stage of alienation and rebellion of world historians and Third World historians against dominant Euro-American perspectives, there ought to be room for dialogue among these fields of historical study. World historians have a great deal to gain from such a dialogue. The literatures on U.S. and European history contain the most detailed, sophisticated, and thoroughly reviewed analyses of the past, and the techniques for research, writing, and teaching in those fields deserve to be tested and exploited in world historical analysis. In addition, the national literatures in U.S. and European history include a great deal of excellent work on social, cultural, and political connections within the nation. By the same token, the literatures on U.S. and modern European history may have much to gain from dialogue with world historians. For one thing, studies in European and U.S. history are restricted to a rather short time frame, from two to as many as five centuries. While world historians too have tended to give disproportionate attention to recent centuries, there exists in the global framework a necessary attention to long-term historical change. Given this long-term framework, historians may identify dynamics that might significantly inflect the interpretation of history in the North Atlantic.

For much of the twentieth century, the place of Europe and North America in studies of world history alternated. At one extreme, Europe and later North America appeared as the central influences in world history; at the other extreme, Europe and especially North America lay outside the scope of world history. The phrase "the West and the rest" served as an ironic reminder of this disjuncture.

The underlying questions about the relationship among these fields of historical study are numerous and conflicting. Does world history encompass the history of Europe and North America? Do the histories and historiographies of the West govern world history? Does world history conflict with the histories of Europe and North America? Do the fields complement each other? What historical works based in U.S. and European history make substantial contributions to the understanding of world history? What contributions show the place of Europe and the United States in world history? While recognizing the importance of these questions, I propose not to face them in this study in order to focus more simply and directly on brief discussion of some recent works on Europe and North America that adopt transregional and thematic approaches to the past.

Historians of Europe and North America have been rethinking their field with a seriousness at least equal to that of world historians and area-studies historians.

67. James Blaut, *The Colonizer's Model of the World: Geographical Diffusionism and Eurocentric History* (New York, 1993); Samir Amin, *Eurocentrism*, trans. Russell Moore ([1988] New York, 1989); Peter Coclanis, "*Drang Nach Osten*: Bernard Bailyn, the World-Island, and the Idea of Atlantic History," *Journal of World History* 13 (2002), 169–82.

In *Telling the Truth about History*, three leading historians reviewed developments in U.S. and European historiography.[68] Gary B. Nash and other participants in the struggle over the U.S. national standards on history wrote a review and analysis of the debates. Ross Dunn's edited collection, "a teacher's companion," sets forth the major approaches within world history.[69] At the turn of the twenty-first century, the Organization of American Historians published a report encouraging more cosmopolitan and connected approaches to history, and the American Historical Association launched a three-year review of graduate education.[70] So far, few studies have been very successful in linking interpretations of Europe and North America in global context.

Some recent studies address the place of Europe and North America in world history directly and with considerable success. Peter Gran, trained as a historian of Europe and of the Middle East, relied on European dynamics as a device for addressing world history in *Beyond Eurocentrism*. In an approach that he described as the application of social history on a world scale, Gran made an original argument based on a structuralist postulate. That is, in the nineteenth and twentieth centuries, elite strata seeking to maintain political dominance and cultural hegemony addressed their general problem with a limited number of models: division of potential opponents by region, ethnicity, class, or gender. As he argued, these may not be the only logical possibilities, but in practice they are the only models in place. Gran's narrative is then the historical evolution of the models themselves, as well as the various national experiences of countries governed by one model or another. It is a global argument.[71]

Among other wide-ranging interpretations by Europeanists, Charles Tilly's analytical study of the past millennium in European politics emphasized the links of cities, states, war, and capital.[72] Felipe Fernández-Armesto, in a narrative approach

68. Appleby, Hunt, and Jacobs 1994. In the years following, two of the co-authors, Joyce Appleby and Lynn Hunt, became presidents of the AHA.

69. Nash, Crabtree, and Dunn 1997. Ross E. Dunn went on to edit Dunn 2000, a compilation of key writings in world history.

70. The Organization of American Historians and the American Historical Association have each undertaken major reviews. The OAH completed a study of internationalizing U.S. history (chaired by Thomas Bender), and the AHA created its Committee on Graduate Education in 2000, with Colin Palmer as chair and Thomas Bender as co-chair. Thomas Bender, *The La Pietra Report: Internationalizing the Study of American History* (New York, 2000)—available online at www.oah.org/activities/lapietra/final.htm; and Philip Katz, "The CGE Hits the Road," *Perspectives* [AHA] (May 2001), 11.

71. Peter Gran, *Beyond Eurocentrism: A New View of Modern World History* (Syracuse, 1996); Stearns 1993b.

72. Peter Linebaugh and Marcus Rediker, *The Many-Headed Hydra: Sailors, Slaves, Commoners, and the Hidden History of the Revolutionary Atlantic* (Boston, 2000); Nicholas Canny, *Europeans on the Move: Studies on European Migration, 1500–1800* (Oxford, 1994); Margaret Strobel, *European Women and the Second British Empire* (Bloomington, Ind., 1991); Charles Tilly, *Coercion, Capital, and European States, AD 990–1990* (Oxford, 1990); Tilly, *Durable Inequality* (Berkeley, 1998).

to the same time period, focused on the changing hegemonies of civilizations.[73] Jack Goldstone's thematic analysis of revolution and rebellion balanced Ottoman and Chinese instances against revolutions in England and France.[74]

Studies of the twentieth century in world history have long tended to be dominated by dense narratives of great-power relations, focused far more on polarities and on power centers than on connections and global patterns. The field of international history, rather closely linked both to diplomatic history and to international relations, nonetheless began to develop a more interpretive and more nuanced approach.[75]

Analyses of the Atlantic world, linking area-studies approaches to Africa, Latin America and the Caribbean, continued to work toward establishing closer links with North Atlantic territories. Michael Gomez's study of African culture and its transformation among African Americans focused on the eighteenth and nineteenth centuries; Richard Powell's study of art and culture in the African diaspora focused on the twentieth century.[76] For the mainland of North America, studies of regional and social interconnection appeared, especially on the colonial era.[77]

The study of medieval European history, long one of the strengths of the historical profession in both methodology and interpretation, showed signs of connecting to studies of world history. Alfred J. Andrea, a historian of the Crusades, teamed up with modern Europeanist James Overfield to develop what rapidly became the most widely adopted documentary collection in world history, with its strength in introduction to and analysis of primary sources. And applying the skills of medievalists to interpretive debate, R. J. Barendse and Stephen Morillo conducted a long and erudite debate on H-WORLD in 1999 that came to be known as the dispute on global feudalism. In it, Barendse asserted that contemporaneous transformations took place in many parts of the eastern hemisphere in the tenth and eleventh centuries, in which subordination of peasantries, development of mounted warriors, and even a cult of Alexander the Great amounted to

73. Felipe Fernández-Armesto, *Millennium: A History of the Last Thousand Years* (New York, 1995); see also James C. Russell, *The Germanization of Early Medieval Christianity: A Sociohistorical Approach to Religious Transformation* (New York, 1994).

74. Goldstone 1991; Linebaugh and Rediker 2000; Canny 1994; and Strobel 1991.

75. Iriye, 1997; William Keylor, *The Twentieth Century World* (Oxford, 1995); Gyan Prakash, ed., *After Colonialism: Imperial Histories and Postcolonial Displacements* (Princeton, 1995); Michael Geyer and Charles Bright, "World History in a Global Age," *American Historical Review* 100 (1995), 1,034–60.

76. Michael A. Gomez, *Exchanging our Country Marks: The Transformation of African Identities in the Colonial and Antebellum South* (Chapel Hill, 1998); Richard J. Powell, *Black Art and Culture in the Twentieth Century* (New York, 1997).

77. Karen Ordahl Kupperman, *Indians and English: Facing off in Early America* (Ithaca, 2000); Carl J. Guarneri, *America Compared: American History in International Perspective*, 2 vols. (Boston, 1997); Joyce E. Chaplin, *Subject Matter: Technology, the Body, and Science on the Anglo-American Frontier, 1500–1676* (Cambridge, Mass., 2001); Donald Worster, *The Wealth of Nature: Environmental History and the Ecological Imagination* (New York, 1993); Gwendolyn Midlo Hall, *Africans in Colonial Louisiana: The Development of Afro-Creole Culture in the Eighteenth Century* (Baton Rouge, 1992).

a widespread development of feudalism. Morillo, in contesting the thesis, argued that such a transition took place, but in a much slower and more discontinuous fashion. The evidence in support of Barendse's thesis was sufficient to make a debate worthwhile, and the controversy suggests that the skills and the documentary analysis developed by medievalist European historians may be relevant in the interpretation of world history.[78]

Viewed through the examples of these works, the boundaries separating European, U.S., and world history do not seem to be very closely guarded.

Conclusion: Promise and Dilemma in World History

At the turn of the twenty-first century, the field of world history expanded in several ways: course enrollments grew, production of books and articles expanded, the institutions for study and teaching developed and strengthened, and the logic of seeking historical connections spread to new fields of study.

Research in world history resulted not only in more publications but in exploration of new issues. Area studies monographs reached the limit of each area and began connecting with adjoining regions, not just with Europe. Thematic studies developed further from work of previous decades. Broad and conceptual studies, previously based in European studies, now drew on area-studies literature to address more of the world. These developments took place not only for those issues that historians have long debated, but especially along the external path, in which historians and scholars in the natural sciences and cultural studies have developed new questions for historical debate.

Historians of Europe and the United States drew on the same intellectual currents, developing more interest in cross-border connections and cross-thematic interactions. Major scholarly journals facilitated discussion on global issues, opening what might optimistically be considered to be a dialogue among Europeanists, Americanists, area-studies historians, and world historians on the direction of historical studies.

---

78. The discussion began with postings from Barendse and Morillo, and continued for a month. R. J. Barendse, "Chinese 'feudalism'," H-WORLD (h-net.msu.edu/~world) 28 October 1999; Stephen Morillo, "Global feudalism," H-WORLD 5 November 1999.

Chapter 6

Narrating World History

In some ways the current generation is making great progress in developing a new, more comprehensive, and more intelligible version of world history. It is certainly a major change to find—within the pages of a single textbook or in the curriculum studied by high school and college students—a survey linking the experiences of people all around the globe for the past two thousand or more years. And it is new to have thematic volumes surveying the ecological and technological history of humankind.[1]

At the same time, the underlying elements and structures of world history are those that have been in place for literate populations for hundreds of years. Today's well-informed world historian may read with profit H. G. Wells's *Outline of History*—published over eighty years ago by an energetic journalist and novelist. The prose and conceptualization of Voltaire in his *Philosophy of History*, written a quarter of a millennium ago, remain relevant to the study of world history.[2]

These two sides of world history—novelty and continuity, the new and the old, the expanded and the continued—need to be set in appropriate balance. The tool for striking that balance will almost surely be the historical narrative. Presenting the past in narrative form, while not unique to the field of history, is nonetheless the most characteristic element of historical studies. The issue, then, is how best to construct and convey world-historical narratives.

In the preceding chapters, I have emphasized continuity more than novelty. Not only is the world itself old, but interpretations of the world are old as well. But in this concluding look at global historiography, since it is so compressed, I shall emphasize novelty and transformation in the interpretation of world history.

To underscore the novelty, I want to argue that world history is not simply another sub-field to be added to the range of historical studies. Instead, because of its focus on connections, it adds a whole dimension to the study of history. The range of studies that could be undertaken in world-historical context is as large as

1. Pacey 1990; J. McNeill 2000.
2. H. G. Wells 1920; Voltaire [1753–1754].

all the studies that have already been undertaken. Since resources for the field are small, it is all the more important to select carefully in undertaking studies and to make an effort to connect the analyses of world history with each other.

The Nature of Narrative

People have been telling stories for many thousands of years, and it is unlikely that the techniques of narrative have advanced much in recent times. To hold the audience, the narrator must offer a problem and a message of interest, presenting a tale that is at once convincing and beguiling. No simple chronicle or list of facts will do. The tale must be conveyed with beauty of word and of image. It must connect the audience to the past and to the topic and characters of the story. It must be an ordered yet not entirely predictable story of a significance that the audience can comprehend. It may relate tales of pain and disaster, yet it must hold somewhere a message of hope. The narrator must introduce an interplay of personality and social forces as well as the mystery of matters beyond human control. The various subplots must sustain their own interest yet converge to make sense of the principal narrative. In short, the narrator faces the challenge of reshaping the story and manipulating the audience, yet leaving that same audience with a memorable impression and with a sense of having taken away a valuable experience.

I was reminded anew of these old lessons in the course of work in creating the Migration CD-ROM.[3] Having read and written many narratives before, I was struck as I wrote by the artfulness and the complexity that must be worked into even the most simple and straightforward historical narrative. The CD-ROM has thirteen sections entitled "narrative," each summarizing a theme over time. But none of the narratives could be a simple chronicle. As author, and with the help of those giving me advice and corrections, I constructed paths for the readers, but these paths repeatedly shifted issues, perspectives, time frame, and mode of analysis. A narrative of families in world history had to include definitions of unfamiliar terms and situations and then descriptions of migrations, marriages, and enslavement. Sometimes these shifts in the narrative were made explicit, but at other times they were included without a word. Writing with the intention of reminding readers of connections between local lives and global patterns, I found it best sometimes to convey clarity and at other times to impart a sense of mystery.

World historians, seeking to develop their own emphases on transnational connections, often contrast their work with the national studies that now dominate historical research and writing. But professional narratives of national and global history are not the only choices. History existed as a field of study long before it became professionalized, and the professionalization of history has not eliminated the other types of history. Family histories have been told by the aged or by self-appointed record keepers, dynastic histories are told by chroniclers, and histories

---

3. Manning et al. 2000. This instructional CD-ROM includes some four hundred documents, thirteen chronological and thematic narratives, and an analysis section with a thousand questions.

of localities and organizations have been told by boosters and by critics. It was straightforward to organize and comprehend the history of such local groups: the documents were not voluminous, and the perspective was easily selected (though debatable). Writing the history of the nation, that great project of the last century or two, was not so easy. It involved a great deal of abstraction and immense quantities of documentation. National history nevertheless became oversimplified—dangerously so—by imposing a single perspective, and creating a master narrative to trace the path of national destiny. Writing the history of the world is different but not much more difficult. The boundaries are less artificial. The documents are more complex, but our techniques for collecting and analyzing them are improving. The most difficult issue is that of constantly shifting perspective.

How then are world historians to tell their stories? Is a tale of world history told like any other story, or does it have certain distinctive qualities? For the balance of this chapter, and assuming that the available techniques of narrative have changed very little over the millennia, let us consider how a specific sort of narrative—the narrative of world history—has changed with time.[4]

Narratives of the World

Who are those who have sought to grasp the history of the world? Why do they assume that the different aspects of the world are connected rather than autonomous from each other? They are people who want to understand their place in the world, but also people who want to change the world. In the latter category, they have included conquerors who desired to control the bodies and loyalties of every person, and religious visionaries who sought comfort and salvation for all human souls. In the former category have been scientists who pursued understanding of the physical heavens and earth, philosophers seeking meaning in all aspects of life, and ordinary persons seeking to find a peaceful place in a complex world. Historians, finally, have sought to tell a tale of the world's past for an audience drawn from the small numbers of persons in these two groups.

Probably most people have wondered about the history of the world at some time in their lives, but it is only in recent times that sizable numbers of people have sought a systematic understanding of world history. Today, in an age of wide literacy and democratic aspirations, the impulse to understand the world—and, in response, to change oneself or the world—has become more broadly based. Even political leaders now encourage, if tentatively, the study of world affairs by the youth in their constituencies.

The problems that world historians have addressed over the ages include the origins and extent of the world and the dynamics of change in the world—in

---

4. Many historians have written narratives of the writing of history, in both abbreviated and extended form, though not usually with this explicit focus on world history. The following narrative corresponds to what E. H. Carr has called "the widening horizon" of history. Carr, *What is History?* (New York, 1961), 177–209. See also R. G. Collingwood, *The Idea of History*, ed. Jan van der Duesen ([1946] Oxford, 1993); Herbert Butterfield, *The Origins of History* (New York, 1981).

social processes, natural processes, and the interventions of supernatural forces. Over time, knowledge about the world has changed and expanded, and concepts for understanding the world have been reformulated and amplified.

Those who first tried to tell tales of the world had no notion of how far back the world went and no way to be sure about how far the world extended. Myths of origin formalized notions of the world, but those who related myths of origin had to speculate about the actual origin of earth and of society. The myths relied on available grains of evidence to describe the past, but they usually conveyed more about contemporary social values than about historical origins.

Written documents redressed the balance somewhat: Herodotus and Sima Qian used a mix of written documents and oral testimony to take them back a few centuries before their own times, but they had no knowledge of what came before, just as they had only sketchy knowledge of lands distant from their own. There was no way to choose between those who argued that the world had been created a few thousand years earlier and those who argued that the world had existed almost forever. Eusebius, the early Christian bishop, emphasized depth as much as span and wrote a history of the world attempting to link many different regions through chronology; he also used the chronology of the Old Testament to estimate a date for the creation of the world.[5] In the work of these and other early authors, their understanding of the world as a whole was limited and flawed, yet they conveyed a sense of context for the histories of their immediate surroundings.

From the time of these early achievements, the basis for understanding world history grew incrementally. With the passage of each year, the known time-perspective of humanity increased. Within that expanding time perspective, the notion of the "other," the person beyond the home community, remained a major problem in world history. On one hand, geographical distance and social difference encouraged authors to demonize distant "others," endowing them with tails and strange customs, or attributing evil designs to nearby "others" who were political enemies. Yet every locality included, in its history, the cases of persons who had come from distant places and became locally influential—as with St. Patrick in Ireland and Omani settlers among the Swahili—so that the notion of interaction was built into world history from the beginning. The concept of the "other" thus immediately became complex, including at once the encounter with unfamiliar persons and the negotiation with those who were familiar but distinctive.

The development of formalized religious traditions surely contributed to the interpretation of world history. For spiritual leaders to think of the fate of all humans, as compared to the gods and the cosmos governing their lives, may raise questions and suggest answers about world history and the interactions of humans with each other. Such global thinking is explicit in the theologies of Buddhism, Christianity, and Islam, and is evident in most religious traditions. By the end of the first millennium C.E., each of these religious traditions had developed a vision of the world in terms of the relations between god and mankind. Under the influence of such traditions, I suggest, world historians of early times

---

5. Eusebius, *The Ecclesiastical History*, trans. Kirsopp Lake ([325] Cambridge, Mass., 1953).

focused their efforts not only on summarizing patterns of human action, but also on understanding patterns in natural history (in order to harmonize one's life with them) and on enabling people to align themselves with the desires of the gods—those forces beyond the knowledge or understanding of mankind.

Beginning shortly after 1200 C.E., two great waves of human encounter dominated a period of five centuries: the Mongol conquests across the Eurasian landmass and the European maritime voyages to most of the coasts of the world. The Mongol conquests of the thirteenth century forced direct military, political, and economic connections among people over most of Afro-Eurasia. Ferociously and effectively, the Mongols created an immense political community. While the community itself broke up into successor states within two centuries, the political effects of encounters among these disparate regions remained in the historical memory.

Less significant socially but more significant conceptually, the news of Magellan's voyage of circumnavigating the earth, verifying that it was a globe and demonstrating the place of the continents on its surface, provided new information for everyone. It was not just Europeans who confirmed that the world was spherical, it was literate and informed people everywhere, of all backgrounds, who could henceforth state with precision the geographical extent and limits of the planet.[6]

These two great efforts to control and understand the affairs of the world reflect a notion of dominance in world history. In every case, however, new encounters and new knowledge showed the limits on human dominance as a strategy for action or as a concept for historical understanding. The Mongols swept all before them in military terms and administered their vast realms with remarkable effectiveness. But the pandemic of bubonic plague, erupting a century after the Mongol conquests and facilitated by the greater social interaction under the Pax Mongolica, disrupted political dominance and verified that politics could not command disease. Similarly, while the circumnavigation of the globe gave humans a clear sense of the limits of the earth and the possibilities of dominating global affairs, it was soon followed by the confirmation that the earth was but one small planet in a huge solar system, a humbling realization for any humans who sought to dominate.

The efforts at domination and exploration, limited though they were by newly discovered realities, brought a wealth of human encounters in the half-millennium after Chinghiz Khan. People learned of the world by traveling or by observing the impact of external influences on their homeland. The old routes of contact remained the main ones—crossing Eurasia, the Mediterranean, the Indian Ocean, the South China Sea, or the African continent, and from point to point within the Americas. But the new maritime routes put people into contact with each other in different ways. African sailors on European vessels, for instance, voyaged not only

6. I am grateful to Adam McKeown for relating to me the story, told today in Malaysia, that the first person to circumnavigate the world was a Malay sailor who worked his way to Europe aboard a Portuguese ship, had joined Magellan's fleet in 1519, and survived the voyage.

to Europe and the Americas, but across the Indian Ocean to Japan and across the Pacific between Acapulco and Manila. What they saw and thought we will never know, but we may be feeling the results of their voyages even as we read.

The experience of knowing the world was not necessarily pleasant. The new connections initially brought more disaster than development, more oppression than enlightenment. The loss of life with the spread of disease, especially for peoples of the Americas, is the easiest way to tell this tale. The warfare and hostility among newly encountering groups—sometimes at first meeting, and sometimes after a generation's acquaintance—brought great cost in loss of life and creation of new prejudices. Yet people also created family and personal ties across the boundaries. The movement of crops, ideas, and people led ultimately to a more dependable interplay among regions of the world and to a global system that began to expand after overcoming the shock of its creation.

Geographers learned how to comprehend the earth as a physical unit long before historians learned to consider it as a social unit. The notion of encounters with those thought to be previously in isolation from each other drew attention in the debate of Sepulveda and Las Casas about slavery in the Spanish Americas and in contemporaneous struggles along the African coastline. But more important in the long run were the renegotiations of relations among those previously in contact. The fifteenth-century accounts of Ma Huan, the chronicler of the Indian Ocean voyages under the Chinese admiral Zheng He, convey the modification of existing relationships among peoples more than the opening of new contacts. The relationships among lord and peasant, Han Chinese and Vietnamese, Hindus and Muslims in the Moghul state, rulers of Songhai and Morocco—all these and more were renegotiated in the context of expanding global contacts.[7]

The encounters and renegotiations of this era from the thirteenth through the seventeenth centuries took place not only in social relations but in the natural world and in the human understanding of the supernatural. Humans explored and acclimated themselves to new geographic spaces and climates. Animals and plants, meanwhile, moved as never before from place to place, each struggling for survival with greater or lesser success. And in a world now shown to be changing and changeable, one had to ask whether the changes were the result of supernatural intervention, or whether the gods were allowing the world to work by its own rules. Though they focused clear attention on these several types of encounters, the historical writings of this era tended to treat them in isolation from one another, and were able to link them only in the sense that Providence and the will of the supernatural guided the events and processes of the world.

Once the issue of the geographic scope of the world was settled, the issue of social perspective became the most difficult problem in world history. Authors, beginning with their own perspective, found it difficult to acknowledge other perspectives. This, along with limits on their documents, restrained their historical

7. Bartolomé de Las Casas, *História de las Indias,* 3 vols. ([1566] Caracas, 1986); Ma Huan, *The Overall Survey of the Ocean's Shore,* ed. Feng Ch'en Chun, trans. J. V. G. Mills ([1433] Cambridge, 1970); Embree 1988; Levtzion 1981.

writing to certain segments of the world. Most authors were enclosed within the limits of their region, languages, religion, and social station. The writings of Voltaire—most accessibly in the charming *Candide*—demonstrate an increasing ability of authors in the eighteenth century to acknowledge and articulate a variety of perspectives on history. But the question of perspective, while engaged, remained unsolved.[8]

Meanwhile, in the eighteenth and nineteenth centuries, historians of the world found themselves able to trace significant change over time, and they labeled that historical change as progress: progress first in their own knowledge, but also progress in the social order and transformation in the natural world. The developing demonstrations of biological and geological evolution did more than anything else to advance notions of progress in human society.

The problem of the time frame of world history was solved, in general if not in detail, during the nineteenth century. The ambitions of philosophers and scientists, as they grew in the eighteenth and nineteenth centuries, led at once to broader generalization and to a more thorough segmentation and classification of issues in the natural and social world. Linnaeus's classification of plant and animal species set the standard for classification. By the mid-nineteenth century, the classification of animals, plants, and geology had demonstrated that the earth, its geologic features, plants, and animals had existed for many millions of years, and humanity had existed for perhaps a million years or more. Yet, while the problem of the extent of time had now been solved, the problem of how to break it up remained. Condorcet, updating an earlier Christian tradition, suggested a series of discrete ages, each with its own character. Buddhist philosophy had earlier suggested cycles at the cosmic level, and Ibn Khaldun had suggested dynastic cycles. It would be some time before historians would learn how to address cycles, transformations, trends, and episodes all at once and put them into a coherent narrative.[9]

The problem of social dynamics in world history appeared to become more complex rather than simpler. Bossuet had presented it as the gradual enactment of God's plan, and Hegel would see it as the struggle for expression of the human spirit. Karl Marx and Friedrich Engels, when they were voraciously reading young philosophers, did as well as anyone in their time at identifying the principle that every element of the world is in change through interaction, at varying rates and because of various conflicts and interactions.[10] Yet the dynamics of world history remained mysterious.

Nineteenth-century analysts of world history sought primarily to address the social issues of their own age. Those issues included the development of new social classes in the wake of economic transformation, the expansion and limiting of slavery, the changing technology and expanding scale of warfare, the complex legal regimes developed to negotiate the social overlaps of expanding empires, and problems of constraint and freedom in commerce and in social relations. In analyzing

8. Voltaire [1756].
9. Headrick 2001: 16–17, 20–27; Condorcet [1795]; Ibn Khaldun [1377].
10. Bossuet [1681]; Hegel [1830]; Karl Marx and Frederick Engels, *Selected Works* (New York, 1968).

these issues, students of broad historical interpretation relied on the social philosophy of the Enlightenment, but they also looked to developments in the natural sciences, particularly the emerging schemes of classification. Out of the classificatory work of the eighteenth and nineteenth centuries, the notion of *civilization* came to be accepted as an all-purpose concept or unit of analysis for world-historical analyses. *Nation* gained even more approval as the putatively homogeneous community, organized into a state, which would be the basis of further human advance. *Class*, a statement of social and economic hierarchy, served at once as an elite affirmation of the necessity of hierarchy and as a rallying cry for the toiling masses who sought to eliminate the privilege of wealth. *Race* and *religion*, one assumed to be inherited and the other adopted, were proposed to mark the great divisions in the human community. All of these terms were conveyed as the legacy of nineteenth-century social science, for twentieth-century historians to adopt or discard in their interpretation of the world. (Only in the late twentieth century did social scientists add the notion of *gender* to this list.)

Meanwhile the development of cultural studies in the eighteenth and nineteenth centuries laid the groundwork for later analysis in world history. In the studies that became known as orientalism, European scholars studied the languages and writings of Asian societies, working especially in Sanskrit, Arabic, Persian, and Chinese languages. In one sense, they were reenacting earlier waves of cultural appropriation, in which scholars had translated works from Arabic to Latin, Greek to Arabic, Sanskrit to Chinese, and Chinese to Japanese. But the literary, scientific, and religious studies of the eighteenth and nineteenth centuries were followed as well by new types of analysis: the development of Indo-European linguistics, the beginnings of academic study of archaeology, and the foundation of the field of anthropology to study ethnic and cultural groups around the world as they fell under colonial domination. These studies were ultimately to link up with studies of elite culture, notably literature, painting, and music, and with the studies of folklore that eventually became the field of popular culture.

At the turn of the twentieth century, writers were finally ready to propose overall statements of world history. They focused principally on a social perspective and on a narrative of expanding and transforming civilizations. H. G. Wells provided a narrative of world history encompassing the full geographic and temporal frame of the world. Oswald Spengler's narrative was more restricted in time and space but fuller in its assertion of a dynamic of civilizational rise and fall. The heritage of the nineteenth century was evident in these syntheses: they relied heavily on notions of dominance and diffusion. The work of Darwin and the advances in biology were impressive enough that Spengler, while contesting Darwinian thought, chose to adopt an organic metaphor for his interpretation of the civilizational dynamic in world history. Spengler argued that the apparent crisis of World War I was instead a natural inflection in a long-term process of growth, maturation, and decline.

The popularity of these syntheses of world history in the 1920s made explicit a dynamic in the formulation of world history that had begun earlier and was to be accentuated with time. First, the audience for world history surged in breadth at times when war, social revolution, political collapse, or economic transformation fed the need for study of world affairs among the reading public. Second, a few writers had anticipated the growing interest and provided books just in time: their analyses addressed current concerns but explored them by providing context ranging far back in time and across wide spaces. Third, the sources for these studies were the historical information collected in the previous era, organized into analytical frameworks from the previous era, including those of philosophy, the natural sciences, and cultural studies. As a result, each wave of studies provided documentation of previously unsuspected or unappreciated dynamics in world history. Fourth, the debate over these restated interpretations of early and recent times—among scholars and in the public arena—led to further study and research.

A remarkable aspect of the dynamic in the development and popularization of world history is the recurring linkage between contemporary issues and evidence on the distant past. New evidence added both clarity and confusion to the patterns of world history. In the early twentieth century, archaeologists unearthed the previously unsuspected cities of Harappa and Mohenjo-Dara in the Indus Valley. The ruins of these ancient societies became a source of pride to many in India: India too had developed ancient river-valley civilizations, a parallel to those of the Yellow River, the Nile, and Mesopotamia. It also became clear that the system of castes had arisen in these early societies, and before the arrival of the Indo-European-speaking Aryans. So it was that the Brahmins, at the peak of the social order, were disappointed to learn that the founders of their caste had not been Aryans.[11] The same evidence, while undermining the elite position of the Brahmins, may have helped develop a sense of Indian national solidarity.

The dynamic of world-historical interpretation brought successive changes throughout the twentieth century. Notions of interaction and relativity arose from new scientific work, especially the captivating theories of Albert Einstein. Such thinking began to challenge principles of diffusion and dominance, and suggested that world history should give greater emphasis to contingency. Fernand Braudel's fame depends in significant part on his advances, at mid-century, in the techniques of world-historical narrative: he assembled a narrative of multiple themes, multiple dynamics, and multiple time frames.[12] The development of social struggles in a time of periodic democratization encouraged the articulation of history through contending perspectives, as in the Cold War, decolonization, and critiques of racial discrimination. Late in the century, clear demonstrations of the pervasiveness of ecological change brought the natural world firmly into the narrative of world history. Further, the discovery of details of human evolution, documenting the waves of human development in Africa and migration to other world regions, restructured the hierarchy of the continents and reaffirmed the importance, in the world-historical narrative, of those early times that had previously been labeled as "prehistory." The expansion of literacy in most areas of the world, plus the development of electronic communications media, greatly widened the audience for debate about the past.

11. Anstey 1929: 49.
12. Braudel [1949].

A fascinating example of debate and contested narrative in world history occupied public arenas around the Atlantic in the enactment of the Columbian Quincentennial in 1992. Early preparations for the anniversary of Columbus's landing in the Americas focused on celebration of the discovery, and included reenactment of the voyage. In the United States, contending voices arose from Native Americans and African Americans, articulating the harm done to their communities as a result of the initial centuries of transatlantic contact; defenders of Columbus and of European expansion responded. As a result, the exhibits at the Smithsonian Institution switched perspective from straightforward celebration to a multivalent debate. Similar battles were fought out in other Atlantic regions. The government of Spain had gained the approval of UNESCO for a commemoration of the quincentennial, expecting a celebration of Columbus. By the time debate settled down, the highlight of the commemoration was a conference in the mid-Atlantic, in the Republic of Cape Verde, at which representatives from African and Latin American nations dominated and gave critical assessments of the Columbian exchange. Representatives in attendance from Spain restrained themselves from any display of imperial triumphalism.[13]

To this expanded arena of public debate was added one more crucial element in the dynamic of interpreting world history: the professionalization of world history at two levels. Teachers faced newly expanded responsibilities for formal teaching of world history in schools and colleges; and scholars in world history, having gained recognition for developing a field of study, began formal interchange with historians in distinct but related fields. Only in the twentieth century did historians begin to grapple with the issue of contending narratives: what should be the subject matter and style of world-historical debate? The debate over Arnold J. Toynbee's magnum opus was mostly wasted in nitpicking at the local level rather than review of his overall analysis.[14] Factual correctness is certainly important, yet the world-historical narrative contains far more than a list of facts. World historians are gradually learning to identify the numerous issues in any narrative or interpretation of global scope—geographic extent and breakdown, temporal extent and breakdown, themes, perspective, dynamics, and interpretation. To assess the dynamics of world history, for instance, we need narratives to trace the development of difference and interdependence, as well as the affirmation of sameness and dominance. There are many stories left to tell.

Conclusion: Expanding our Understanding

World history consists of studies of the past working from the postulate of connection rather than the postulate of autonomy. In every era, analysts of society

---

13. I was privileged to attend this conference. I arrived, as it happened, having just left Los Angeles at the time of the devastating riots following the acquittal of police officers who had beaten black motorist Rodney King in the course of his arrest a year earlier; much of the conference discussion focused on the tensions of urban life worldwide.
14. W. McNeill 1989.

have worked from the problems of their age, with the knowledge base of their age, and with available procedures for analysis. The particular approach of world historians has been to address the problems of their era by seeking to establish the context in space and time and locate the constituent elements of each problem. In so doing, world historians have tended to relativize their objects of study. In time, world historians seek out the limits of time relevant to any process, then investigate the dynamics of change over time within those limits. In space, world historians locate the widest geographic space relevant to any historical issue, then explore the categories and dynamics of regional interaction. World historians have worked in most detail on the social sphere, focusing especially on politics, warfare, commerce, and the rise and fall of states. The natural world, about which so much has been learned in the past two centuries, has recently become a significant focus of world historical studies. Studies of culture are now undergoing an extraordinary process of development, and the effects of this new knowledge are already beginning to show up in studies of cultural issues at the global level. The relations between mankind and god, between the known and unknown, have been constricted as the power and knowledge of mankind have expanded. Yet the old questions of supernatural forces and human spirituality reappear in new fashions: Comte in his time (and Toynbee a century later) wrote volumes of comprehensive and pragmatic explanation of the world, yet changed in later years to adopt an emphasis on spirituality, in hopes of improving the path of history through an exercise of the will. One can be certain that the issues of religion and spirituality will not disappear from studies of world history.

A world-historical narrative written today has much in common with a narrative written a century ago, in underlying form and technique. Yet new dimensions of the narrative have emerged and will continue to emerge as the task of world history develops and clarifies. To locate the relevant new techniques, historians should look elsewhere in our societies for possibilities. For instance: television soap operas, now popular in many languages, are a rhetorical form allowing for multiple narratives and multiple perspectives, all contained within one tale or at least one program. With the help of this or other devices, historians will learn how to tell a tale that is an envelope of many tales—an overall story but not a simplified one in that it allows the variety of underlying stories to show through. In addition, world historians must learn to convey their theory and analytical perspective at the same time as telling the story or stories. World history can move ahead as a form of analysis and narrative only if the authors and the audiences learn how to adopt and utilize multiple and shifting perspectives. We may also expect, given current changes in technology, that future narratives of world history will be presented in multiple media.

Finally, a world-historical narrative must provide a basis for the audience to respond. Each reader or listener needs to be enabled by the narrative itself to make independent statements about the world. World history as indoctrination will doubtless be attempted again, as it has been in the past, based on the argument that the facts of history lead to an inevitable interpretive conclusion. But the logic of world history, while reliant on the facts as they are known, leads inevitably instead to a multiplicity of interpretations. Thus, writers a century ago chose to focus on "civilization" as the basic concept in world history, and attempted to write master narratives focused on this concept. But as the scope of world history expanded, "civilization" proved too small a concept to contain the developing narratives of ecological change, migration, technology, family, and expressive culture. By the opening of the twenty-first century, civilization had ceased to be an absolute standard. It maintained its significance, but, like everything else in world history, civilization had to be relativized.

Part II

Revolution in Historical Studies

History is not what it used to be. The study of the past has undergone dramatic transformation, especially in the late twentieth century, and as a result it seems poised for a great expansion and renewal. The chapters in this section focus on three great axes of change—new or transformed disciplines, the rise of area studies, and the rise of global studies. The results are expanding the frontiers of historical study, strengthening the tool kit and broadening collaborative responsibilities of historians. Every field of history has changed in response to this expanded definition of history, but the study of world history has changed and benefited most of all. World history, newly enabled by the expanded methods, differs from previous historical studies in addressing a wider range of topics, specifying previously neglected connections among arenas of human experience, tracing broad patterns in the past, and clarifying relationships among different scales of the world's events and processes.

In using the term *revolution in method* to characterize these multiple and compounded changes in scholarly method, I must emphasize the two edges of this revolution. First, the changes in each discipline, while incremental in their own terms, added up to a massive reorganization and potent advance of the scholarly apparatus. At the same time, I think it is important to retain the second and more negative connotation of the term *revolution*: the debates, clashes, destruction, and enmities that accompanied the process. In particular, world history, while emerging from the revolution in method as an increasingly respected new field, has also been stigmatized in past academic debates as superficial speculation at ethereal levels of generalization, as undermining the advances in area studies, and as hostile to the history and historiography of the West. Similarly, struggles have emerged among the practitioners of world history: between supporters and opponents of the study of civilization, over the meanings of "global" and "world" history, and between those focusing on recent and earlier times in history. The revolution in method has neither been easy nor entirely pleasant, but it leaves world historians with immense potential for achievement.

In each of three chapters I trace the panoply of new information, techniques, perspectives, and theories that developed and entered the purview of historical studies. Scholars working along the two methodological paths to world history— the historians' path and the path of other specialists that I have labeled the "scientific-cultural" path—responded to the new developments somewhat differently, so that they brought different types of new results to world history. The scientific-cultural path, for instance, has led to major changes in the understanding of long-term processes in history, but also to important and unsuspected short-term changes in cultural patterns.

Of the new developments in method, it has been easiest for historians to utilize or appropriate the results of the new data and new techniques, especially through a growing reliance on computers. New perspectives have brought benefits at a somewhat slower pace. For historians of the scientific-cultural path, shifting perspectives has meant especially shifting among the standpoints of various disciplines: from cell biology to zoology to cultural anthropology to ethnomusicology. For those working along the historians' path, the shifts in perspective have been principally from one social standpoint to another: history as seen from the standpoint of gender categories, of regional or class perspectives. National perspectives and national histories remain significant, even among world historians, but it has now become almost automatic for world historians to view "the nation" as only one of numerous social perspectives and, in viewing the nation, to consider it from any of several disciplinary perspectives.

It is in the realm of theory, however, that historians still tend to drag their feet in appropriating the benefits of the revolution in method. Historians of many earlier generations became adept at voicing the terms of each new theoretical framework (such as "relativity" in the wake of Einstein's theories) without learning, applying, or formally modifying the substance of the theories. Historians, being what they are, will always subordinate theory to empirical data. But the prominence of many sorts of theoretical formulations in the rapidly transforming disciplines leaves historians with the clear need to learn and implement a selection of the available theories as part of their historical work.

Chapter 7

Disciplines

In universities of the United States today, the Department of History is sometimes included among the humanities, and at other times with the social sciences. Historians, Janus-like, have sought at once to privilege both narrative and analysis. The gradual move of history departments from the humanities and toward the social sciences results in part from changing preferences of historians, but also from more general shifts in university structures. Yet if the discipline of history has an essence, it lies beyond any single region of the human experience: not in the focus of the humanities on consciousness or in the focus of the social sciences on institutions—or in the focus of the natural sciences on the physical and biological environment for human existence.[1] The essence of historical study goes beyond the specifics of subject matter and centers on its approach to posing and analyzing questions of change over time.

But concern with time was never a monopoly of historians, and it became even less so in the late twentieth century. Analysts in fields from anthropology to literature to zoology found that certain of their results made more sense when expressed in terms of change over time. Without giving up their own identity, they began to crowd into the field of history. The resulting overlaps of approaches to change over time led, on one hand, to apparent confusion and fragmentation in historical studies but also, on the other hand, to great opportunities for linking changes and connections in the past.

---

1. Historical writing is associated more closely with some social sciences and humanities than with others. In the social sciences, history is close to politics, sociology, and economics, but not to geography, psychology, demography, anthropology, or archaeology. ("Psychohistory" was briefly popular.) In the humanities, history is close to literature, material culture, and law, but not to linguistics, philosophy, visual art, or music. Similarly, of the natural sciences, history has been closer to some than to others.

    Hayden White notes that historians have relied on a somewhat devious tactic to minimize criticism of their work: when criticized by social scientists for "softness" of method, historians respond that history does not claim the status of a pure science; when criticized for unwillingness to explore the depths of consciousness and modes of literary representation, historians argue that history is a "*semi*-science," requiring reliance on historical documentation. White, *Tropics of Discourse: Essays in Cultural Criticism* (Baltimore, 1978), 27–28.

The Revolution in Historical Studies since ca. 1960

Peter Novick, in a 1988 book entitled *That Noble Dream*, made innovative use of the traditional techniques of the historian—leafing through documentary archives—to reveal some unheralded patterns of change in historical studies. He studied the letters, speeches, book reviews, and notes of historians working in the United States over the century from the founding of the American Historical Association in 1884. He examined the twists and turns in the debate about whether history could be an objective science or whether it was dominated by the subjective impressions of historical authors. In the concluding chapters to that study, Novick noted the collapse, from the 1960s, of an apparent consensus among historians, a consensus that had been dominated by agreement on the main lines of political history. Instead, as he noted in a chapter entitled "There was no king in Israel," the contending perspectives of women's history, black history, and social history led to an impression of the fragmentation of historical studies.[2]

I prefer, however, to interpret the last forty years of historical studies not in terms of fragmentation of a consensus, but as a methodological and theoretical revolution, accompanied by rapid expansion of the scope (geographical, thematic, and temporal) of historical studies. Biology had its revolution with the breaking of the DNA code; physics and chemistry had earlier revolutions with the development of quantum mechanics; economics had its revolution with the development of macroeconomics. History has had not one but several such innovations concurrently. My purpose here is to call attention to the changes in historical methods and their implications for world history and for historical study generally.[3]

2. Novick's detailed and subtle analysis is a major contribution to the history of history. Through it, for instance, entering graduate students can learn more easily where they fit into the range of debates and specializations in historical studies. Novick 1988: 573–629. For other important studies of changes in the historical profession, see Appleby, Hunt, and Jacobs 1994; and Smith 1998. See also Dorothy Ross, *The Origins of American Social Science* (Cambridge, 1991); and Page Smith, *Killing the Spirit: Higher Education in America* (New York, 1990).

3. I have chosen the term *revolution* to emphasize the breadth and thoroughness of change in historical studies, more than its rapidity. This notion of sudden change remains controversial. Interpretation of the industrial revolution has gone from the notion of sudden change to gradual transformation. Stephen Jay Gould has emphasized the notion of "punctuated equilibrium" to argue that episodes of rapid evolutionary change lead to species development. But evolutionists generally have not adopted the notion. In evolution of *Homo sapiens*, recent work emphasizes long and gradual development (in Africa) rather than sudden appearance of a new species with advanced culture (in Europe). Jan DeVries, "The Industrious Revolution and the Industrial Revolution," *Journal of Economic History* 54 (1994), 249–70; Stephen Jay Gould, *The Panda's Thumb* (New York, 1980); Sally McBrearty and Alison S. Brooks, "The Revolution that Wasn't: A New Interpretation of the Origin of Modern Human Behavior," *Journal of Human Evolution* 39 (2000), 533–34; K. Greene, "V. Gordon Childe and the Vocabulary of Revolutionary Change," *Antiquity* 73 (1999), 97–109.

Change in historical studies has come from every direction and gives the appearance of resulting from multiple causes. The very pervasiveness of change is perhaps the reason why so few historians have used the term *revolution* to describe the changed nature of the field: rather than massive change in one area, it is incremental change in every area.[4] In the following discussion, I will alternate among the following epistemological categories or types of new knowledge that have come to influence the study of history: perspectives, techniques, data, theories, and linkages.

Among the perspectives that developed substantially new influence in history are history "from the bottom up" (popular and working-class history), history by gender (especially women's history), and history by world region (with the development of substantial area-studies literatures in history, and history from the vantage point of each region). Additional novelties in perspective include history by various time frames (for instance, with emphasis on the Braudelian *longue durée*), history by philosophical outlook (with increasingly explicit contrasts among radical, liberal, and conservative outlooks), and the expanded inclusion of environmental considerations in historical studies. In these cases the changing currents of thought and debate seem to be key in developing new interpretations.[5] Only in the last case can one argue that an exogenous determinant of thought, the threat of environmental crisis, brought the new perspective suddenly into existence.

The new technique of greatest importance to history has been the electronic storage of data. This made far more usable the many questionnaires, surveys, electoral results, and text documents awaiting attention in archives. With computers came the advance of quantitative analysis and statistics in history, but the same technology also brought advances in textual storage, annotation, and criticism. Radiocarbon dating techniques added precision to dates of human remains, and advanced techniques of protein analysis linked traits among populations and even suggested evidence on human evolution. In these instances new technology or new artisanal techniques, rather than new logical systems, have caused an expansion in historical knowledge.

New data, emerging out of archives and analysis, have led to substantial changes in historical thinking. We now have data, unavailable to previous generations, on the pace of population change across history, the many campaigns for recognition of national identities in the past two centuries, the episodic reduction in the world's forest cover, the dimensions of slave trade across the Atlantic and in the Old World, the decimation of populations with periodic epidemics, and great migrations that can be traced by their linguistic remnants. New facts, once they can be verified, stimulate new hypotheses and even whole fields of study. Not least of the developments is the growing realization that historical data are not simply retrieved from the past, they are actually created by processing materials from the

---

4. A further peculiarity of the revolution in historical studies is its uneven impact—it has been felt to a great degree in new research, to a much smaller degree in undergraduate and high school teaching, and almost not at all in the training of graduate students as researchers or teachers.

5. Abu-Lughod 1989: vii–x.

past with new techniques.[6] New empirical evidence, from whatever source, has itself spurred developments in historical thinking.

New theories have emerged in discipline after discipline, and previously existing theories have been applied in new ways. To list briefly some of the theories that will be discussed in more detail later in this chapter: microeconomic and macroeconomic theory created substantial historical literatures, sociological theory (of Parsonian, Weberian, and Marxian variants) brought numerous historical studies, two branches of linguistics (linguistic philosophy and historical linguistics) each had great influence on historical study, theories of gender roles led to historical analyses, and so forth. One might mention literary theory, plate tectonic theory, kinship theory in anthropology, and evolutionary theory as influenced by empirical studies in biology. In these and other arenas of study, new arrays of variables, dynamics, and predictions have stimulated new interpretations.

Finally, but least widely noted, a wave of new linkages among disciplines has been a central element in the revolution in historical methods: ecological studies have emerged through the confluence of geography, geology, biology, medicine, and other fields; gender studies have developed out of a mix of linguistics, psychiatry, sociology, and psychology; and social history draws on sociology, economics, demography, and more.[7]

The timing and interaction of these disciplinary innovations—the successive moments of discovery and of predominance for each approach—is worthy of much closer and more nuanced study than I can provide in this rapid review. Instead, I will explore the range of innovations, discipline by discipline, moving from those most closely associated with history in the past to those for which the connections to history are more recent.

Economic history and economics. The development of macroeconomic theory and national income accounting led to great efforts, beginning in the 1950s, to

---

6. Examples of the creation of new historical data are the reconstitution of family structures from parish records and the estimation of past temperatures through study of polar ice cores. I am indebted for this point to Marjorie Murphy, who teased me in earlier years about making up data for simulating the demography of Atlantic slave trade.

7. Peter Burke has written a book-length assessment of the interaction of history and other social sciences, focusing not only on the recent and dramatic changes in historical studies but on the past two centuries of change in the relations between history, sociology, and other fields. Burke's first edition, in 1980, concentrates on sociology; his more recent edition addresses social sciences and cultural studies more broadly. Peter Burke, *Sociology and History* (London, 1980); Burke, *History and Social Theory* (Ithaca, 1992). Other reflections on interactions among the disciplines have appeared recently, especially for studies beyond Europe ankd North America. See, for instance, Bates, Mudimbe, and O'Barr, 1993. For a collection of studies on disciplinary change in early modern times, inspired by the more recent changes, see Donald R. Kelley, ed., *History and the Disciplines: The Reclassification of Knowledge in Early Modern Times* (Rochester, N.Y., 1997).

develop historical estimates of national income and output for most countries in years for which adequate statistics were available. For such countries as the United States, Sweden, and England, it was possible to develop estimates of national income as far back as the eighteenth century. These historical studies of national income fit into a great theoretical and empirical emphasis on economic growth.[8] Most of this work was done by economists, and the field of economic history therefore grew considerably.

In the wake of these studies in economic growth, economists began turning microeconomic theory, along with the logic of hypothesis-testing through statistics, to historical purposes. The canonical article is the 1958 study of Alfred Conrad and John Meyer on the profitability and viability of slavery in the American South, in which historical questions were formally quantified, and alternative hypotheses were tested against each other. While the quantitative and theoretical work was esoteric and arcane, the effective point was that historians usually failed to specify their variables or to test their conclusions systematically. Robert William Fogel, an ex-Marxist who became a neoclassical economist, assumed leadership in the field through two major studies. In *Railroads and American Economic Growth,* he contrasted the actual railroad system of the nineteenth century with a "counterfactual"—the canal system as it might have been—as a way of estimating the contribution of railroads to economic growth. In *Time on the Cross,* along with Stanley L. Engerman, Fogel focused on emphasizing the economic rationality and productivity of antebellum slavery in the United States.[9]

The 1993 Nobel prizes awarded to Robert Fogel and Douglass North were for their theory-based work in quantitative economic history conducted in the 1960s and 1970s. Andre Gunder Frank—who, also in the 1960s, set forth the thesis of Latin American underdevelopment that sparked an important debate on interaction in modern world history—spoke up after the Nobel Prize award to criticize the work of both Fogel and North for interpreting narrowly within the limits of the United States or Western Europe.[10] That is, we still have much to do in working out the mutual implications of the many new developments in historical studies.

For all of this new work, the logic of hypothesis-testing was central. The analyst is expected to define a set of variables under study and to postulate a set of relationships among them, with one or more variables as the dependent or

8. Simon Kuznets, *National Income and Its Composition, 1919–1938,* Vol. 1 (New York, 1941).

9. Alfred H. Conrad and John R. Meyer, "The Economics of Slavery in the Ante Bellum South," *Journal of Political Economy* 66 (1958), 95–130; Robert William Fogel, *Railroads and American Economic Growth: Essays in Econometric History* (Baltimore, 1964); Douglass C. North, *Growth and Welfare in the American Past: A New Economic History* (Englewood Cliffs, N.J., 1966); Robert William Fogel and Stanley L. Engerman, *Time on the Cross: The Economics of American Negro Slavery,* 1 vol. and supplement (Boston, 1974); Gavin Wright, *The Political Economy of the Cotton South* (New York, 1978).

10. Frank 1967; Frank 1993.

resulting variable and the others as independent or causal variables. The hypothesis is then assessed according to its consistency with the available data. (Often enough, the available data do not fit exactly with the theoretical variable, so the analyst uses the available data as a "proxy" for the desired data—that is, if one does not know the number of religious believers, one can estimate it with the number of church members.) But in assessing the hypothesis, one requires a standard by which to assess it. This standard is provided by the "alternative hypothesis," a contrary interpretation. In the statistical analysis, one must find observations on enough individual cases to know, with a given level of confidence, whether they give more support to the hypothesis or the alternative hypothesis.[11]

In the United States, this "new economic history" was studied more in economics departments than in history departments. The *Journal of Economic History* and *Explorations in Entrepreneurial History* became the main organs for publication and review of quantitative economic history. In Britain, in contrast, departments of economic history continued work that was less theoretical and more institutional, and this work appeared in the *Economic History Review*.

Meanwhile, the expansion of radical perspectives and Marxian theory in the 1960s and 1970s gained a significant foothold within the economics profession in the United States and elsewhere. *Review of Radical Political Economy* emerged as a major journal for this perspective, which focused on labor economics and on macroeconomic perspectives.

Social history and sociology. The "new social history" developed at much the same time as the "new economic history." Community studies by Stephan Thernstrom on the Massachusetts towns of Newburyport and Boston laid the groundwork for studies that were census-based, quantitative, and focused on differences in ethnicity, occupation, and religion. Some studies in social history looked at demography and the structure of families, as well as rates of birth, death, and population growth.[12]

For social history, the radical critique had an earlier and more pervasive influence than in economic history. This included work inspired by E. P. Thompson's 1964 *Making of the English Working Class*, a study of "history from the bottom up" that privileged the viewpoint of artisans and wage workers and, in the hands of other authors, the crowd and peasants.[13] Works in African American history gained wider attention in the era of the U.S. Civil Rights movement. The new social history arose in a parallel that was equally quantitative and sometimes

---

11. For more on hypothesis-testing in history, see chapter 17.
12. Stephan Thernstrom, *Poverty and Progress: Social Mobility in a Nineteenth Century City* (New York, 1969); Thernstrom, *The Other Bostonians: Poverty and Progress in the American Metropolis, 1870–1970* (Cambridge, Mass., 1973); John Blassingame, *The Slave Community* (New York, 1972); Louise Tilly and Joan Scott, *Women, Work and Family* (New York, 1978).
13. Thompson 1964; George Rudé, *The Crowd in History: A Study of Popular Disturbances in France and England, 1730–1848* (New York, 1964); Charles Tilly, *The Vendée* (Cambridge, Mass., 1964).

equally theoretical with the new economic history. Feminism brought a new critique to history and an expanded literature on women's history. Overall, the social-science dimension of history brought adoption into history of formal methodology (especially quantitative techniques), formal theory (neoclassical and Marxian economics, Parsonian and Marxian sociology, psychoanalysis and feminism), and explicit identification of standpoints: working-class history, feminist outlooks, African American perspectives, Third World viewpoints.

While the logic of hypothesis-testing was similar in economic and social history, the actual data in the two fields differed enough to create a huge gulf in the practice of their work. Economic historians worked almost entirely with interval data—with prices and incomes that could be added and divided. For these data, economic historians used the statistical procedure of linear regression to test their hypotheses. Social historians, in contrast, worked mostly with nominal (or categorical) and ordinal data—discrete categories such as male and female or ethnic groupings, or ordinal categories such as grade levels in school.[14] Such data can sometimes be ranked, but they cannot be added or divided. Social historians thus used a range of statistical procedures appropriate to their data for testing their hypotheses. (They included linear regression as one of their techniques, but they did not use it in such depth as did economic historians.) Thus the new economic and social historians, while unified by the general nature of their projects, were separated by the statistical nature of their quantitative data.[15]

As quantitative studies in history progressed, changing computer technology influenced the path and speed of change. In the 1960s most work was on manual desk calculators and a few mainframe computers. Small-scale linear regressions and the simplest of nominal statistical tests could be performed in this way. By the 1970s, mainframe computers and statistical packages enabled social history projects to expand greatly in size.[16] With microcomputers in the 1980s and 1990s, large-scale data analysis came within the range of any scholar, and new openings in cultural history emerged with the development of relational data sets and, at a few major universities, electronic text centers. With the quantitative analysis came journals focusing on publication of the new work.[17]

---

14. To convey the new methods to historians, Richard J. Jensen directed the Newberry Library Summer Institute in Quantitative History in the 1970s and early 1980s. This institute, supported by the National Endowment for the Humanities, introduced as many as fifty historians each year to statistics, computer analysis, and to the literature in social, political, and economic history. In the 1990s Jensen launched another major venture with NEH support: the H-Net discussion lists in history.

15. See chapter 17 for further discussion of these techniques.

16. Statistical Package for the Social Sciences (SPSS) and Statistical Analysis System (SAS).

17. The *Journal of Interdisciplinary History* and *Social Science History* became major journals, and the *Journal of Social History* came to focus significantly on theory-based quantitative studies. *Radical History Review*, *Radical America*, and *History Workshop Journal*, and later *International Labor and Working-Class History* applied detailed analysis of sources and the approach of "history from the bottom up" to a wide range of social issues in recent centuries.

For all the importance of quantitative analysis, with its focus on hypothesis-testing based on positivistic theory, its application has been largely restricted to the social sciences and to local and national studies in history. Hypothesis-testing has had little impact in world history, with the exception of some cross-sectional, transhistorical studies such as those of Frederic Pryor on the origins of the economy and Orlando Patterson on slavery. Both of these studies relied on the Human Relations Area Files, a great ethnographic data set constructed through the efforts of the anthropologist George Peter Murdock, and both used the technique of linear regression to locate correlations among variables, with individual societies as the case studies, to propose and test historical hypotheses.[18]

Neo-Marxian social history had a wider range of approaches than did neo-Marxian political economy. While the works of sociologist Erik Olin Wright on the quantification of class were to have a certain influence in historical studies, the artisanal and eclectic approach of E. P. Thompson was to have greater influence. Thompson was closer empathetically to the young Marx, with his focus on the problem of human alienation, than to the old Marx who sought to analyze social forces. His study gave voice to English artisans and their struggle to maintain old identities and gain new ones as their lives were reshaped by the pressures of industrialization.[19] Thompson's great success was in linking many small snippets into a broad narrative showing the voice and the agency of those whose lives underwent social transformation, and the reflections of his techniques were to be seen in works of many perspectives in the generations thereafter.

For most of the work that I have labeled as neo-Marxian, there was an emphasis on developing a broad and logically consistent set of theoretical propositions, and on linking them to empirical evidence. This approach to scholarship did not, however, put as much effort into hypothesis-testing or other formalized ways of validating the research results as did the neoclassical approaches. Whereas the new quantitative historians focused specifically on measuring the key variables within their purview, the New Left historians focused broadly on linking economic, social, political, and cultural factors. Their analysis focused particularly on identifying distinctions in social class based on occupation and relating this factor to the other factors included in social-science theories. In addition, while concentrating on class, the new Marxian historians also focused particularly on community, and thus on neighborhood, ethnicity, and occupational groups, and on oral history and discourse analysis as techniques of study. Verification of the conclusion was generally conducted by inspection rather than through a statistical test.

Political economy. The intense social conflict centered on the year 1968 provides a convenient marker for the emergence of a wave of studies in political economy,

18. Frederic Pryor, *The Origins of the Economy: A Comparative Study of Distribution in Primitive and Peasant Economies* (New York, 1977); Orlando Patterson, *Slavery and Social Death: A Comparative Study* (Cambridge, Mass., 1982); Murdock 1959.
19. E. Thompson 1964; E. P. Thompson, *Poverty of Theory and Other Essays* (New York, 1978); Erik Olin Wright, *Classes* (London, 1985).

and neo-Marxism is a convenient label for the approach, though in fact this body of scholarship was much broader ideologically and more gradual in its emergence. The term political economy, in use for two centuries, was appropriated by those emphasizing that their analysis of economic history included social class and other power relations as part of the theory, rather than treating them as border conditions outside the analysis.

The most widely recognized statement within this frame of reference was that of Immanuel Wallerstein on the modern world-system. Christopher Chase-Dunn and Thomas Hall extended the study of world-systems to a much longer time frame, developing both quantitative and qualitative comparisons.[20] Within the field of economics, the Union of Radical Political Economists and its journal, *Review of Radical Political Economy*, led a parallel development of intellectual effort.

Political history. Political history too became more theoretical and more quantitative: the work of Ronald Formisano led in applying statistical techniques to historical data.[21] Political history took on quantitative work as in electoral studies, and political historians rethought their field in the light of new work in social and economic history. The analysis of political history remained centered in history departments, where micro studies explored voter behavior, policymaking, and public opinion, and where macro studies explored constitutional law, empire, and revolution.

Sociology. The long tradition of sociological investigation of historical topics continued through this period. Talcott Parsons's emphasis on modernization in the immediate postwar years gave way to a renewed interest in Marx, Weber, and Durkheim, whose contributions were reviewed repeatedly by Anthony Giddens. Immanuel Wallerstein's explorations in political economy remained based in the field of sociology. Barrington Moore, Theda Skocpol, and Charles Tilly were prominent among those producing macrosocietal analyses of social change.[22]

History and literature. Several developments in the study of history and literature came to have wider implications in the study of history. Hayden White's view of language and linguistics, developed out of his exploration of Giambattista Vico's earlier attempts to explore language as the main expression of consciousness,

---

20. On Wallerstein, see chapter 5. Christopher Chase-Dunn and Thomas D. Hall, *Rise and Demise: Comparing World Systems* (Boulder, 1997); Samir Amin, *Accumulation on a World Scale: A Critique of the Theory of Underdevelopment*, trans. Brian Pearce ([1970] New York, 1974); Samir Amin, *Unequal Development: An Essay on the Social Formations of Peripheral Capitalism*, trans. Brian Pearce ([1973] New York, 1976).
21. Ronald P. Formisano, *The Birth of Mass Political Parties, Michigan, 1827–1861* (Princeton, 1971).
22. Moore 1966; Anthony Giddens, *Central Problems in Social Theory* (London, 1979); Skocpol 1979; C. Tilly 1984.

resulted in his analysis of tropes in historical writing, in which he studied European historical writers of the nineteenth century.[23] French academic life in the 1960s led to the emergence of a number of schools of thought that had great impact on the historical literature. Michel Foucault's studies of social deviance led him beyond documentary studies of history to seek out the sources of broad transformations in society, most notably through his history of sexuality. While he focused significantly on the changing history of institutions, his primary inquiry was on the complexity and indeterminacy of social connections and on individual agency and the interplay of power and knowledge.[24]

In what became known as "the linguistic turn," scholars in several disciplines came to study language and its role in mediating social forces. (The term *linguistic turn* is tricky, in that it refers not so much to a turn toward the field of linguistics—though linguists contributed to the discussions—but to a general focus on the role of language.) One important result of the linguistic turn was the emergence of a formal literary theory, based on the notion of "deconstruction."[25] Rather than focus simply on the work itself, literary scholars now turned to analyze the author and his or her perspective. While deconstruction became unpopular with those who sought simply to enjoy their literature rather than dissect it, the development of this formal mechanism for identifying the perspective of authors added a formidable arrow to the scholarly quiver. Edward Said's *Orientalism*, published in 1978, was an application of this methodology to the field of Middle East studies.[26] Then in the "New Historicism" in literature, literary scholars turned to an emphasis on the historical influences on the authors of major works.

Gender studies. The rise of a new feminist movement in the 1960s soon had its impact in academic life, first with the elaboration of feminist theory and the creation of programs in women's studies, and later with the development of gender studies. Feminist theory, in turn, relied heavily on developments in psychiatry and psychology, as well as literary theory.[27] The results of this work led, at the most basic level, to widespread efforts of scholars and teachers to locate women in the past where their presence had previously been ignored. On a more sophisticated level, feminist studies led to the elaboration and critique of gendered concepts of society and historical change and to the identification and critique of the notion

23. Hayden White, *Metahistory: The Historical Imagination in Nineteenth-Century Europe* (Baltimore, 1973).
24. Michel Foucault, *The History of Sexuality*, 3 vols., trans. R. Hurley ([1976–1984] Harmondsworth, 1978–1986).
25. Terry Eagleton, *Literary Theory: An Introduction* (Oxford, 1983); Fredric Jameson, *The Political Unconscious: Narrative as a Socially Symbolic Act* (Ithaca, 1981).
26. Edward Said, *Orientalism* (New York, 1978).
27. Shulamith Firestone, *The Dialectic of Sex: The Case for Feminist Revolution* (New York, 1970). See also Sheila Rowbotham, *Women, Resistance and Revolution: A History of Women and Revolution in the Modern World* (New York, 1972).

of the "master narrative," the idea that a single narrative could sum up the main lines of human history, either within national or global context.[28]

Cultural history. Several strands of work in cultural history developed in the late twentieth century. American Studies arose as an interdisciplinary linking of history and literature; in England, Raymond Williams led in the development of a movement known as "cultural studies," in which Marxist analysts of literature and popular culture sought to link culture to social structure.[29] In a similar but more eclectic fashion, the studies of popular culture, reinforced by the new social history, and principally for the United States but increasingly extending their scope to other regions, are laying the groundwork for a significant advance in historical studies of culture.[30]

Literary studies provide a reminder that the development of historical studies will rely substantially on the continuing evolution of electronic technology. The availability of texts and images via the internet and the World Wide Web expanded the availability of these documents. The development of mark-up languages and other software advances led to the creation of relational data sets and electronic text centers, which helped sustain advances in cultural history, as through the creation of elaborately annotated manuscripts online.[31] We cannot know what precise developments to expect next, but it seems certain that electronic technology will permit the creation and analysis of more complex data sets, and that these technological possibilities are likely to permit advances in the study of world history.

Revolutions at the Edge of Historical Studies

The changes brought to history from fields beyond the previously established topics of historical discourse are those most clearly indicating that history has gained new boundaries and new conceptual tools, not just new perspectives on old issues. Knowledge about the dramatic changes in fields of social science, humanities, and natural science reached historians in various ways: the announcement of individual discoveries, circulation of fragments of texts from outstanding analysts, the realization that scholars outside history were writing histories, and the entry of

28. Joan Scott, ed., *Feminism and History* (New York, 1996); White 1973.
29. The leading journal in American Studies, *The American Quarterly*, was founded in 1949; Raymond Williams, *Keywords: A Vocabulary of Culture and Society*, revised ed. ([1976] New York, 1983).
30. On literary theory, studies in popular culture, and their combination in the historical literature, see Eagleton 1983; Herbert Gans, *Popular Culture and High Culture: An Analysis and Evaluation of Taste* (New York, 1974); David Sabean, *Power in the Blood: Popular Culture and Village Discourse in Early Modern Germany* (Cambridge, 1984). The *Journal of Popular Culture* appeared in 1967.
31. Various specific mark-up languages are based on SGML, Standard Generalized Markup Language. Of the many electronic text centers now in existence, the center at the University of Virginia is outstanding: www.etext.lib.virginia.edu/eng-on.html.

new personages and topics into the historical literature. In the short run, these developments seemed to add spice to what was already a cosmopolitan field of study. In the longer run, however, the impact was to broaden significantly the frontiers of history and reshape the relations among historical sub-fields. The same influences, acting over a longer time period, might even reorganize the basic structure of historical institutions and inquiry. This review of disciplinary change in fields at the edge of historical studies begins with social sciences, then goes on to humanities and natural sciences.

Anthropology. Area-studies historians, as part of their interdisciplinary training, gained an introduction to the anthropological literature. Anthropology introduced these historians to social and cultural details of each region, but also to global comparisons, because the field of anthropology is organized more by theme than by region. Area-studies historians, through this introduction to anthropological paradigms, thus gained a view of social and cultural affairs that differed substantially from historians of Europe and the United States, where sociologists dominated social analysis and anthropologists were hard to find. Nevertheless the work of anthropologist Clifford Geertz, based on field work in Morocco and Indonesia, gained attention among historians of Europe and the United States as his emphasis on the complexity of social interaction and his technique of "thick description" for revealing complexity facilitated an expanded interest in cultural history.[32]

The fields of social and cultural anthropology, meanwhile, underwent fierce controversy as the established discipline, based on the study of kinship and debates on cultural evolution, met new critiques. The rise of neo-Marxian analysis and especially the impact of decolonization made clear to what a great degree anthropology had served as a research agency for colonial rulers seeking to maintain social control. New journals emerged—*Dialectical Anthropology* among them—and established journals such as *Current Anthropology* focused heavily on the debates. Kinship theory in particular was revealed to be deeply problematic, and anthropologists became particularly adept at debates over the standpoint and perspective of analysis, in which the categories of "neutral observer" and "participant-observer" lost their apparent simplicity.[33]

Demography. Demographers had long focused on the regions with the best demographic data and hence on Europe and North America. They had also abstracted from the complications of migration. But from the 1970s a combination of social concerns and technical advances brought demographers to the serious study of many world regions and to migration and other complexities in

---

32. Clifford Geertz, *The Interpretation of Cultures: Selected Essays* (New York, 1973). For an example of area-studies history relying heavily on anthropology, see Jan Vansina, *Kingdoms of the Savanna* (Madison, 1966).
33. Johannes Fabian, *Time and the Other: How Anthropology Makes its Object* (New York, 1983); Kuper 1988; Henrika Kuklick, *The Savage Within: The Social History of British Anthropology, 1885–1945* (Cambridge, 1991).

demographic analysis. Concerns with famine and birth control brought investment in long-term demographic surveys in Asian and Latin American countries, the development of electronic spreadsheets made easy the calculations of demographic tables, and advances in theoretical notation facilitated the representation of migration and changes in status. As a result, it became easier to study populations all around the world, and easier to trace the links among populations. Meanwhile, the continuing work on European demography led to such analytical landmarks as the estimation of England's national population from 1581 to 1841 and a comprehensive review of the complexities in the nineteenth-century decline in European fertility.[34] Major centers at Cambridge University, the Ecole des Hautes Etudes en Sciences Sociales, Princeton University, University of Pennsylvania, University of California–Berkeley, and Harvard University sustained much of this new work.

Archaeology. The field of archaeology continued its multidisciplinary tradition, linking studies from human biology, radiocarbon dating, plant physiology, art history and more into a comprehensive attempt to reconstruct pieces of early human history, site by site. While endlessly restrained on one side by a shortage of funds for excavation and analysis, and by pilfering of sites on another side, archaeologists nonetheless made striking discoveries in every area of the world, that provided a growing basis for adding a long-term perspective to the study of history. These pressed back the dates of earliest known human activities on many fronts: the stages of human evolution, the settlement of Australia and the Americas, the beginnings of kingdoms in China, the use of iron in Africa, cosmopolitan connections along the Nile Valley, and many others.

Geography. In the United States, geography almost disappeared as a teaching field, but thanks to the National Geographic Society, it continued to be influential at the level of research and in the eye of the general public. A major technical development, the Geographic Information System (GIS), created an electronic system for storing information linked to precise coordinates, and it led to great advances in mapping and in correlating information by geographic distribution. Geography, like history, is a highly interdisciplinary field, though the two fields have remained only episodically connected. In research, geographers have been important in environmental studies. The expanded teaching in world history and global studies, because they require basic geographic information of students, is providing an opportunity for geographers to connect to teaching as they have connected to interdisciplinary research. The National Geographic Society supports global studies centers at colleges and universities throughout the

34. E. A. Wrigley and Roger S. Schofield, *Population History of England, 1581–1841* (Cambridge, Mass., 1981); Ansley J. Coale and Susan Cotts Watkins, eds., *The Decline of Fertility in Europe* (Princeton, 1986); Peter Laslett, *The World we have Lost* (New York, 1966). See also Michael Gordon, ed., *The American Family in Social-Historical Perspective* (New York, 1973); and Dennis D. Cordell and Joel Gregory, eds., *African Population and Capitalism: Historical Studies* (Boulder, 1987).

United States, and Martin Lewis and Karen Wigen made a stir among history teachers with their volume criticizing the reliance on continents as the pieces of global geography.[35]

Philosophy. The field of philosophy is complex, filled with countercurrents, and difficult to summarize in itself, because it overlaps so much with other fields. The broad categories of positivism, Marxism, and postmodernism are insufficient to capture the range of contending ideas that philosophers debated in the late twentieth century. Here are two of the many influential lines of argument emanating from philosophy. In Marxian philosophy, with its emphasis on social change, the philosophy of Antonio Gramsci came to have wide influence. Gramsci, the Italian Communist whose prison notebooks of the 1920s and 1930s form the basis of his influence, focused in the intellectual influences in class conflict and developed such concepts as "hegemony" and "organic intellectual," which have influenced many studies in social history.[36] A later coalescence of critical ideas, centering in France during the 1960s, gave birth to postmodernist philosophy. In a mixture of philosophy, psychotherapy, and literary criticism, Jacques Derrida and Jacques Lacan formulated the idea of "deconstruction," according to which any artifact is a text and in which the analyst is to "read" the many intersecting texts of cultural life.[37] The work of the analyst cannot become definitive, however, in that the "intertextuality" gains a life of its own, so that the author of any text must share its meaning with the reader, and so forth.[38] These ideas, though transformed greatly in their wider use, gave substantial emphasis to the importance of language and provided a basis for new theories in literature, feminism, and other fields.

Historical linguistics. The initial strength in the field of historical linguistics came in nineteenth-century studies of Indo-European languages. Research and controversies in that field continue to add new information about long-term patterns in cultural history and on the origins and dispersion of populations speaking Indo-European languages. African languages became a major focus for successful work in the mid-twentieth century, with Joseph Greenberg's classification of African languages into four groups, each roughly equivalent to Indo-European languages, and his identification of southeast Nigeria as the point of origin for the migrations that ultimately covered the southern third of the continent with Bantu languages. Further study of African languages has led, on one

35. M. Lewis and Wigen 1997.
36. Antonio Gramsci, *Prison Notebooks*, ed. and trans. Joseph A. Buttigieg, 2 vols. ([1975] New York, 1996).
37. Jacques Derrida, *Of Grammatology*, trans. Gayatri Charavorty Spivak ([1967] Baltimore, 1976); Jacques Lacan, *Language of the Self*, trans. Anthony Wilden ([1969] Baltimore, 1981).
38. For a useful discussion of postmodernism as it has influenced historical studies, see Appleby, Hunt, and Davis, 1994: 198–237; see also David Harvey, *The Condition of Postmodernity* (Oxford, 1989).

hand, to detailed information on direction and timing of migrations and, on the other hand, to detailed studies of loan words and hence on the ideas and artifacts passed among populations. Greenberg classified the languages of the Americas into three groups, reflecting successive migrations from Asia into North America; and he and others focusing on Eurasian languages have laid out preliminary classifications of all the world's languages.[39]

Art history. In the later stages of this disciplinary revolution, several fields of cultural studies came to influence historical studies. Such fields as visual art and music, relying as they do on complex nuances and on the individual talents of the creator and performer, are not so easily formulated in theoretical terms as is economics. But if one speaks of a discipline in artisanal terms, then the practiced hand of the specialist has much to contribute to the explication of visual art, music, and of popular or elite culture in general. The field of art history, long focused on Renaissance and Early Modern European painting and sculpture, and with an analysis centered far more on aesthetics than on social context, began in the late twentieth century to expand beyond its traditional base. Art historians developed both a practical and theoretical basis for studying different periods of time and regions outside of Europe. One innovative and transdisciplinary initiative was the transference of the concepts of "pidgins" and "creolization" from linguistics to art history in the study of twentieth-century art in colonized areas of the world.[40]

Music history. The field of musical studies is of particular interest, since it has evolved in recent years into a joint project of work from the traditional musicologist approach (treating the musical work itself as the object of study) and the approach of ethnomusicology (treating the social context of the music and the perspective of musicians as part of the object of study). The study of music, as reflected in its principal U.S. journal, *Musical Quarterly*, is now global and interdisciplinary.[41]

39. Joseph H. Greenberg, *The Languages of Africa* (Bloomington, Ind., 1966); Greenberg, 1987; Peter Bellwood, "The Austronesian Expansion and the Origin of Languages," *Scientific American* 265 (July 1991), 88–93; J. Greenberg and M. Ruhlen, "Linguistic Origins of Native Americans," *Scientific American* 267 (November 1992), 94–99; Greenberg, *Indo-European and Its Closest Relatives: The Eurasiatic Language Family*, Vol. 1, *Grammar* (Stanford, 2000).
40. Paula Ben-Amos, "Pidgin Languages and Tourist Arts," *Studies in the Anthropology of Visual Communication* 4 (1977), 128–139; Jan Vansina, *Art History in Africa: An Introduction to Method* (London, 1984); W. McAllister Johnson, *Art History, Its Use and Abuse* (Toronto, 1988); 41.
41. Stephen Blum, Philip V. Bohlman, and Daniel Neuman, eds., *Ethnomusicology and Modern Music History* (Urbana, 1991). In recent years the *Musical Quarterly*, founded in 1915, has published a substantial number of articles written from an ethnomusicological perspective.

Geology. The acceptance of theories of continental drift in the mid-twentieth century and the elaboration of detailed analyses of the mechanisms of plate tectonics linked geology more tightly than ever to the evolutionary biology of animals and plants. For more recent times, the excavations of geologists into the sea floor and polar ice caps have provided information on the history of climatic change.

Biological and environmental history. Yet another dimension to the change in historical studies involved biological and environmental history. Studies in nutrition, disease, and other aspects of biological history began to be conducted in greater numbers. Beyond human biology, historians also undertook study of other elements of the environment—plants, animals, land, and the atmosphere.[42]

Medicine. Medical history developed as a close companion to medical practice to keep track of individual illnesses and larger social patterns in illness. Out of epidemiology grew studies of major waves of infectuous disease, that now have become a regular part of world-historical analysis. Through studies of the genetic components of illness have emerged such patterns as the link between sickle-cell trait and endemic malaria in West Africa. From public-health investigations a substantial literature developed on techniques for handling drinking water and refuse that became central to the history of cities.[43]

Perspectives, Theories, and Training in Historical Analysis

All of this new theory and artisanal practice, in its multitude of perspectives, has poured incommensurately into historical research and historical discourse. The impact is at once exciting and distressing. It expands the possibilities for history, yet makes interpretation more difficult. Sorting out the new methods and approaches, not to mention applying them, makes it clear that a comprehensive graduate course on historical methods today would be a far cry from one forty years ago. The field of history requires an updated conceptualization. This new view of the past should account for the new lenses 'on the past—the perspectives and frameworks through which we view history—and for the theories and metaphors with which we model the dynamics of past life.

It is well established in general that history is written from varying perspectives. For this reason, for the historian to accommodate to new perspectives is mostly a matter of devoting sufficient energy to understanding the background for each outlook. Nevertheless, the practices of historians explicitly identifying

---

42. On biological history, see Kenneth F. Kiple and Virginia Himmelsteib King, *Another Dimension to the Black Diaspora: Diet, Disease, and Racism* (Cambridge, 1981). On ecological history, see Cronon 1983; and Crosby 1986. The American Society for Environmental History was founded in 1976.
43. Kiple and King 1981.

their own perspective and considering the past from a variety of perspectives have expanded considerably in recent years. The question of why so many new perspectives should enter historical discourse is too vast to pursue in detail here, although it is certainly the case that the admission to advanced levels of the academy of people from a wide range of social backgrounds is a part of the change.[44] The practical task is that of identifying the legitimacy of each perspective (though not necessarily approving it) and learning enough detail to be able to recognize and articulate each perspective.

Addressing new theory is more complex; it involves learning the formal structure of each analytical system. In addition to studying individual theories, historians need to consider how the inclusion of formal theory changes the nature of historical study. In practice, the discussion has moved ahead only irregularly. Historians are not habituated to theoretical discussion. In earlier times, economic historians ignored economic theory, social historians ignored sociological theory, and cultural historians ignored anthropological theory. Historians learned only enough of the disciplines on which they were drawing to make some basic distinctions. When a new form of analysis becomes popular, historians have tended not to debate the concepts of a new analysis, but merely to adopt its keywords (deconstruction, gender, productivity, and so forth).[45]

Historians must find some techniques for formulating and navigating theoretical discussions, or they will simply waste the opportunity for clarifying the past brought by the confluence of so many disciplines in historical studies. Perhaps some agreement can be achieved for the meaning of the term *theory*. As defined in natural and social sciences, a theory is a set of assumptions and logical propositions relating a set of variables to each other. The theory yields predictions that may be compared to the observed patterns. Within the framework of the theory, one may propose hypotheses on particular variables and verify the hypotheses if they fit the observed data. More broadly, if the empirical data show patterns that are consistently at variance with the theory, a new theory may be proposed. This comprehensive vision of "theory" works best for theories that are focused on quantitative (usually interval) data, and for which one can perform experiments to collect data. On this basis, "theory" may be distinguished from such terms as hypothesis, conclusion, interpretation, model, metaphor, and so forth.

This notion of "theory" becomes more problematic when the "variables" are qualitative rather than quantitative (e.g., social class, religious belief, political power), and when the patterns under study are in the past, where they cannot be replicated.[46] And it becomes even more problematic for a field such as literature,

---

44. More generally, social standpoint has much to do with analytical or ideological viewpoint, among both current analysts and the historical figures they analyze, though standpoint and viewpoint definitely do not correlate precisely.

45. And by now the field of history has engaged so many different theories that a historian can claim it is impossible to know all of them and thus avoid theoretical discussion again, if on a higher level.

46. Note the difference between experimental fields, where new data may be collected according to experimental criteria, and fields collecting historical data, where the

in which the actual number of variable factors is immense, so that the distinctions made in literary theory are mainly the definition of various categories, a level of analysis that would be called "typology" rather than "theory" in other fields. Nevertheless, while a uniform definition of "theory" for historians seems out of reach, perhaps it can be agreed that the term should be restricted to the most formal statements of relationships in any discipline.[47]

The great expansion of new disciplines into history and the accompanying expansion of analytical approaches within history seemed for a time to displace the centrality of narrative in historical writing. This tendency was reinforced even with the introduction of more sophisticated literary analysis into history, because deconstructionists were exploding the notion that a narrative could be neutral.

In fact, the new methodological strength of history provided the tools with which to become clearer about the meaning of narrative. Most basically, historians need to remind themselves about the difference between their methods of analysis and their methods of presentation. Narrative is a method of presentation. It presumes an analysis that has been conducted in whatever form by the author and is now to be explicated for the reader. Narrative, clearly, is more than chronology, more than a simple ordering of all the evidence so that the documents will speak for themselves. The historical narrative, as with the literary narrative, cannot be independent of the perspective of the author.[48]

In the 1980s, conservative historians began to clamor for the return of narrative, as they objected to the focus on analysis and worried that it was alienating historians from the general public.[49] One way or another, narrative has worked its way back into prominence in historical writing, but no longer as the straightforward story of the past. The deconstruction of a narrative reveals the way in which authors lead the reader, dropping topics here, avoiding questions there, and shifting from macro- to micro-analysis and then back. The narrative is not so much the whole story as it is a work of art, in which the author selects parts of the story to tell and decides how to tell them. The critique of the narrative in a work of history is a project distinct from the critique of the analysis. The topic is

Continued

observer can retrieve but not set the conditions for the data. Geology, archaeology, astronomy, zoology, and physiology are among the latter, along with history. In these fields analyzing historical data, however, new data may be created through development of new techniques of analysis.

47. Rather than rely on formal theory to structure their analyses, historians have tended to rely on metaphor. For more detailed discussions of metaphor, comparative method, and the logic of systems, see chapter 16.

48. See the section on methods of analysis and presentation in chapter 17.

49. Lawrence Stone, "The Revival of Narrative: Reflections on a New Old History," *Past and Present*, No. 85 (1979), 3–24; Hayden White, "The Question of Narrative in History," *History and Theory* 32 (1984), 1–33; Marjorie Murphy, "Telling Stories, Telling Tales: Literary Theory, Ideology, and Narrative History," *Radical History Review*, No. 31 (1984), 33–38.

important for world historians, in that there is not yet a recognized and accepted form for presenting world-historical narratives.

Meanwhile, as the many new trends unfolded in each field of research and analysis in history, graduate study in history entered a crisis. In the United States, the employment of new Ph.D. historians reached a peak of 1,200 in 1975, then fell suddenly to 600, and began to climb slowly from the lower level only after a decade. Demographically, historical studies had undergone a boom during the 1960s as colleges and universities were constructed rapidly. Undergraduate student enrollment peaked in about 1970 and then declined, partly as the baby boomers passed through college age, but also as greater flexibility in the curriculum led to relaxation of the traditional requirements for U.S. history and Western Civ that had employed so many college history teachers. Many new doctoral programs had opened up in the 1960s, and new Ph.D.s focused particularly in social history. Yet from 1975, a new Ph.D. historian had a 50 percent chance or less of finding a job in history.[50]

Ironically, then, at the moment of greatest creativity and advance in historical research, demoralization came to dominate graduate education. Faculty members, seeking to avoid the prospect of training students who would never find work, and exhausted by the strain of reading two hundred dossiers for each replacement of a retiring colleague, put little energy into graduate programs. Undergraduate programs suffered less, but in an era when book publishers focused mainly on competing versions of shiny textbooks, the logic of product differentiation dominated: moving the chrome strips was a safer tactic than marketing a whole new design, much less introducing an Edsel. The emphasis in teaching remained centered on synthesis of established facts rather than on presentation of new research results. Thus, while historical research was developing and changing dramatically, the institutions for the study and the teaching of history changed very little. In reality the philosophy of historical studies at the level of research was under debate and in flux, but it was rare for undergraduate or graduate students to perceive this excitement.

As a result, graduate study in history in the late twentieth century fell out of touch with most of the new research techniques and results in historical studies. Instead, graduate training for historians went ahead at a reduced level within the bunkers of the approaches of earlier generations. The inherited system was characterized by students working with individual mentors in a single department to specialize in studies of a chosen region with a chosen methodology. According to this model, the only way in which broad and transregional studies are developed is for the historian to do an apprenticeship in local studies and broaden out to consider wider areas at mid-career and beyond.

50. Contractions in various sub-fields of history proceeded at different paces: African and Middle East history had tightened up in earlier years, but new and replacement positions continued to be offered in African American history. Robert B. Townsend, "Sharp Increase in Number of History PhDs Awarded in 1997," *Perspectives* (October 1999), 3–5.

World History and the Disciplines

What will be the place of history in a redrawn disciplinary map? The discipline of history has a dynamic of its own, as seen recently in the ups and downs of demand for courses, the rise of public history, and the interactions among fields (e.g., quantitative social history and popular culture).[51] But historical scholarship also reflects the dynamics of the disciplines with which it is closely associated—social science and humanities disciplines, as well as environmental and biological studies. Each of these shows a development, in recent times, toward broader and more interconnected styles of inquiry, highlighted by a rising importance of theory and databases.

History as it used to be threatens to be swallowed up in the transformations of the disciplines surrounding it. My guess, however, is that in wake of this ongoing reorganization of intellectual and academic life, the discipline of history will reemerge with a recognizable approach and character. I argue, not surprisingly, from my own experience: I was trained in the 1960s both as an Africanist and as a cliometrician, a new economic historian. I watched as the field of economic history moved from history departments to economics departments. For a time it seemed that hypothesis-testing would be the only way to do economic history. Indeed, hypothesis-testing remains central in that field. But I watched as the economists who stayed with the subject gradually became more like historians: their writing style improved; they began to season their bold and decisive analyses and to pause on nuances, on the specificities of one situation or another, on the ironies of timing.[52] Consider again the case of Douglass North, who began his career in quantitative analysis on American materials in which institutions were only constraints at the edge of his system. He then moved to a study of European development in which institutions became the key to growth, and then to a study that is really the philosophy of history.[53] His Nobel Prize was awarded actually for the first stage of his career, but the long-term development of his thinking is more illustrative of the interplay we might expect between the innovative introduction of social science theory into history and the longer-term adjustments in history and its adjoining fields.

The new disciplinary frontiers will be different from and more permeable than the old: we will read journals across what were once disciplinary lines, use each others' research techniques, and apply each others' theories. But I think that when the dust settles from this particular set of transformations, the study of history will still be, recognizably, the offspring of historical studies in the past.

One side of the historian's task will be a continuation of the traditional role of guardian and synthesizer of the evidence and teller of nuanced tales of the past.

51. Robert B. Townsend, "New Data Reveals a Homogeneous but Changing History Profession," *Perspectives* (January 2002), 15–17.

52. G. Wright 1978; Roger L. Ransom and Richard Sutch, *One Kind of Freedom: The Economic Consequences of Emancipation* (New York, 1977).

53. North 1966; North 1973; North 1981; North, *Institutions, institutional change and economic performance* (Cambridge, 1990).

There will still be narrative history, and historians will remain the specialists at combining diverse categories of evidence into stories constructed with a focus on the passage of time.

The new side to the historian's role will be that of synthesizer of methodology. Historians taking on this new function will address their topics by mediating among the theoretical and methodological alternatives and combining them or alternating among them artfully in interpreting the historical record to provide a comprehensive and, one hopes, realistic view of the past. Historians may be masters of few of the academic trades they will ply but journeymen at many of them. And as history in the past was tied closely to the traditional library, so will history in the future be tied to many dimensions of the transformed library, with its multimedia collections.

The migration of disciplines from social science, humanities, arts, and natural sciences into historical studies is a movement with tremendous potential for advancing the understanding of world history. Rather little of that potential has been realized so far. World historians have generally worked with established tools of their field and have made their contribution more by expanding their scope rather than by methodological innovation. On the internal path toward world history, world historians have drawn selectively on new perspectives, and they have been changed more by perspectives than by theory. World historians have drawn heavily on area-studies perspectives and on the critique of Eurocentrism. World historians have drawn on the late-twentieth-century atmosphere of religious ecumenism and political decolonization and the expanding literature in global political economy. The shifts in social-science perspectives to focus substantial attention on class, race, gender, and ethnicity have had more effect on studies in national history than on world history.

World historians have drawn more selectively on new theory. Instead of taking up macroeconomic or microeconomic theory or social-historical analysis, world historians have concentrated on histories of commerce and institutions that can be traced back to earlier innovations. The global studies of slave trade and other migrations, however, are an area in which new demographic theory and social analyses have contributed significantly.

Along the external path to world history, the innovations have come to history not only in the form of new perspectives, but also as new data and new theory. Perhaps the best examples of new perspectives come in the field of anthropology, where the decolonization of the field led to fundamental challenge of the notion of the "other" and the "primitive contemporary" on which so much of the anthropological literature had been based. This revolution in anthropological perspective also led to changes in theory, as conceptualization of kinship and ethnicity changed in the late twentieth century to give a great deal more emphasis to situational choices and less emphasis to inherited tradition in explaining the social structures and decisions of the populations under study.

New data and new theory have been outstanding in archaeology, in which the reconstruction of human evolution has become dramatically more detailed, and in which those results have reinforced human monogenesis at unexpected levels. In studies of the physical environment, the development of systems approaches to the geosphere, atmosphere, and biosphere have now established long-term frameworks for human existence and at the some time have shown human impact on all aspects of the environment.

Cultural studies have developed in manifold fashion but have yet to bring substantial change to the literature on world history. The global patterns and changes in culture, while actively under study, have yet to be summarized in ways that will sharply inflect interpretations of world history. If the patterns in other fields are followed, scholars in disciplines beyond history will begin to offer historical analyses of cultural transformations, and historians will join in the discussion subsequently.

With so many new fields to draw on, where will world historians gain adequate training in the handling of perspectives, theories and models, and in the specifics of multiple disciplines? Alfred Crosby's studies of environmental history provide an encouraging example; Crosby drew on old and new findings from several disciplines to present a bold picture at the planetary level. He completed his global studies as an individual scholar rather than as part of an interdisciplinary group, though he benefited from academic leaves that put him in close contact with medical and biological scholars.[54] In Crosby's case individual energy and imagination were sufficient to span the range of evidence and ideas. But for other fields, the materials may not so easily be assembled and digested by a single scholar, or there may not appear such enterprising and energetic individuals. The task of developing scholars to match the breadth of emerging connections in world history is daunting.

This chapter has emphasized the role of historians as recipients of new ideas from other fields of study. Along the internal path this is the familiar interplay of historians with sociologists and literary analysts; along the external path it is the newer and sometimes jarring encounter of historians with biologists and mathematical theorists of chaos. Yet I suggest that the result of the expanded interaction of disciplines will elevate historians to a new position as the synthesizers of methodology. Is there any evidence that as the scope of history expands, historians are becoming creators and exporters of frameworks and theories in addition to their established role as artisans applying the conceptions of others to historical data? I think it will be uncommon for historians to make the theoretical discoveries that come with deep disciplinary specialization, though historians rank high among those who discover new evidence that is itself thought-provoking. But historians have been especially attentive to the operation of large systems and have pointed out the inconsistency and incompleteness of previous studies in a way that adds clarity to our understanding of the system. One example is the literature on slavery and slave trade, in which a network of studies since the 1960s has broken down the earlier pattern of episodic and isolated national studies to reveal a worldwide pattern of the expansion of forced labor into the early nineteenth century and its gradual dissolution and transformation in the century thereafter. In contrast to earlier views of slavery as a backward and exceptional

54. Crosby 1986.

institution, slavery is now emerging as a central though problematic pillar in the development of the modern world. Another example is the literature on environmental history, in which historians have linked the many individual discoveries in environmental studies into a model of the global environmental system, showing the interactions among human and other environmental influences, from region to region, and in the distant past as well as in recent times. The ideas being created by historians are complex rather than simple, but they are nonetheless important for an understanding of the world they address.

Conclusion: New Tools

Historians live in an academic world that is now populated by many disciplines. Each of them addresses fascinating issues in human society and the natural world. And there are so many compelling and relevant combinations of disciplines. Faced with this intellectual feast, historians will find the choices difficult: it will take a substantial expense of energy to achieve depth in disciplines, connections among disciplines, and critique of disciplines all at once.

History, though recognized as a discipline in itself because of its emphasis on change over time, has never been able to maintain independence from other disciplines. In earlier times, history was most closely tied to studies of politics and war, then to studies of commerce and literary culture, and in the twentieth century to sociology. But as knowledge and theory advanced in every discipline during the late twentieth century, history became implicated in all the changes. Groups of social historians gathered to assess the new developments in social history; individual scholars assessed the possibilities in environmental and cultural history.

The point that is only now becoming clear is that there is more to this phenomenon than a series of parallel changes in different arenas of historical studies: the changes in each segment of history are changing the field as a whole. No longer do historians simply develop an additional specialization in an adjoining field.[55] Now they learn the theories and study the interplay of several historical disciplines at once. The graduate students I work with are immersed in a complex mix of intellectual, social, cultural, environmental, and political history.

Thus the two changes brought to historians by the disciplinary revolution are that historians must now learn theory in their field of specialization, and that they must also understand groups of theories and how to combine them. World historians need to specialize, generalize, and balance the two tendencies. Thirdly, world historians need to maintain their critical sense and scrutinize the work done in each discipline.

World historians also need to know when enough is enough. As we will be reminded in the chapters to follow, in addition to working in multiple disciplines, world historians have taken on the problems of working across large and diverse

---

55. In my case it was economic history in graduate school, then social and demographic history in later years.

spaces and often over long periods of time. The exciting new prospect is that the revolution in historical methods gives us wonderful new tools and techniques to work with. The problem in facing this methodological feast is finding a way to achieve depth of analysis without sacrificing breadth or global connection. But the very breadth of interdisciplinary work presents certain advantages; one can view problems from differing perspectives, and one may see the parallels and linkages from one discipline to another. There are global patterns to be discovered in the linkage of academic disciplines, just as there are global patterns to be located in the history of the world.

Chapter 8

Area Studies

A rea-studies scholarship is revolutionary in a reformist sort of way. It is not fundamentally distinctive in its analysis: instead, it is the application of the established sort of academic work to regions beyond the established terrains of study. The move to establish parallel disciplines and equal standards for scholarship on for each area of the world did, however, conflict directly with the inequalities of the age of imperial, colonial, and racially discriminatory scholarship. The rise of area-studies scholarship thus brought about an intellectual decolonization and democratization paralleling, in some measure, the contemporary transformations in global politics. In addition, and precisely because the area-studies scholars were organizing their fields at a time of methodological innovation, they were able to take advantage of some new approaches more readily than their colleagues focusing on Europe and North America, and thus make up for more of their deficit.

At the same time, there remained some distinctiveness in the organization and approach of area-studies scholarship. First, it was interdisciplinary: specialists in politics, economics, geography, sociology, anthropology, and languages worked closely together, sharing insights and methods. Second, history played a central part in area-studies ventures. In each case, scholars for each region needed to emphasize that the region had a history and a deeply rooted identity. Even in these cases, the distinctiveness may have been a function of these changes having come earlier to area-studies scholarship than to social sciences focusing on Europe and North America. Area-studies scholars, while they were sometimes treated as insurrectionists by metropolitan leaders of their fields, more often than not worked with the objective of acceptance into their disciplinary folds. If they were social revolutionaries, they were not academic revolutionaries.

Yet the ultimate impact of area-studies scholarship was to contribute forcefully to the revolution in historical studies along two major axes. First, by applying skilled social-scientific research to areas of the world previously neglected, area-studies scholars tested and advanced the existing scholarship on patterns at local, national, and regional levels. Second, the move toward equalizing scholarship across regions allowed for the recognition of global patterns that were not previously visible. In this survey of area studies and its impact on historical studies, I address the emergence of area studies, the character of area-studies scholarship, the interdisciplinary character of area studies, area studies as a model for developing new scholarly fields, some particular insights from African studies, and world history as Third World history.

The Emergence of Area Studies

Universities expanded everywhere in the world in the aftermath of World War II, but they expanded with particular force in the United States. Not only did American universities take responsibility for educating a larger portion of the population and expanding into new disciplines, but they attempted to broaden knowledge about every area of the world. Given strong support from the federal government, especially in major universities, the postwar era brought the expansion of historical studies for each of the recognized world regions and a renegotiation, at the intellectual level, of their relations with each other. The result was a revolution in perspective within history as Latin American history was reaffirmed and as historical studies were organized for East Asia, the Middle East, South Asia, Slavic studies, and Africa.

Studies of British history in the United States had risen to a relative peak in the years after World War II, and some of these British historians concentrated on the British Empire. These persons, in turn, were key figures in the early years of area-studies programs, as they were able to focus, through the imperial angle, on the history of India, Southeast Asia, Africa, the Middle East, Latin America, and so on. Meanwhile the American-based approach to area studies was more like earlier German approaches than like the British approach: U.S. libraries and instructional programs expressed interest in all areas of the world, not only those to which national political ties were the strongest.[1]

As the 1950s progressed, the U.S. federal government began funding area-studies centers. These centers provided support for new faculty members, library acquisitions, administration, and language study. The language study was supported by the National Defense Education Act and the National Defense Foreign Languages Act. This legislation, on the premise that it was in the interest of U.S. national defense to develop academic specialists with strong knowledge of languages relevant to their study, underwrote the language training of many area-studies doctoral students.[2] The major area-studies programs built up faculties to carry on multidisciplinary work in history, anthropology, political science, sociology, economics, and languages. At a later stage literature and the arts were added to some programs. In addition to their degrees, graduate students received interdisciplinary certificates of study in their area. Area-studies programs were also required

1. As two examples among many, Graham Irwin in the United States moved from British Empire to African studies, and Henri Brunschwig in France moved from French Empire to African studies.
2. Ravi Arvind Palat gives particular emphasis to the Cold-War origins of area studies scholarship. Palat, "Fragmented Visions: Excavating the Future of Area Studies in a Post-American World," in Neil L. Waters, ed., *Beyond the Area Studies Wars: Toward a New International Studies* (Hanover, N.H., 2000), 64–66.

by their federal grants to conduct outreach programs aimed at conveying new knowledge on each area to public school teachers and to the public generally.

The expansion of area-studies programs relied not only on governmental support, but on substantial support from private philanthropic foundations. The Ford and Carnegie Foundations were leaders in area-studies scholarship, and especially in the creation of new programs for interdisciplinary study, notably in the social sciences and history. The American Council of Learned Societies and the Social Science Research Council, each of which had existed for some decades, assumed roles as coordinating bodies for area studies.[3]

The dates of creation of area-studies journals and associations provide a good outline for the emergence of this organization of academic work. Major area-studies journals were founded during and after World War II: thus, *Slavic Review* (1941), *Journal of Asian Studies* (1941), and *Middle East Journal* (1947). The main journal in Latin American history, *Hispanic American Historical Review*, was launched at the end of World War I, in 1921, while the *Journal of African History* was founded relatively late (and in Britain), in 1960. The Latin American journal is historical because it was founded before area studies; the African journal is historical because of the importance of affirming the existence of history in the study of Africa.

The African Studies Association was formed in 1958, the last of the major area-studies associations to be established in the United States. Prior to it were formed the Association of Asian Studies, the Middle East Studies Association, the American Association for the Advancement of Slavic Studies, and the Latin American Studies Association. These interdisciplinary organizations became the main professional focus for historians in these fields. Since area-studies historians tended to attend meetings of their respective area-studies organization rather than the American Historical Association, the AHA remained primarily a forum for historians of Western Europe and the United States. Thus, while the historians of Asia, Africa, and Eastern Europe expanded greatly in number and produced rising quantities of new work, the reflection of this work at the level of the main organization of historians in the United States, the AHA, was meager. Historians of Europe and the United States might thus seriously underestimate the quantity and quality of work by area-studies historians even into the 1990s. The exception to this rule was the historians of Latin America. Latin American history was organized as a field at the end of World War I, long before the rise of area-studies programs, and the Conference of Latin American Historians (CLAH) was formed in the interwar years as an affiliate of the AHA, organizing its own segment of the annual program. Even with the rise of the Latin American Studies Association, historians continued to participate actively at AHA meetings through CLAH. Meanwhile, occasional moves to establish equivalent conferences for historians of Africa or Asia, to increase their participation at the AHA, came to naught. (In Canada, in contrast, most all of the learned societies met together each year, so that historians of Europe and North America were more likely to be aware of the work of area-studies historians.)

3. The ACLS (www.acls.org) was founded in 1919, the SSRC (www.ssrc.org) in 1923.

The major universities of the United States each came to specialize in one or two areas. African studies became particularly strong at Wisconsin, Northwestern, UCLA, and Indiana. For Latin American studies, Berkeley and Texas were major centers; Harvard developed strong centers in East Asian studies and in Soviet and East European studies; Chicago was a major center for Middle East studies and South Asian studies.[4] The regions were covered unequally: American academia gave relatively little emphasis to South Asia or Southeast Asia.

Scholarship in Area Studies

The objective of many area-studies scholars, in the competitive mind-frame of the time, was to establish that the scholarship and the historical experiences of "their" area were on a par with those of Europe and North America. In particular, area-studies scholars explored and justified the nationalist movements of the areas they studied and gradually found ways to show that the national experiences of Mexico, Nigeria, India, and Turkey paralleled those of France and Germany. Area studies thus served to replicate national and civilizational paradigms for the study of history in every part of the world. Latin American studies focused on the main nations of the region, Mexico and Brazil, as South Asian studies focused on India. Or to present the framework with a different twist, since we have a long tradition of teaching African history in the same semesters that we teach French history, students in the United States came to think of Africa as a country rather than a continent composed of many countries.

Within the area-studies framework, comparative analysis has been central in the methodology of transnational studies. The logic of the research design is that comparison among cases provides a way of getting variance in historical data and transformations, thereby enabling the researcher to identify the most important elements within each case. Such a comparison, while often effective in analyzing individual cases, does not easily lead to explaining the operation of larger systems. To return to the example with which I opened chapter 1, local and comparative studies of the rise of racist ideology and racial segregation can succeed in posing the question of why racial segregation spread across the world with such force at the turn of the twentieth century, but they do not get far in providing an answer.[5]

The scholarly accomplishments of area studies are considerable: two generations of area-studies scholarship have refined important new techniques of historical research, collected an immense volume of new evidence, and fundamentally shifted the regional balance of historical discourse. In methodological terms, one may note the formalization of study of oral tradition, especially for Africa but for every other region as well. The reaffirmation of field work in

4. For the area-studies programs at the University of Wisconsin and their impact on studies in world history, see pages 327–328.
5. Michael Adas has become one of the leading advocates and practitioners of comparative approaches to world history. Michael Adas, *Prophets of Rebellion: Millenarian Protest Movements Against the European Colonial Order* (Chapel Hill, 1979).

historical research was so successful that a "library dissertation" became a term of opprobrium and the meaning of "document" was expanded from diplomatic correspondence to include oral "texts," archaeological remains, and popular music. While much of the emphasis of area studies has been on interdisciplinary study, it is equally true that the researches of area-studies historians have uncovered major new written sources.[6]

The debates within area studies took on a character of their own. The progress of decolonization brought the field of anthropology everywhere into an encounter of its complicity with colonialism, and to a wave of self-critique within the discipline. For Middle East studies, a parallel debate led to an extensive critique of Orientalism, the philologically based studies of Arabic and other Islamic texts that had been associated with a Christian-based critique of Islam. In much shorter time-perspective Southeast Asian studies rose and fell with the Vietnam War. Russian and Eastern European studies remained well funded yet mired in Cold War politics, until they fell into sharp decline in the 1990s. East Asian Studies centered on the encounter with the West, particularly in the nineteenth century. Studies of Africa focused heavily on critique of colonialism and racism.[7] For each of these areas, scholarship linked the region to Europe or the United States more than to the world in general. For Africa and Latin America, for instance, studies on slavery and slave trade brought an eventual link of research to U.S. and European literatures.[8]

Among the topics of historical literature for which bookshelves became more heavily weighted as a result of area studies scholarship were slavery (especially for Africa and the Americas), peasantry (for every region), regional and global commerce, economic innovation and growth, family and social change, empire and colonialism, resistance to colonialism, and nationalism and national identity.[9]

6. Research during the 1960s, for instance, located Arabic-language documents written in the eighteenth and nineteenth centuries for the courts of the kingdoms of Gonja and Asante in modern Ghana—held in the royal library of Denmark. Ivor Wilks, *Asante in the Nineteenth Century: The Structure and Evolution of a Political Order* (Cambridge, 1975), 347–348.

7. On the Middle East, see Bernard Lewis, *The Emergence of Modern Turkey* (London, 1965); Lewis, *Islam in History: Ideas, Men and Events in the Middle East* (New York, 1973); Lewis, *The Muslim Discovery of Europe* (New York, 1982); and Said 1978. On Russia and Eastern Europe, see Richard Pipes, *The Formation of the Soviet Union: Communism and Nationalism, 1917–1923* (Cambridge, Mass., 1954); Pipes, *Property and Freedom* (New York, 1999); Pipes, *Communism: A History* (New York, 2001). On East Asia, see John K. Fairbank, *Trade and Diplomacy on the China Coast: The Opening of the Treaty Ports, 1842–1854* (Cambridge, Mass., 1953); Ssu-yu Teng and John K. Fairbank, *China's Response to the West: A Documentary Survey, 1839–1923* (Cambridge, Mass., 1954); John K. Fairbank and Edwin O. Reischauer, *China: Tradition and Transformation* (Boston, 1978).

8. E. Williams [1944]; Frank 1966; David Brion Davis, *The Problem of Slavery in the Age of Revolution, 1770–1823* (Ithaca, 1975).

9. For bibliographic surveys on slavery, see Joseph C. Miller, *Slavery and Slaving in World History: A Bibliography, 1900–1991* (Millwood, N.H., 1993), and Patrick Manning, "Introduction," in Manning, ed., *Slave Trades, 1500–1800: Globalization of Forced Labour* (Aldershot, U.K., 1996), xv–xxxiv.

The time perspective of area-studies research has focused heavily on modern times, particularly on the nineteenth and twentieth centuries. Still, area-studies research is associated with important discoveries on early times—the fields of archaeology and linguistics benefited from the rise of area studies and filled in new information on ancient times.[10]

The energy with which each of these area-studies enterprises expanded meant that scholars concentrated heavily on their own region. If there was substantial exchange of information among scholars of different disciplines working on Southeast Asia, there was precious little exchange of those same scholars with others working on the Middle East. Until the 1980s, the area-studies groupings remained largely segregated from each other.

Nonetheless, some ideas did pass among the various developing area-studies traditions. The rise and transmission of the modernization paradigm provides a striking example of the dynamics of connections among area-studies scholars. This formalization of the division of societies and outlooks into the traditional and modern, with its universalization of the modern and its teleological anointing of those who would lead the struggle to become modern, developed first as an explanation of the conflicts in social change within the Middle East. The model was soon taken up by scholars in African studies, but within a few years it was abandoned in all aspects except for the occasional use of the term. However heartening as an image of the future, the paradigm was oversimplified: it abstracted from too many important factors, such as divisions of social class, global economic relations, and the complexity of beliefs, so that it predicted little. Nonetheless, scholars working on Latin America and East and South Asia took up the modernization paradigm as it was being set aside by their predecessors. Europeanist and Americanist scholars tried out the paradigm last and went through the same steps of adoption, exploration and rejection—though they might have learned from their area-studies colleagues.[11] Better communication among scholars working on different areas might have resulted in considerably less analytical waste. Perhaps the problem was that area-studies scholars, preoccupied with practical problems of creating literatures for their fields, were still ready to acknowledge the intellectual and theoretical leadership of their Europeanist colleagues.

Area-studies approaches to history, in contrast to the literatures in U.S. and European history, tend to be multinational and decentralized, crossing many lines of language and culture. This has been most true for studies of Africa, for which scholars are employed in African nations, in Western European nations, in the United States, in Latin American and Caribbean countries, and beyond. For Latin America, scholars based in the United States and Europe interact with those based in the various Latin American nations. For the Middle East, scholars working in

10. Mair 1998; Greenberg 1966; Greenberg 1987.
11. Parsons and Smelser 1956; Almond and Coleman 1960; David E. Apter, *The Political Kingdom in Uganda: A Study in Bureaucratic Nationalism* (Princeton, 1961); Manfred Halpern, *The Politics of Social Change in the Middle East and North Africa* (Princeton, 1963).

the various languages of the region interact with those from Europe and North America who have taken up specialization on the Middle East. For East and South Asia, the variety of perspectives is somewhat reduced by the strength of the academic traditions in China, Japan, and India. For India and South Asia generally, most scholarship is produced in the English language. For China, the differing perspectives among scholars based in the People's Republic, Taiwan, Hong Kong, and overseas gradually changed from ideological confrontation to a critical dialogue. Studies of Eastern Europe and the Soviet Union have been something of an exception; they tended to be highly ideological and conformist on both sides of the political divide during the Cold War era. In general, scholars in area-studies fields are ideologically diverse, and their debates challenge a wider range of presumptions than might otherwise be the case. In addition, the literatures in area studies for both ancient and modern times have tended to include a focus on interaction in human affairs, if only because of concern for the impact of the West on each region.

An additional emphasis of interest is that studies of Africa, Latin America, the Middle East, and Eastern Europe have entailed frequent references not just to local or national units but to the region as a whole and to its diaspora communities. Southeast Asia has come somewhat more slowly to be seen in the scholarly literature as a coherent region. The big nations of East and South Asia tend to be treated on their own, but studies of Chinese, Indian, and Japanese diasporas add a global dimension to regional study. Thus each of the area-studies traditions provides its scholars with experience in thinking beyond the national level to consider broader, regional commonalities and interactions.[12]

At present, and outside of East and South Asia, it is probably true that more research is being carried out by scholars based in North America than on any other continent. At the peak of the area-studies movement, the large programs with government-funded programs were the sites of most research—Columbia University, for instance, maintained such programs for study of Africa, Latin America, the Middle East, East Asia, and Eastern Europe. With the passage of time, and with the graduation and employment of doctoral students from the major university programs, a much larger number of small-scale programs has emerged, so that area-studies research in North America is now spread across a wide range of institutions. Meanwhile, universities throughout the world have developed expanded programs for historical study. In Africa this trend has been particularly striking; only a tiny number of universities existed at the end of the colonial era, while now most nations have at least one university, and Nigeria has over thirty-five. Yet for Africa, as for so much of the world, universities have experienced periodic closure because of political upheavals, and budget constraints from periodic recessions or from the external dictates of World Bank structural adjustment programs have limited the ability of university faculty members to conduct and publish their research.

---

12. Colin Palmer, "Defining and Studying the Modern African Diaspora," *Perspectives* (September 1998), 1, 22–25.

The point is that area-studies scholars contribute to broad regional literatures or a series of regional literatures in several languages. The diversity of cultural and national viewpoints among area-studies scholars is impressive enough, and the full range of diversity is quite striking. We may contrast this diversity with studies of history and social science of the United States, Britain, France, or Japan, where virtually all the leading scholars come from a single national tradition, and where one or two universities sometimes achieve a hegemonic position in directing research and interpretation.

In the United States, Britain and Canada, annual conferences gather scholars of most disciplines in the various area studies. In France, periodic interdisciplinary conferences organized by the national research center, CNRS, bring area-studies scholars together. Various research councils sustain contacts among scholars within each of the regions: in Africa, CODESRIA (Council for the Development of Economic and Social Research in Africa) has done a remarkable job of supporting research and publication, working especially to link Anglophone and Francophone traditions. In addition, UNESCO has provided essential support for conferences, research and publication for scholars from Africa, the Caribbean, Latin America, and the Middle East. The coverage of these meetings is less than comprehensive, and there remain many separations and some antagonistic divisions among groups of scholars. Still, the structures of area studies are more remarkable for their inclusiveness than for exclusiveness. Reviews of the literature have helped reaffirm the inclusiveness of area studies, both in empirical studies and in analytical approach. In the United States, the Social Science Research Council commissioned a set of research overviews on areas of scholarship within African studies during the 1980s. These reviews did an excellent job of integrating the continental literature.[13]

Area studies histories in the postwar era began as political histories and went through the transformation to inclusion and even dominance of social history, as happened in European and U.S. history. But the transition was eased for area studies; historians gained some formal instruction in other disciplines as part of their graduate study. Even in the study of political history, area studies brought a difference: area-studies historians became acquainted with the field of political science and added to their narratives the conceptual rubrics and development schemes then current in political science.[14]

Area-studies scholars also studied anthropology, and therefore participated actively in the wave of peasant studies of the 1960s and 1970s—more actively, it might be argued, than their Europeanist colleagues.[15] Area-studies scholars

13. Research overviews were published in the *African Studies Review* each year from 1981 to 1990.
14. Apter 1961; Apter, *The Politics of Modernization* (Chicago, 1965); Halpern 1963; Gabriel A. Almond and Sidney Verba, *The Civic Culture: Political Attitudes and Democracy in Five Nations* (Princeton, 1963). But see Rupert Emerson, *From Empire to Nation: The Rise of Self-Assertion of Asian and African Peoples* (Cambridge, Mass., 1960); and James Smoot Coleman, *Nigeria: Background to Nationalism* (Berkeley, 1958).
15. Eric R. Wolf, *Peasant Wars of the Twentieth Century* (New York, 1969).

received substantial training in the language and literature of their regions and, to a lesser degree, training in linguistics and psychological studies. This led to such interpretive emphases as that on dependency for Latin America, social and cultural anthropology and historical linguistics for Africa, geography for East Asia, and political science for the Middle East.

But the changes were problematic as well. In economic history, economists gained control for the United States, but elsewhere economic history remained under the leadership of historians, and often historians with little or no economic training. Thus, at times the links to other disciplines made area studies historians more broadly interdisciplinary in their training and thinking, leading to works of new sophistication. At other times, the links to other disciplines caused historians simply to become more audacious, and willing to speculate in areas beyond political history without training themselves in those fields.[16]

The interdisciplinary and transregional approaches of area studies, in sum, brought great new strengths to the historical analysis of large areas of the world, and were absolutely requisite to developing cosmopolitan approaches to world history. Yet the distinctions as well as the commonalities of area studies and world history retain their importance: for instance, area-studies history has come to rely rather heavily on the anthropological literature, while world historical studies remain virtually devoid of anthropological emphasis.[17]

Area-Studies Models for Creating Academic Industries

The rise of area studies brought major changes to the structure of the historical profession. Prior to 1965, few members of history departments in the United States conducted research on areas outside the United States and Europe, but by 1980 most departments of history had area-studies faculty lines. Similarly, in many other academic fields, area-studies specialists gained a significant number of positions. Beyond gaining jobs, with the passage of time, area-studies scholars rose to be chairs of departments, heads of search committees, and deans and presidents of colleges. Area-studies scholars rose to the peak of the elected and appointed positions within the historical profession and achieved substantial recognition in such influential bodies as the SSRC and ACLS.[18]

The central achievement of the area-studies project has been not simply that of creating employment, but rather of creating knowledge. Within a few decades— through creation and exploration of archives, publication of books and articles,

16. For instance, the *Indian Economic and Social History Review* and *African Economic History* include few articles by scholars with training in economics, and the latter journal is not indexed in the *Journal of Economic Literature*.

17. For instance, works in anthropology are rarely reviewed in journals on U.S. or European history but are commonly reviewed in area-studies journals and in historical journals addressing regions outside the North Atlantic.

18. Sandria Freitag, a historian of South Asia, served as executive director of the American Historical Association in the mid-1990s; during her term Joseph C. Miller, a historian of Africa, served as AHA president.

definition and resolution of debates—the scattered elements of scholarly studies addressing most of the world's area and population had been caused to coalesce into well-organized fields of study. The area-studies approach put substantial resources into the hands of leading scholars in comprehensive university programs, backed by effective professional associations able to coordinate efforts and review the research agenda. The results of this research led to significant changes in public policy, public education, and the popular images of the regions addressed in this new research.

A further strength of area-studies scholarship is that it is inherently international and has various built-in structures to ensure that scholarly discussion will include a number of national and ideological perspectives. In this respect, area studies scholarship has an advantage over that, for instance, of the consensus-based scholarship of the United States during the 1950s and 1960s. There a homogeneous social background and common ideological outlook of scholars meant that despite their high technical proficiency, they were unlikely to debate or challenge many basic assumptions. For the history of the Middle East, in contrast, scholars from the United States and Europe must carry on debate with those from the various countries of the region itself and of surrounding countries—scholars of various religious and secular backgrounds within Islam, Christianity, and Judaism. In this case, perhaps too much time is spent challenging assumptions, but there is little danger of an orthodoxy gaining uncritical approval.

Each of these fields of study required not only addition of new knowledge within existing frameworks, but critique of preexisting frameworks and creation of new paradigms and new categories of knowledge. In African studies, the identification of racialistic and imperial biases in data and interpretation was an important part of analysis. For Middle East studies, the identification of anti-Muslim biases in orientalist studies was a major issue. In East Asian studies, scholars worked against stereotypes as well. Slavic studies was perhaps the most problematic of the area-studies projects, in that it developed precisely in the era of Cold War confrontation between the United States and the Soviet Union, so that scholars were divided between an adopted loyalty to the region under study and a political project of overthrowing its current leadership.

This experience suggests that for construction of a substantial field of study of global history, area-studies scholarship provides a tested and successful model on which to build a new field. The lesson goes beyond the area studies concerned. It shows what results can be achieved by placing significant effort on the development of new fields of study, under conditions requiring interdisciplinary study, intensive language training, and research in distant locations. To exaggerate only a little, we know how much to spend, how many students to support, and how to organize the journals and conferences. It is surely the case that establishing global historical studies will have to follow a slightly different pattern, accounting for previous developments. But world historians need not start from scratch.

At the same time, the area-studies project has had its limitations and its failures, and these should be kept in mind in proposing an analogy between area studies and any project in global historical studies. While area studies research centers in the United States are more broadly dispersed than in the United Kingdom or

France, where Cambridge, Oxford, London, and Paris remain the great centers of scholarship, the unevenness of resources is a great problem. Area-studies scholars can carry out their work at the major centers, but those in institutions beyond those centers are restricted in their ability to participate.

The problem of uneven resources, while notable within the United States, is far more serious at an international level. While it was a common expectation of area-studies scholars, in their early days, that academic leadership would soon be assumed by scholars and universities in the home countries of Africa, Asia, and Latin America, this has rarely turned out to be the case. For reasons having both to do with academic politics and with global political economy, universities outside Europe and North America have been sharply restricted in resources, unable to build library collections, incapable of supporting large programs of graduate study, and unable to support active publication programs.[19] Scholars flee their home countries to find work in the United States, thus broadening the discourse in the United States but narrowing it elsewhere. The main academic journals, based at metropolitan universities, accept few submissions from overseas, arguing that the authors are not sufficiently well read in the recent literature or have not consulted relevant documents. As a palliative to this growing gulf, travel funds have been located to bring scholars from Pakistan or Brazil to area-studies meetings in the United States, but this practice does little to strengthen scholarship in Pakistan or Brazil. The gulf is a gulf in language and race as well as in location, for the English-language literature is encouraged to grow at the expense of publications in Urdu or Portuguese, and the academic meetings in the United States, while always cosmopolitan, remain dominantly white in complexion.

A second limitation of area studies, from the viewpoint of global studies, is that area studies serve to reaffirm regional solidarity and intraregional contacts, but not connections with other regions. The positive side of this organizational emphasis is the development of cosmopolitan ties and consciousness within the region. Certainly the enterprise of African studies has served to develop ties between East and West Africa and between Anglophone and Francophone Africa—though the tendency to emphasize national units means that Nigeria, South Africa, Kenya, and Senegal get a disproportionate amount of attention. But the negative intellectual result of this regional solidarity is parochialism and exceptionalism—parochialism in that Africanists know about Africa but not other regions, and in that Africanists who know their own corner of Africa tend to assume that it is representative of the continent.

The problem of exceptionalism in area studies is more severe. It shows up particularly when an area of study finds itself on the defensive, as is the case for Russian and East European studies in the aftermath of the Cold War. To sustain study within the region, scholars may be tempted to argue for that region's uniqueness. Thus, rather than engage scholars working on other parts of the world to seek explanations of the reasons for the distinctive features of Russian life, one is tempted to obviate any comparison and perhaps any analysis by declaring that Russia—or Japan—is unique. The mutual loyalty of area-studies scholars

19. Jan Vansina, *Living with Africa* (Madison, 1994).

to each other and to the topic of their study may conceal a degree of exceptionalism that will slow substantially the development of a global perspective on the past.

African Worlds and Global Analysis

While I have just argued against regional exceptionalism in principle, I can hardly argue against the distinctiveness of the historical experience of each region. The distinction between exceptionalism ("you can't compare your area to mine") and distinctiveness ("the difference between your area and mine is . . .") may be small, but it is fundamental. Each region has its specific characteristics, and it is out of those specific characteristics that analysts may hope gain in understanding of the world as a whole.

Because of my own regional specialization and the benefits I see in it, I propose the history of Africa as a useful model for world historians.[20] Africa is a large portion of the world, with 20 percent of the earth's land and a population that is now one tenth of all humanity and that was a larger portion in earlier times. The fact that Africa's role in world history is often neglected or minimized can be used, for these purposes, as a way to elucidate comparisons with the study of world history, which has suffered a neglect of its own. African studies have both contributed to and benefited from global thinking.

African history, first, provides a successful example of the rapid creation of a major academic enterprise. In 1950 African history did not exist as an organized field of study, but by 2000 it had major research centers, an impressive monographic literature, academic appointments in college and university history departments throughout the world, and an established place in the community and discourse of historians.

African historians have conveyed new concepts, methods, and interpretations to historians generally. The concept of diaspora as used today by historians was elaborated by Africanist scholars working with students of the Americas.[21] The literature on slavery has tied together the history of the continent and has shown its links to other regions. The methods of historical linguistics and oral history have developed substantially through the work of Africanists, as has the linkage of history with social and cultural anthropology. For recent times, scholarship on Africa has led to a substantial rethinking of imperialism, colonialism, and nationhood; for early times, studies of Africa have led to great revisions in the history of human evolution, plant domestication, animal husbandry, and metallurgy.[22] African historians have also shown how to apply established historical methods to new problems and new evidence. They have created archives and documentary

20. This section draws on Patrick Manning, "African History, World History: The Production of History on a Global Scale," a paper presented to the Program of African Studies, Northwestern University, 25 April 1991.
21. Harris 1982.
22. Greenberg 1966; Cooper 1996; Herbert 1984.

collections and set up the usual sub-disciplines in economic, social, cultural, political, intellectual, and environmental history.[23] They have struck an effective balance between the need to collect a great amount of empirical information on the continent and to apply appropriate theories and paradigms to process it. The debates among African historians on the paradigms of modernization, world-systems, neo-Marxian political economy, new social history, and Afrocentricity are as enlightening (and as inconclusive) as the equivalent debates in other fields of history.

African history is international, interdisciplinary, and connected. It is international in the sense that the scholarly field is based in numerous African nations and several European and American nations, and also in the sense that the scholars, wherever they work, were themselves born and educated in a wide range of societies. This provides African history with the wide range of perspectives that keep it from becoming a narrow and conformist field of study.[24] African history is interdisciplinary in that the training and research of many practitioners relies heavily on social-science and humanities disciplines, such as anthropology, sociology, linguistics, political science, economics, and archaeology.[25] African history is connected in that its specialists work in interaction with historians of other fields. In the formative era of African history, Africanist scholars worked closely with historians of Europe and European empires, in reconsidering the history of colonial Africa. As the field of slavery studies expanded, historians of Africa worked with others specializing on slavery in the Americas. In social, cultural, and political studies of the African diaspora, Africanists have worked with scholars focusing on the Americas. In another growing connection, scholars working on Islamic Africa work with scholars on the Middle East and North Africa.[26]

23. The Cooperative Africana Microform Project, with its headquarters at the University of Chicago, is an exemplary case of international cooperation in building research facilities.
24. African history is an international field of study, with scholars holding appointments in the United States, Britain, France, Canada, Germany, Belgium, and in African countries (especially Nigeria, South Africa, Egypt, Morocco, Kenya, Senegal, and Congo). Other fields of African studies include these countries plus Japan, Russia, Brazil, Italy, Poland, and several Caribbean countries. Major journals include *Journal of African History* (Cambridge), *International Journal of African Historical Studies* (Boston), and *Cahiers d'Etudes Africaines* (Paris).
25. As in other fields of study, interdisciplinary work in African history has tended to focus on recent centuries. For an example of work on earlier times with particular strength on ethnology and linguistics, respectively, see Derek Nurse and Thomas Spear, *The Swahili: Reconstructing the History and Language of an African Society, 800–1500* (Philadelphia, 1985); Ehret 1998.
26. On colonial politics, see Ruth Schachter Morgenthau, *Political Parties in French-Speaking West Africa* (Oxford, 1964); on slavery studies, see Serge Daget, ed., *De la traite à l'esclavage*, 2 vols. (Nantes, 1988); on the African diaspora, see Robert Farris Thompson, *Flash of the Spirit: African and Afro-American Art and Philosophy* (New York, 1983); on links of African and Middle East studies, see John O. Voll, *Islam: Continuity and Change in the Modern World* (Boulder, 1982); and Lidwein Kapteijns, *Mahdist faith and Sudanic tradition: the history of the Masalit Sultanate, 1870–1930* (London, 1985).

The historical study of Africa has developed some characteristic dynamics. One is that the region as a whole is not readily summarized by reference to sub-regions. East Asia is often summarized by reference to China, South Asia is summarized by reference to India, Europe by reference to Britain and France, and Latin America by reference to Mexico and Brazil. But Africa can not so easily be summarized by reference to Nigeria and South Africa. In a second characteristic, African studies generally explore the outside as well as the inside of the unit of study. Europe is often studied on its own, but Africa is generally studied with reference to Europe, the Americas, or the Middle East. For both these reasons, the historical literature on Africa gives particular attention to interregional connections and interactions. A third characteristic of the study of Africa is that scholars have had to confront systematically the heritage of race, racial discrimination, and racialized interpretation of the past in the historical record of Africa.

None of these advances in historical study of Africa could have taken place without an adequate institutional base. History is today a professionalized field of study, and it is produced in institutions. Many of the advances in the historical study of Africa came because of the creation of strong and well-funded institutions for research and teaching. World history will not gain a secure place in the academy until an equivalent set of institutions is created. While individual scholars and writers have made key contributions by dint of individual effort, the field as a whole cannot progress without research centers, graduate training, teacher preparation, library development, and more.

In the United States, African history gained its strength through its participation in multidisciplinary area-studies programs. Scholarship was conducted especially in a few large programs (Wisconsin, Indiana, Michigan State, UCLA, and several others), including centers for study of African languages. The large programs have become less dominant than they were in earlier years, as federal funds for area studies have become limited in the post-Cold War era, and smaller programs have grown as African history has become widely accepted as a specialization. European universities, once great centers of African studies, have allowed their programs to shrink steadily. African and Caribbean universities have growing numbers of students, but they work with such limited resources that little of their scholarship is published. Latin American universities, especially those of Brazil, have become significant in African studies. Overall, African studies are at once centralized (in the United States) and decentralized. Language study fellowships came from the federal government through the National Defense Foreign Language program, and fellowships for doctoral field work came from the Ford Foundation and agencies of the Social Science Research Council. An initial wave of journals created in the 1950s and 1960s was supplemented by additional journals founded thereafter. Professional associations in the United States, Canada, Britain, and Africa brought scholars together annually.[27] Of particular importance

27. Conferences of African and diaspora studies include, in the United States, meetings of the interdisciplinary African Studies Association, the National Council of Black Studies, and disciplinary groups such as the Association for the Study of Afro-American Life and History. African studies associations meet in Canada, in the United Kingdom, and

for maintaining the participation of African-based scholars in the field has been UNESCO, which has supported numerous conferences in Africa and publications including the work of African scholars.

At present, it is probably true that more research is being carried out by scholars based in North America than on any other continent. A few large programs—at UCLA, Indiana, Florida, Wisconsin, and York—concentrate much of the work, but there are many Africanists working in smaller programs or on their own, and the large programs are less dominant than they once were. Meanwhile, African universities have developed some impressive programs of research and training, particularly in Nigeria and South Africa, but also in Senegal, Zaire, Kenya, and Cameroon, and even in such small countries as Benin and Botswana. European programs have lost their earlier dominance in African studies, especially in Britain where universities have undergone such cuts, but African studies programs remain important in Britain, France, Germany, and also in the Netherlands, Belgium, Sweden, Portugal, and Italy. Latin American and Caribbean universities have significant programs in African studies, for instance in Haiti, Jamaica, and Brazil. The work of Japanese Africanists is of remarkable depth; in addition, African studies programs exist in the Philippines and elsewhere in Asia.

The inclusiveness of Africanist analysis—in which small villages and poor people become subjects of history, metaphors for larger processes, and even causal factors in historical change—may have something to offer to global analysts. Thus, whether I work on a village, a region, a nation, or all of Francophone Africa, my work is accepted as "African" history. (This is pointedly not the case for planetary studies, in which work on a village or a nation is generally not seen as helping make sense of the globe.) The existence of pan-African identity underscores the inclusive dimension of African studies: virtually no Africans are left out of history on the grounds of their being "uncivilized." Since African identity has been developed primarily in response to the experiences of slavery, colonialism, and racism—largely imposed from outside the community—African studies give as much emphasis to interaction among communities as to autonomous development within communities. In addition, since the sense of common African identity cannot deny the immense diversity among peoples of Africa and the diaspora, there has developed among black people a sense of unity in diversity, in which the very range of differences is seen as a measure of underlying unity.

Philip Curtin, in his significant contribution to establishing African history as a full and equal field in historical studies, consistently justified African history by presenting it as part of world history. African studies have now developed in their substance and recognition to the point at which the reasoning can be inverted: the achievements of African studies can now be used to elucidate and to justify

Continued

to a lesser degree in France. In Africa, CODESRIA (Council for the Development of Economic and Social Research in Africa) holds seminars drawing scholars from many parts of the continent, and UNESCO provides support for occasional conferences and volumes on key issues. The Association of Caribbean Historians meets in the Caribbean and draws on all the language traditions of the region.

studies of world history. Some examples of recent work show several ways in which those working on Africa and the diaspora are developing global interpretations of a sort whose logic might be applied to studies of the entire planet. John Iliffe's recent work on poverty in Africa showed that it is possible on this key issue, to do continent-wide work that is both monographic and synthetic. To perform a historical study of poverty and the planet would not be many orders of magnitude more complex. Other recent studies of continental scope which combine monographic and synthetic work are John Thornton's book on Africans throughout the Atlantic world in the early modern period, and my study of slavery and slave trade throughout Africa and in the diaspora. An exemplary and global documentary project is the Marcus Garvey and UNIA papers project, which reconstructs, through a very broad search for documents, not just a man or an organization but a social movement in its many dimensions on four continents.[28]

World History as Third World History

The expansion of area studies created one of the main paths toward studies of world history. To put it in the simplest terms, as the area-studies literatures grew larger and stronger, scholars grew in curiosity about comparisons and linkages of one area with another. Their initial connections developed into a framework for seeing world history as Third World history.[29] The occasional (and usually unrepeated) joint meetings of area-studies associations (for instance, Africa and Latin America, Africa and the Middle East) reflected this growing interest in tracing the links joining major regions to each other.

Which path did the increasingly active historians of the Third World pursue? Can their work be best characterized as the further development of established fields of historical study, or as the introduction of new types of knowledge to history? It was surely some of each. The famous denunciation of African history by H. R. Trevor-Roper, the critique of the field's reliance on oral history, and historians' disdain for Africanist reliance on anthropology would suggest that this field was widely treated as being the work of "other specialists" rather than fitting into historical studies.[30] But most scholars in the field saw themselves as on the internal path, expanding conventional historical studies to new terrain. The success of Africanist and other area-studies scholars in creating lines for appointments

---

28. Iliffe 1987; John Thornton, *Africa and Africans in the Making of the Atlantic World, 1400–1680* (New York, 1992); Manning 1990; Robert A. Hill, ed., *The Marcus Garvey and Universal Negro Improvement Association Papers*, 9 vols. (Berkeley, 1983–1996). While the Garvey movement in the United States has been the center of previous interpretations of the movement, the African and West Indian series in the collection shows these regions to be central rather than peripheral to the overall story and will surely lead to a redefinition of the movement's history.
29. Stavrianos 1981.
30. Trevor-Roper wrote, in the BBC *Listener* (28 November 1963), 871, that African history traced "the unrewarding gyrations of barbarous tribes in picturesque but irrelevant corners of the globe." See Vansina 1994: 123.

within history departments indicates that they were successful. The topical emphases of area-studies historians tended to confirm that interpretation: they worked principally on politics, trade, and elite culture.

This thread of thinking led to the vision of world history as Third World history. World history interpreted from this perspective was a series of regional histories, with interesting comparisons among case studies in continent after continent. Commonly, therefore, courses intended to convey "the world" focused on regional comparisons, as among Mexico, Nigeria, and China in the nineteenth and twentieth centuries.[31] Europe was at once the center of discussion and the invisible participant—the analysis focused on the response of each region to the forces of modernization seen to be emanating from Europe.[32] Only with time did this approach include connection to the United States, especially with the growing political presence of multiculturalism.[33]

At an intellectual and conceptual level, area-studies work adds not only evidence but specifically regional perspectives on the interpretation of world history. The area-studies framework provides an alternative to global analyses that focus either on great-power relations or on the division between rich and poor. The African continent, for example, has been treated as marginal in both types of global analysis. Africa, lacking in great powers, drops out of sight in the great-power vision of the world. In the rich-versus-poor vision of the world, the poor side is often made to appear as an undifferentiated bloc, and African examples are most often displaced by those from Asia or Latin America.

It is unwise, however, to simplify in this fashion and allow the marginalization of Africa (or "poor countries") in global debates. It is wrong in the sense that modern notions of social equality ought to give Africans a full voice simply on the basis of moral principle. It is foolish, because in neglecting Africa on the grounds that its people are poor and weak, one also assumes that we can neglect the mechanisms of impoverishment and disempowerment that brought Africans to that state. Such reasoning tends to assume, narrowly, that interaction in human affairs consists of no more than one scenario and two steps: the diffusion of influences from the powerful to the weak, and the acceptance or rejection of those influences by the weak. A focus on Africa in the world spurs the analyst to focus not simply on dominance but on the ways in which those who are dominated nonetheless participate in and influence the operation of the global system.

31. At Stanford, such a course emerged in 1984, with support from the Mellon Foundation and the National Endowment for the Humanities, as a stage in that institution's long debate over requirements in Western Civilization and the non-Western world. This course may have been influential in the evolution of David Abernethy's view of European empires. James Lance and Richard Roberts, " 'The World Outside the West' Course Sequence at Stanford University," *Perspectives* (March 1991), 18, 22–24; Abernethy 2000; Allardyce 1982.

32. Stavrianos 1981.

33. Joan Nordquist, compiler, *The Multicultural Education Debate in the University: A Bibliography.* Contemporary Social Issues: A Bibliographical Series, No. 25 (Santa Cruz, 1992).

Conclusion: New Places

Area-studies research has formally addressed one of the great inequities in scholarship—the privileging of the study of Europe and North America and the separation of studies of "the West" from analysis of the rest of the world. In addition, area-studies scholarship has faced the choice on whether to replicate the organization of social science and humanities study for Europe and North America or to develop distinctive approaches. The primary emphasis has been on replication (or equalizing), so that the history of Southeast Asia comes to be written by the same standard as that for Western Europe. But area-studies work has its distinctiveness, for three main reasons: first is the adoption of a different mix of disciplinary emphases than for Europe and North America (with more emphasis on anthropology, for instance); second is the influence of the intellectual traditions of each region, as the institutions and ideas of Africa and the Middle East continue to influence scholarship on those regions; and third is the work of area-studies scholarship in the languages of each area. This third part of the area-studies tradition is spreading literacy and improving access to documents in major languages, and it is assisting some communities in preservation and documentation of their languages: it provides a major counterweight to the English-only approach that appears to be gaining ground in scholarship and in public discourse. In sum, the expansion of area-studies scholarship has enriched greatly the scope of the historical record and the interpretation of history. One may hope for continued vitality in this important category of learning.

Yet area-studies scholarship brings with it a parochialism that parallels the inward-looking essentialism of national histories written in the early twentieth century. A significant number of area-studies scholars have begun to explore connections within and beyond the inherited limits of their field, but they remain outnumbered by those who continue to emphasize segmented, microlevel analyses of small regions, typified by the anthropological field study of a village. The villagers under study just as often are migrating great distances to work or school, buying items produced far away, and debating world affairs. The villagers worry about being marginalized in the world, just as the area-studies scholars worry about being marginalized by global studies. My hope is that more area-studies scholars will take the approach of the villagers and go out to encounter the world they worry about.

Chapter 9

Global Studies

G lobal studies began to be formalized as a framework for scholarly analysis during the 1990s. Of course the idea of global studies had long existed at some level, since programs in international studies and international relations were in place at a number of universities. But the late-twentieth-century period of intense global interconnection—in economics, politics, communication, migration, and culture—spurred new global thinking. Beyond this immediate and cosmopolitan experience, the late twentieth century brought a more general philosophical turn in analytical priorities, so that thinkers in many fields found themselves focusing more on connections among systems rather than on subdividing issues and analyzing the pieces separately.

Globalization brought a revolutionary insight—the recognition that issues in social and cultural affairs can be analyzed successfully in a framework crossing the frontiers of regions, time periods, and themes. Not yet, however, has there been enough time or study to turn the global insight into secure global knowledge. The long habits of social scientists and humanists, thinking within constructed and reified social boundaries, are unlikely to be set aside by the power of a few brilliant cross-cultural observations.

This chapter begins by reviewing the wave of globalization of the 1980s and 1990s as it generated a heightened global consciousness and brought into existence programs of teaching and research in global studies. Then, for the social sciences and history, I review recent problems of setting priorities among global studies and area studies. Expanding this issue to a more general level, I review the logic of global studies, or rather the competing versions of the logic of study at the global level. Then I return to the specifics of world history and consider how best to organize this profession and link it to other realms of global studies.

A Wave of Globalization

Globalization took many forms, and global consciousness responded to each of the changes. International trade was a large and growing portion of national output for most countries, and most consumers were buying some goods created far away. Corporations had long operated in multiple countries, but now their governing boards as well as their work forces were multinational. The international flow of capital was reflected in the development of stock markets and commodities markets on every continent. The great powers met regularly, now using the term G-7 to refer to the industrial giants. They replaced the General Agreement on Tariffs and Trade, created in the aftermath of World War II, with a streamlined World Trade Organization, which in turn attracted opposition from those critical of big business. Small powers too had their economic globalization, as with the continuing influence of the Oil Producing and Exporting Countries cartel, and the expanding influence of the regional and national economies that came to be known as the Asian Tigers and the broad coalition of poor nations in international economic discourse, the Group of 77.

In politics, the global conflict of capitalism and communism (or capitalism and socialism) had dominated much of the twentieth century. This conflict was expressed in terms of conflicts among states and their political alliances, among political parties, and among social classes. The rise of widespread movements for democratization, the overthrow of apartheid in South Africa and the collapse of the Soviet Union led to political seismic waves around the world.

Over a somewhat longer period of time, international organizations had become a major element in the global scene. While such organizations emerged as early as the Red Cross in the 1850s, Standard Oil in the 1890s, and the World Court shortly after 1900, it was in the era after World War II that international organizations in public, private, and nonprofit sectors came to restructure world affairs—governmental organizations such as the United Nations and the World Bank; corporate organizations such as Nestlé and General Motors; relief organizations such as Oxfam International.[1] These international organizations were part of the phenomenon of globalization and became both subjects and directors of works in global studies.

The broadcast communication of television and radio, especially through satellite hookups, enabled governments and corporations to spread their messages further, and the individual communication through fax and cell phone created new elements of autonomy. The Internet and the World Wide Web emerged as a hybrid, enabling both broadcast and individual communications to reach new levels. Ecological issues entered global consciousness as localized problems of waste disposal and toxic chemicals emerged in many regions. Generalizing those concerns, people in all regions have learned to worry about global warming, weakening of the ozone layer, deforestation, and loss of species on land and in water. Culture became globalized as musicians and their music moved from region to region, exchanging sounds and dance forms. Culture contact included the export of items in popular culture from industrial centers, the less well-funded yet culturally potent spread of Third World cultural practice, and the development of domestic multiculturalism within both great and small powers.

---

1. Ida M. Tarbell, *The History of the Standard Oil Company* (New York, 1925); John F. Hutchinson, *Champions of Charity: War and the Rise of the Red Cross* (Boulder, 1996); Howard N. Meyer, *The World Court in Action: Judging among the Nations* (Lanham, Md., 2001).

Another aspect of globalization in the United States was the expansion of global studies in classrooms at all levels. The National Geographic Society continued its campaign to reestablish the basic instruction in geography that had been lost decades earlier and supported the creation of global studies curricula including world geography. The expanding public support for standards-based education in public schools led to the adoption, in most states, of standards requiring global studies or world history in some form. Teachers began to seek out workshops to strengthen their work in the globalizing curriculum. In the college classroom, international studies programs arose in the 1990s, in which undergraduate majors took an eclectic range of courses drawn from area-studies offerings. The courses were focused on the social sciences plus language study, but they did not generally include studies in cultural or ecological issues. These students, some of whom were able to study abroad, tended to continue into law school and policy studies.

"Globalization is the catch-word of the day." So argued A. G. Hopkins in introducing a recent collection of studies on globalization in history.[2] Yet, as Hopkins went on to argue, this was not the first wave of globalization or the first development of global consciousness. This wave was different, however, in that the global consciousness flowered in peacetime rather than in the shadow of war. The parallel between the meanings and use of the term *globalization* in the 1990s and those of *modernization* in the 1950s is intriguing.[3] Each served at once as slogan and scholarly model. Each focuses on the short term, devaluing the past as dead weight and dismissing its influence on the future. Yet there are differences between the two: modernization theory relied on a Parsonian sociology, while globalization commonly relies on an unspecified, apocalyptic vision.[4] And while the modernization paradigm was initially welcomed in area studies (though later discarded), the vision of globalization and global studies met with suspicion among many leaders in area studies.

2. A. G. Hopkins, "Introduction: Globalization—An Agenda for Historians," in Hopkins, ed., *Globalization in World History* (London, 2002), 1.
3. I am thankful to Jeffrey Sommers for emphasizing the parallel in paradigms of modernization and globalization. In the dissertation within which he developed these ideas, he focused on the dynamics of opinion management, in the context of periodic economic expansions and contractions, especially in the United States. Chapters address various corners of the modern world-system, including regions that sought to gain autonomy from pressures for incorporation, but center especially on such cases as the campaign to develop support for U.S. involvement in World War I and the development of conservative think-tanks in the 1970s. Jeffrey W. Sommers, "The Entropy of Order: Democracy and Governability in the Age of Liberalism" (Ph.D. dissertation, Northeastern University, 2001).
4. Roland Robertson, perhaps the earliest serious theorist of globalization, seeks to apply Norbert Elias's notion of the civilizing process, arguing that it is now elevated to a global level. Robertson is still left with a "search for fundamentals" in globalization. Parsons 1937; Roland Robertson, *Globalization: Social Theory and Global Culture* (London, 1992); Elias [1939].

Research Priorities: Global Studies vs. Area Studies

In the social sciences, the expansion of global studies brought a collision with area studies. The area-studies movement, at the peak of its achievements, ran into two sets of difficulties in the 1990s. Especially in the United States, the end of the Cold War brought a restriction in funds allocated to combat communism and, therefore, to area studies. Second, the rise and labeling of "globalization" suggested to some that regional studies might now be outdated. Area studies thus faced a challenge to the relevance of particular regions of concentration (most seriously in Russian and East European studies), and a general challenge to relevance of regional rather than global studies. These two challenges to area studies were made more serious by the more general limits on the growth of the academy, in the United States and elsewhere, in the 1990s.

One major battleground of area studies and global studies was in the Social Science Research Council, the agency that provided many of the fellowships for study by graduate students and junior faculty members in the social sciences. (The SSRC gained support from the American Council of Learned Societies and funding in particular from the Ford Foundation.) In 1994 the SSRC restructured its program of graduate fellowships, combining many of the area-studies programs into a broad International Dissertation Research Fellowship (IDRF), which was to provide support for both global and regional work in international studies. From the point of view of area-studies scholars, their own area-studies programs were now thrown against those of other world regions, and especially against global studies, and with a smaller pot of funds to draw on.

Nonetheless, the leaders of area-studies programs were able to protect themselves. Within the first few years of the IDRF, the awards went from an initial expansion in support for global and comparative work to an emphasis on doctoral work addressing one place or comparing two adjacent places. The process of reviewing applications, relying on faculty members with expertise in international studies, ended up reproducing the old area-studies priorities rather than encouraging transnational research.[5]

My impression is that there was little chance for foundation support of dissertation research in global historical studies. The proponents of global studies placed historical dimensions of globalization at a low priority and did not seek out historians as allies. Area-studies scholars viewed world history as an accumulation of regional narratives and assumed that the prior research patterns within area studies would provide adequately for the development of world-historical studies. World historians, working as individuals rather than in teams, focusing on

---

5. I was able to observe this process in detail, since I served as a reader of applications in 1995 and 1996, and since I was the advisor of several Northeastern doctoral students who applied for fellowships. I concluded that the program, despite its title and mission statement, was providing support for dissertations that were area-studies specializations rather than global or transregional analyses. I wrote letters of inquiry and concern to SSRC, arguing that two separate programs should be set up—one for global studies and one for localized area studies. The proposal was debated for a time, but then rejected.

teaching rather than on research, and functioning within a professional organization in its early stages of development, made almost no attempt to participate in
the debate. As a result, the logic of research in world history played virtually no
part in the major realignment of social science research funding during the 1990s.

The World History Association, in contrast to area-studies associations,
remained small, charged a low level of membership dues, and relied on academics working overtime rather than on a professional staff. Area-studies associations
(with a thirty-year start) have become professionalized, have high dues, meet with
each other at ACLS and SSRC meetings, lobby the federal government, hold large
conferences, have institutional bases at major universities, and have designated
fellowship programs supported by foundations. The World History Association,
though supported energetically from its grassroots, lacked a strong, organized
constituency.[6] For lack of resources, it carried on almost no program of communication or lobbying with other professional organizations, philanthropic
foundations, or government agencies.

The situation is distinctive, and perhaps paradoxical. The rise of world history
clearly represents the biggest area of change in historical studies at present—it
is a sort of bandwagon onto which many are climbing. Yet world historians and
the World History Association stand alone both organizationally and intellectually. The intellectual distinction is that world historians are the principal
scholarly group that claims to identify long-term continuities at the global level
and that approaches globalization today less as a radical disjuncture than as a set
of recent nuances in a global pattern of fluctuations and gradual transformations.
Individual researchers from other disciplines have contributed actively to this
vision of world-historical continuity, but the disciplines in general have little use
for world history. As a result, the WHA has few allies.

Those supporting global analysis in the social sciences were poorly organized
in terms of academic politics, and they were poorly organized in terms of intellectual content. For academic politics, I mean that they did not have established
programs or representatives to contact and lobby major funding organizations.
For intellectual content, I mean that they did not have programs providing
coherent global analysis in social sciences, so that the research proposals were

6. Membership, as reported on the website of each organization in 2002, was 15,000 for
    the American Historical Association, 11,000 for the Organization of American
    Historians, 7,500 for the Association of Asian Studies, 5,500 for the Latin American
    Studies Association, 5,000 for the American Association for the Advancement of Slavic
    Studies, 3,000 for the African Studies Association, and 2,600 for the Middle East Studies
    Association. Reports to the World History Association Executive Council give a
    membership of 1,400. In 2002 the WHA set up a headquarters at the University of
    Hawaii.
       The U.S. Department of Education, in 2002, supported 114 National Resource
    Centers for area and international studies under Title VI, but none of them addressed
    world history. They included 33 for Asia (East Asia 15, South Asia 9, Southeast Asia 8,
    and Inner Asia 1); 19 for Latin America, 16 for Eastern Europe, 14 for the Middle East,
    11 for Africa, 10 for Western Europe, 2 for Canada, and 1 for the Pacific; there were
    8 centers for International Studies, but none emphasized history.

appropriate and well argued. Even if the supporters of global analysis had been well organized politically and intellectually, they might well have lost just as badly to the defenders of area studies.

I thought and still think that it is a shame to set up global studies and area studies in the social sciences as a battle and a zero-sum game. The need for continued expansion of area-studies knowledge remains great. But the need to develop knowledge for social-science issues at a global level is now becoming apparent, and it is equally compelling. The approaches to work are significantly different, though by no means directly opposed. A rational system would allocate research funds for each approach and award funds according to criteria appropriate within each category.

World historians fall between the move for globalization and the programs of area studies. What should be the structure of the academic enterprise of world history? Should it organize the world as another "area"? Should there be government-funded centers for global studies?[7] The lack of a clear voice of research-oriented world historians during the 1990s meant that these questions were never debated at the decision-making levels of government, universities, or foundations.

The area-studies organizations rely on tight alliances of historians, anthropologists, language scholars, economists, and others focusing on the region in question. In contrast, the field of world history has no links to such fields as "world anthropology," "world sociology," or "world philosophy." Sociologists and anthropologists focusing on Latin America are generally willing to acknowledge that the phenomena they study are historically rooted, so that historians and other social scientists make common cause, at least in organizational politics. Africanists and Latin Americanists could even study "modernization" in their areas without breaking from historians.

There exists no well-established Global Studies Association uniting historians and social scientists in common study of the earth, on the shared assumption that there are historically rooted global patterns. Some activists in international studies programs assume that they will be able to develop a cross-disciplinary coalition to resolve the conflict between area studies and global studies and perhaps develop new space for the study of world history.[8] So far, however, they propose to build this coalition at the level of textbooks and undergraduate teaching rather than at the level of research.[9] In contrast, the high-profile globalization institutes

7. There was some public discussion of these issues among world historians during the 1990s: see chapter 19 for a review of it.
8. As Neil Waters notes in his review of the field of international studies, it benefited from the 1993–1994 tilt by foundations toward global and transcultural studies, though it did not include doctoral training within its purview. Waters, "Introduction," in Waters 2000.
9. Ravi Arvind Palat criticizes area-studies research as a triumph of microperspectives that reifies the areas and isolates them from studies of the West, but argues that international studies might be able to close the gap. Similarly Ian Barrow argues that world history arises as a threat to area studies history, yet proposes that both levels of study share the location of change in such "disembodied entities" as civilizations and nation states. He proposes that by disaggregating agency into the agent, instrument, and patient, it will be possible to open a discourse between area studies and world historians. Palat 2000; Ian J. Barrow, "Agency in the New World History," in Waters 2000: 190–212.

recently established at Yale and Columbia focus on international politics, assuming that they can plan the future of the world without regard for its past. Historians have not been part of this sort of global-studies movement largely because they decline to accept the ahistorical premise that underlies globalization.[10]

Articulating Global Perspectives

Global studies form not a discipline, but rather a framework for disciplinary studies, parallel in a sense to the framework of area studies. The categories for area studies already existed, having been created by the logic of civilizations, continents, and nations. The category of the world also existed already, but data had not been organized on a worldwide basis, at least not in the social sciences or in cultural studies. The will to conduct global analyses has now been expressed, but the process of articulating the approaches and techniques of organized global study is just getting launched.

In the natural sciences, research and interpretation have been primarily at the global level. That is, local phenomena were known and respected, but there were thought to be no major boundaries between localities and the global or universal level. For the natural sciences, globalization and global studies meant a reaffirmation of the dominant outlook. The dramatic announcements of potential global warming, deforestation, and ozone depletion had an impact because these were new research results in ecology, but not because ecologists had suddenly adopted a global perspective. For the social sciences and cultural studies, in contrast, the central beliefs of scholars emphasized the distinctiveness of existences within national, civilizational, and religious boundaries. Scholars who maintained global perspectives were held at the margins of these fields. Globalization and the global studies movement provided an incentive to reorganize some basic assumptions, but there was no way to make instant revisions of fundamental assumptions in response to this demand.

Global studies naturally began by assembling existing scholarly groups. This assembly took place informally and eclectically—there was no war effort or political confrontation to cause governments to convene scholars for a general assessment of the potential of global studies. Global studies naturally focused on topics that stirred the widest interest and immediate policy concerns. Short-term economic concerns, international relations, and ecological issues requiring policy decisions became the center of global-studies programs.

10. Yale Center for the Study of Globalization (www.ycsg.yale.edu); the Columbia Earth Institute (www.earth.columbia.edu); UCLA Center for Globalization and Policy Research (www.sppsr.ucla.edu/cgpr/); Berkeley Center for Globalization and Information Technology (www.bcgit.berkeley.edu/). The Columbia Earth Institute lists social scientists along with other scholars included in its activities, but the undergraduate and graduate programs all focus on natural sciences, with an occasional reference to issues of health and economics. The Globalization Research Center at the University of Hawaii (www.globalhawaii.org) does include an affiliated historian, world historian Jerry Bentley.

And the gulf separating old from new perspectives was large. In area studies, after more than a generation of scholarly activity, there remain many scholars who prefer to focus on the ideas of colonial rulers than on the actions of local populations in Asia or elsewhere. For this reason, we can be sure that not every historian will soon abandon the national perspective to take up global studies. Nor is it a matter of emulating natural scientists and simply focusing on the commonality of phenomena throughout the world. The social and cultural boundaries within which we have worked are real. This generation of globalists will learn more about how to cross and surmount those boundaries, but it will not get far by denying their existence.

The rise of global studies brought revolutionary implications in that it encouraged analysis beyond localized categories study in the social sciences. In one sense, global thinking was nothing new: leaders of great powers and great corporations had long-term experience in thinking at the planetary scale because they were able to exert some control at that level. In another sense, global thinking was new indeed, as it reflected improved skills in seeing connections across systems: the various logics of interaction are now being articulated with increasing specificity.

Parochial Globalisms

Global studies, in its current and still early incarnations, results from the application of partial and occasionally lopsided approaches to global analysis. As I argue here, most programs of global studies, while proposing to conduct broad analyses, impose substantial but rarely explicit constraints on their breadth, so that much of their of analytical logic and practice is parochial rather than global. The different ways that global studies are constructed result in conflict and confusion among their approaches. What might at best be the launching of a healthy variety of approaches threatens instead to become a cacophony of competing globalisms, each trumpeting success and courting sources of new resources, but none conducting a comprehensive, global analysis.

I define global studies, for the social sciences and cultural studies, as interactive studies of wide scope. By "wide scope" I mean studies extending their scope to large geographical regions, wide slices of time, and a broad range of human and natural phenomena. By "interactive studies" I mean analyses considering the dynamics and interactions among the various regions, phenomena, and time periods under study. Of course one cannot study everything at once. But the point of global studies is that one does not automatically limit, *ex ante*, the regions, phenomena, times, or the connections among them that will be considered. Let me call this a *maximal* or *general* definition of global studies.

In practice, however, my maximal definition exceeds greatly the reality of "global studies" insofar as such studies have been undertaken. While everyone accepts that "global" means big analysis, not everyone agrees on how it should be done. Instead, various *partial* definitions of global studies have been implemented and debated. As I seek to show, there are several sorts of constraints that are placed on the meaning of *global studies* by various groups participating in or commenting on these studies. Some see "big" as meaning a wide space but not a wide slice of time, and not a broad range of themes. Others see the scope of global studies as the whole world at once: that is, they decline to address global phenomena and local phenomena at the same time. To summarize a complex situation, I offer five main ways in which a global framework of analysis may be constrained or narrowed.

Privileging certain spatial frames. What regions are considered in global analysis, and how do we break them down? One approach is to assume that the world as a whole is the unit of study and to neglect any smaller units. Such an approach might seem relevant for such ecological issues as the atmospheric ozone layer, though observations have to be made at specific points, not for the earth as a whole.

Another approach is to assume that the nation is the basic or most essential unit of which the world is made up. The idea that national experiences are the "building blocks" of the global experience reveals the assumption that the nation is more basic and even more real than the regions within it or the regions enclosing it.[11] The approach may begin with the practical point that data are available for nations and not for other units. In this case, the world is seen as the summation of a group of nations. A parallel approach is to assume that the world is made up of continents. This vision, which is expanding in some approaches to world history and geography, not only reifies continents but neglects islands and oceans, and it neglects close relations of places that are nearby each other by on separate continents. In natural sciences, some research provides an approximation of being global in scope: the GIS system of global positioning works by coordinates, not by political units.[12] But the global analysis of economists and international relations specialists continues to work with national units, so that the United States, China, Belgium, and Sri Lanka each serve as national observations in their calculations, though their areas and populations differ by orders of magnitude. It would be at least as logical, in order to obtain a global understanding, to break the United States and China into units more like the size of Belgium and Sri Lanka.

Privileging certain temporal frames. What times are addressed in global analysis, and how do we break them down? Particularly in economic and political studies of globalization, there is a tendency to privilege short-run change and to assume that the current situation results from a sudden, recent transformation, rather than from a long-term process of change. A more thorough approach would consider both long-run and short-run dynamics and consider the possibility of gradual evolution or rapid change within each of these time frames.

Global studies programs have rarely included history. One exception has been the group of scholars led by Bruce Mazlish, who sought to define "global history"

---

11. On the notion of "building blocks" or essential units in a hierarchy of structures, see the additional discussion in chapter 15.
12. For the Global Information System, see www.gisresearch.com.

in contrast to "world history," and to link global history to the contemporary concerns of globalization.[13] In economic studies, the expanded discussion of the global economy led some economic historians to apply global approaches to times before the twentieth century, and in environmental studies, students of the twentieth-century environment joined a small number of historians to explore global environmental history.[14] These exceptions aside, the initial impact of "globalization" thinking was to marginalize history, and to assume that a brave new world was in creation, which would have few connections to the lives and experiences of those who had gone before.

Privileging certain thematic frames. What themes are addressed in global analysis, and how do we break them down? Studies of politics and economics at the global level gain particular attention, in large part because these issues are well documented and because governments seek to exert policy influence on them. Environmental or biological issues can be studied as specializations in themselves, or they can be studied in association with the various societal issues that can connect with them. International-studies programs focus on social sciences, not much on cultural studies.

Privileging certain dynamic frames. What dynamics are addressed in global analysis, and how do we break them down? The analyses in global studies have focused very much on the current era and rather heavily on wealthy nations. Studies of economic growth and depression, the arms race, and industrial pollution have centered on the big powers. So too have studies of ideological conflict and even human rights issues. To a lesser degree, global concerns have focused on disparities and conflicts between wealthy nations and poor nations: in levels of debt, social conditions, violations of human rights, and aspects of the environment. Here the rich and poor—but especially the poor—are often made to appear as undifferentiated blocs. To summarize, the analysis of global issues has tended to focus on diffusionist dynamics in the short term, assuming causes to be located in centers of wealth and power, with effects appearing (perhaps with a delay) worldwide. Global studies ought to provide attention to feedback processes and the possible confluence of short-term and long-term dynamics. It may be that distinct dynamics govern processes at various levels of space, time, and thematic breadth, so that we may ultimately learn to identify specifically local, national, civilizational, and global phenomena and the more complex mixtures of them.

Privileging certain organization of study. I have been emphasizing that in global studies, there must be a global object of study, but more localized objects of study should also be included in the analysis. A related question is whether the

13. Mazlish 1993b. See also Mazlish, *The Uncertain Sciences* (New Haven, 1998).
14. Kevin H. O'Rourke and Jeffrey G. Williamson, *Globalization and History: The Evolution of a Nineteenth-Century Atlantic Economy* (Cambridge, Mass., 1999).

organization of study should be local or global. In the natural sciences, in which it is assumed that the phenomena under study are the same anywhere on earth, the organization of study is largely global. In the social sciences and cultural studies, in which it has been assumed that the phenomena are localized to the nation or the language community, the organization of study is mostly local.[15] The rise of global studies amounts to a reconsideration of that assumption. For the time being we remain reliant, ironically, on national studies of the global order.

The Profession of World Historian: Organization and Links

The World History Association is not the only organization representing the common professional interests of global historians, but it is the largest and best-organized such grouping.[16] So I will center on it in this discussion of the strategic choices open to world historians. Perhaps because the term *world historian* has the ring of an ambitious, even arrogant claim of credentials, rather few people are ready to identify themselves as such. Deborah Smith Johnson, in interviews with over seventy well-established teachers and professors associated with world history, found that less than half were willing to label themselves as world historians. Though all the interviewees were enthusiasts for the study and teaching of world history, some thought that their own reading and understanding of history were not broad enough, and others, particularly teachers, were happy to identify themselves as globalists but would not accept the label "historian" because they were not active in research.[17] I will be a bit more free in applying the label "world historian" than these interviewees have been.

Some 1,400 dues-paying members have kept the World History Association functioning in recent years. The WHA has carried out substantial efforts to increase membership in recent years, though without much success. The organization includes a core of active members, and the annual conference in June commonly brings 250 registrants. A substantial dues increase, beginning in the year 2001, has provided the basis for the establishment of a headquarters at the University of Hawaii. Depending on the priority that potential members see in the organization, it has the potential to grow, and to grow internationally.

15. I do not wish to argue that natural scientists are beyond reproach in their design of global research or that social scientists lack good sense. The current opportunity presented by global studies in social and cultural affairs is to learn about global patterns without neglecting the local and regional patterns that have been the focus of study in the past.

16. The International Society for the Comparative Study of Civilizations (www.iscsc.net), established in 1961 in Austria, moved to the U.S. in 1970; the Forum of European Expansion and Global Interaction, established in the U.S. in 1994, meets periodically; the Political Economy of World Systems section of the American Sociological Association (www.acad.depauw.edu~thall/pewshp.htm) meets annually.

17. Deborah Smith Johnson, "Rethinking World History: Criteria for the World History Survey" (Ph.D. dissertation, Northeastern University, 2003).

Meanwhile H-WORLD, the online discussion group, has 1,500 active subscribers; other readers choose to consult the archives without maintaining a subscription. I estimate that fewer than a third of H-WORLD subscribers are WHA members, and vice versa. In addition to these adherents of world-historical organizations, I think there are at least as many additional scholars and teachers who maintain an active involvement in the study of world history.

The list of interest groups with which world historians might enter into formal and informal contact is long enough to present daunting choices. It begins with history teachers and administrators in elementary and secondary schools, especially through the National Council of Social Studies and its affiliates. The list continues with organizations of professional historians at the national and international level.[18] In the United States, the American Historical Association has numerous affiliated organizations, many of them of interest to world historians. Next come the area-studies associations, organized at national and sometimes international levels.[19] Disciplinary associations, organized at both national and international levels, include those for natural sciences, social sciences, and the many fields of cultural studies.[20] Some of these disciplinary groups will have their attention centered on current affairs rather than historical issues, and there arises the question of whether to seek common ground between them and historians. For all of these interest groups, one might say that the basic relationship is that of colleagues—teachers and researchers working in parallel fields of academia.

At another level, world historians may be interested in connections with coordinating groups for scholarly activities such as the Social Science Research Council and the American Council of Learned Societies in the United States.[21] Philanthropic foundations are important both in the determination of research priorities and in the allocation of research awards.[22] Government agencies, at national levels but also at regional and at supranational levels, are important determinants of research priorities and sources of research funds.[23]

18. The International Congress of Historical Sciences; the American Historical Association. The more than fifty affiliates of the AHA include the American Society of Environmental Historians, Economic History Association, the Social Science History Association, and H-Net (Humanities On-Line).

19. Area-studies organizations for the United States include the African Studies Association, the American Association for the Advancement of Slavic Studies, the Association of Asian Studies, the Latin American Studies Association, and the Middle East Studies Association. UNESCO might be considered to be an area-studies association at a global level; CODESRIA is an area-studies association for Africa.

20. In the natural sciences, the American Chemical Society; in the social sciences, the American Anthropological Association and the International Sociological Association; in cultural studies, the Modern Language Association and the College Art Association.

21. Equivalent groups exist in many other countries.

22. Foundations based in the United States include Ford, Rockefeller, Carnegie, MacArthur, Mellon, and Lilly.

23. In the United States, the National Science Foundation, National Endowment for the Humanities, National Endowment for the Arts, and Department of Education. The European Union provides substantial funding for social science research. In Canada

Involvement of world historians in organizational work and alliances only makes sense if there are important purposes to be served by such activity. Simple desire to establish of a stronger community of world historians, in research and teaching, is surely what has held the WHA (and latterly H-WORLD) together. In addition, a desire for formal interdisciplinary linkage might connect historians to globalists in other disciplines. The desire to govern and expand the employment of world historians, in schools and in colleges, is surely of professional interest. World historians may wish to support creation of major programs in the preparation of teachers, and to encourage widespread availability of additional courses and professional development institutes for practicing teachers. Similarly, world historians may seek support for training of graduate students entering the field, for research on dissertations, and for postdoctoral research projects. Further objectives might include the establishment of research centers and the organization of conferences to define and review global historical research. Finally, world historians may wish to combine with other scholars to organize large projects for data collection and analysis at the global level.

The patterns by which world historians will be employed in colleges and universities remain undetermined.[24] The contrast with the employment of area-studies historians seems to be rather sharp. In the past two generations, area-studies scholars have gained employment in history departments through the argument that they were covering additional nations (or civilizations). The success of this approach is near complete, and it has reinforced the tradition that historians are appointed by geographical region rather than by thematic specialty.[25] World historians seeking employment can position themselves in various ways: as specialists on a given nation or area who also study world history; as specialists on world history treating the world as one more area; or as specialists in world history treating the world as

Continued

and the United Kingdom, government-supported Social Science Research Councils exist, and in France the Centre National de Recherche Scientifique includes sections focusing on social sciences.

24. In the early days of world history, most scholars were self-trained, and that has remained the case almost to the present. Positions in world history were defined at mid-level state universities and at liberal arts colleges, though in most cases these added world history to the responsibility of an existing area-studies position. Major universities made almost no appointments in world history. There were virtually no candidacies for college or university tenure based on expertise in world history. In contrast to the numerous monographs and general works of area studies, world history became focused on the undergraduate textbook. Even at present, the research literature on world history is dominated by the work of senior scholars, virtually all self-trained, who have undertaken work at the global level.

25. In the early days of area studies, many of the scholars were self-trained, but the rise of area-studies centers meant that research-oriented specialists came rapidly to the fore. New area-studies Ph.D. recipients in history gained jobs in major universities until these positions filled, though they also went to public colleges sensitive to multicultural constituencies. They taught survey courses and upper level courses in their area and wrote area-studies monographs to gain tenure.

a thematic concentration in history. The first approach leaves the organization of history departments undisturbed but gives the appointee double duty and perhaps little encouragement to do research in world history. The second approach adopts area-studies logic and would seem to leave undisturbed the nature of appointments in history. The third approach raises the possibility of reorganizing history departments and making appointments by theme as well as by region.

How should future work in world historical research best be conceived? How are we to find or create data at a global level to match the global historical questions we now explore? While I address these questions throughout this volume, I want in this paragraph to emphasize our heritage of three types of transregional analysis by historians as they have concentrated on civilizations, empires, and diasporas. Works on civilizations range from the ancient to the modern, from Mesopotamian to American, and are defined in political, cultural, or religious terms.[26] Studies of empires overlap greatly with civilizational studies for early times, but especially for the second millennium C.E., empires are treated as distinctive structures. The study of British Empire history, as it developed in the nineteenth century and especially in the early twentieth century, was thus a significant predecessor for studies of modern world history, and the data collected and analyzed by British Empire historians remain central to studies of world history.[27] The history of diasporas has been incorporated into world history only recently, but already studies of Jewish, African, and Chinese diasporas have become substantial.[28]

A task that I find to be of particular importance is to encourage the development of worldwide collections of data. While most of this work remains to be done, some past projects give indications on what such work might look like. To begin with, some national governments have collected and published series of historical data.[29] B. R. Mitchell's remarkable work has developed parallel series of national data for many countries.[30] The U.S. National Bureau of Economic

26. Martin Bernal, *Black Athena: The Afroasiatic Roots of Classical Civilization*, vol. 1 (New Brunswick, N.J., 1987). The ISCSC group has explored civilizations in transhistorical context, and Chase-Dunn and Hall (1991 and 1997) have brought this approach into macrosociology.

27. Hancock 1937–1942; see also the British Parliamentary Papers, House of Commons Sessional Papers. For a major study of a global phenomenon explored through records of the British empire, see David Eltis, *Economic Growth and the Ending of the Transatlantic Slave Trade* (New York, 1987). William Roger Louis, ed., *The Oxford History of the British Empire*, 5 vols. (Oxford, 1998–1999); P. J. Cain and A. G. Hopkins, *British Imperialism, 1688–1990*, 2 vols. (London, 1993).

28. For a fuller range of research projects in world history, see chapters 10–13. On diasporas, see Cohen 1997; Wang 1997; McKeown 2001; Harris 1982. Jahnheinz Jahn, trans. Marjorie Grene, *Muntu, The New African Culture* ([1958] New York, 1961); R. Thompson 1983.

29. See, for example, United States Bureau of the Census, *Historical Statistics of the United States: Colonial Times to 1970*, 2 vols. (Washington, 1975).

30. B. R. Mitchell, *Abstract of British Historical Statistics* (Cambridge, 1962); Mitchell, *European Historical Statistics* (New York, 1975); Mitchell, *International Historical*

Research, under the leadership of Simon Kuznets, developed estimates of national product for the United States and other nations for the twentieth century and in some cases for earlier times.[31] Quite a different global collection of data resulted from the efforts of George Peter Murdock to establish a global survey of ethnography. The Human Relations Area Files resulted in compilation and coding of a large amount of ethnographic data, in cross-sectional perspective.[32] Another major collection of data is in the form of an electronic archive, the Inter-University Consortium for Political and Social Research, housed at the University of Michigan.[33]

These data structures and archives, created a generation and more ago, have provided immense service to social-scientific research but remain constrained by the national, ethnic, and disciplinary paradigms in which they were created. Now, in the era of globalization—with new problems, new theories, new data structures, newly available data, and a developing global perspective—it would be wise to invest in creating the institutions to study social-science issues at the global level.

Turning now to the question of strategies that organized world historians might pursue, I want first to urge a strategy of activism rather than patience. While some individuals have worked tirelessly to build world history as an

Continued

Statistics: Europe, 1750–1988 (New York, 1992); Mitchell, *International Historical Statistics: The Americas 1750–1988* (New York, 1993); Mitchell, *International Historical Statistics: Africa, Asia & Oceania, 1750–1988* (New York, 1995). These are designed and presented as parallel sets of national statistics and thus reflect the gaps in data and inconsistencies in unit size that are the inevitable result of assembling national histories. But the repository of Mitchell's data and the accumulation of techniques he used in collecting and transforming them represent a major advance in global organization of historical data.

For a compilation of national historical statistics intended to show the feasibility of constructing national historical statistics even when a century's history included three colonial regimes and an independent state, see Patrick Manning, "African Economic Growth and the Public Sector: Lessons from Historical Statistics of Cameroon," *African Economic History*, No. 19 (1990–1991), 135–70.

31. The National Bureau of Economic Research (www.nber.org), founded 1920, established its importance firmly with the comprehensive report of Kuznets 1941.

32. Human Relations Area Files (www.yale.edu/hraf/), founded 1949. Murdock and his associates divided the world into ethnic groups and subgroups and, working with a standard coding system, processed thousands of ethnographic reports. The results emphasize the summary data for each ethnic group, but also provide the data input from each report. These data, recorded in quantitative form, have been used for regression analysis of cross-societal correlations and for historical interpretations based on such correlations. Murdock 1959 is a study of Africa based on data collected for the files. Studies based on the files include Patterson 1982 and Pryor 1977.

33. The Inter-University Consortium for Political and Social Research (www.icpsr.umich.edu) became a valuable electronic archive for local and national studies performed in various social sciences. While the archive makes no effort to establish comparability among the data it houses, it does require that each data set be thoroughly described.

organized field of study, the wider world of teaching and research has followed a laissez-faire approach to world history, so that many opportunities for development of the field are thoughtlessly passed by.[34] At base, the evolution of world history will be very gradual unless world historians develop a stronger constituency and a wider set of alliances. To create a stronger constituency of world historians is no easy matter, as world historians are not born into any recognizable ethnic or regional category but are self-selected out of interest. In any case, the cause of expanding the organizational strength of world historians can be expressed in terms of three choices.

First, what should be the geographic scope of the WHA? Should it become an international organization or a national organization? At present, the organization is balanced between the two options. In formal terms, the WHA exists as a global organization, though in practice it remains predominantly based in the United States. Will WHA grow as one organization with local affiliates around the world? Or will separate organizations grow in various countries and an international federation grow out of it? So far, the WHA has avoided creating any national structures and substructures, but has affiliates that are sometimes subnational and sometimes supranational. In part, this question will be settled by the approach the WHA takes to its relationship with the American Historical Association and with the International Congress of Historical Studies. The latter, while formally global, is a largely European organization, heavily bureaucratic and without much space for rapid change.

Second, what should be the relations of organized world historians with other disciplines? In practice, the WHA has opened almost no discussions with groups of scholars outside the field of history, although individual scholars from other fields have participated actively in WHA meetings. Should the WHA conduct interdisciplinary academic diplomacy on a bilateral basis? This approach would be for the WHA to set up direct relations with the associations of sociologists, anthropologists, political scientists, economists, language associations, other associations in cultural studies, and with area-studies associations. Or should the WHA seek to form a multidisciplinary global studies association, parallel to area-studies associations? This would require organizational meetings in association with globalists from other disciplines, then the launching of the Global Studies Association, and thereafter the work to ensure that historical studies are given adequate attention within the expanded association.

Third, should the WHA maintain its current form as a teacher–researcher coalition, or should it become two organizations? The mix of teachers and professors in the membership and leadership of the WHA makes it distinctive among professional organizations. I am among those who believe that this alliance has brought great benefits; professors have learned about both teaching and research from teachers, while teachers have learned about both teaching and research from

34. In labeling the overall approach of academia to world history as one of laissez-faire, I do not want to ignore the work of leading activists in world history. The full list is long and distinguished, but I think that four names would be on every version of the list: Kevin Reilly, Ross Dunn, Heidi Roupp, and Jerry Bentley.

professors. Nonetheless, teachers remain a distinct minority in the WHA, though they might reasonably be expected to be a majority. At some level, the WHA has not become a relevant organization for most teachers of world history.

In evaluating these strategic choices, world historians should study the earlier experience of other expanding fields of study—in the United States, these include the growth of black studies programs from the 1960s and women's studies programs from the 1970s.[35] I do favor opening of formal communication among world historians, area-studies associations, and disciplinary associations. While the call for such communication may begin to sound like an assignment for many additional meetings and e-mail messages for scholars and teachers who are already pulled in several directions, there do exist some institutions that provide support for such connections. In the U.S.-based Social Science History Association, "networks" have formed around numerous topical and disciplinary interest groups, and these networks organize the sessions at the annual meeting. The networks create and maintain scholarly connections that ease communication among the various other groupings to which members belong.

In practical terms, it is the area-studies teachers and scholars who have shown the most interest in the global-historical project—as has been exemplified in the collaborative World History Symposia held in Boston in recent years.[36] Nonetheless, world history is different from area-studies history. In ideological terms, and especially when on the defensive, area-studies scholars tend to focus on exceptionalism—on the specificity of their region and on its incomparability with other regions. Globalists may readily admit to regional specificity, but they are loath to admit to exceptionalism—to a regional uniqueness that is impervious to external influence. The discussions of world historians with area-studies historians (and with globalists in other disciplines) will surely be rocky at times, but it is out of such connections and debate that learning takes place.

Conclusion: Thinking Big

Many types of big thinking have appeared recently on the intellectual horizon. Scholars, teachers, and funding agencies are currently at the stage of exploratory tinkering, trying out eclectic approaches to the creation and utilization of knowledge in global and interactive frameworks. Which are the most important global issues? How are we to use the range of disciplines, the range of area-studies knowledge, and our awareness of global patterns to advance our understanding of our world? How are we to sort out the priorities for the best research and teaching?

World history, because of its interpretive focus on connections among regions, topics, and time periods, has the potential to become a scholarly nexus linking many fields of study. While it can by no means represent the totality of the new work in the regional and disciplinary sections of the historical literature, it might

35. Nordquist 1992.
36. See pages 358–359 for details.

develop a pattern of reflecting the main new trends in each field and their interactions with each other. For world history to become a scholarly meeting ground
and not just another group of specialists, energy and resources that are not
yet available will be required. World historians will have to take on major tasks in
academic diplomacy, in addition to the demands of their work in research and
teaching. The World History Association would have to grow substantially in
membership, financial resources, and institutional activities. Whether such an
expansion and coherence of world history will take place, making it into a leading
research field, is now impossible to predict. Perhaps there are other forces that will
restrict the development of world history as a field. Perhaps these same forces will
enable global insights and global connections to be studied adequately without
developing a strong group of world history specialists. Perhaps world historians
will become part of the interdisciplinary scholarly crowd rather than leaders in
communication among regional studies and among the disciplines. But it strikes
me that the wealth of new ideas developing in historical connections will confront
increasingly the national paradigm that continues to constrain most studies in
humanities and social sciences. One significant possibility for negotiating this
paradigmatic confrontation is the expansion and flourishing of world history as a
major field of study.

Part III

Results of Recent Research

The five chapters of part III summarize major results of recent study in world history. They show how historians and other scholars have put to work the methods of disciplines, area studies, and global studies, applying them to practical questions in world history. The results comprise what Ross Dunn has labeled "the new world history."

This new world history comes in numerous flavors. The days of world history as a single, all-encompassing field of global synthesis are gone, if they ever existed. The range of methods and perspectives of historians and the range of human experiences combine to create many sorts of world history rather than just one. I have chosen to identify sub-fields of study, organized according to familiar groupings of academic disciplines, within the field of world history. These sub-fields—political history, economic history, social history, and so forth—are treated as identifiable and relatively autonomous arenas of human activity and experience. My emphasis in part III thus shifts somewhat as compared with earlier chapters, and especially in chapter 4, in which I used the term *theme* to describe a focus on specific relationships or processes within history generally. While there is a close relationship between the themes of life and the disciplines through which we study them, I find it to be clearer, in the more specific chapters of part III, to categorize the works considered in terms of disciplinary fields and sub-fields instead of by themes.

In the chapters of part III, I present seven sub-fields in four chapters. I give considerable emphasis in each chapter to the *research agenda,* defined in chapter 10 to mean the historical questions under study for each sub-field. For the bulk of each chapter, I review recent monographic studies. For political and economic history, I focus on problems of governance and on the production, consumption, and exchange of goods and services. For social history, I review research on the many dimensions of family and community. For technological history I center on human devices for control of nature. In ecological history I address at once the influence of nature on human society and the impact of humanity on the environment. The review of history of health focuses on problems of illness and healing. For cultural history, I review studies seeking to address the full range of human representations of their experience and understandings. Following these reviews within disciplinary limits, I address the main techniques and issues for debate among world historians. This discussion of debate completes the circle of interpretation, going from questions to answers to arguments about the answers and back to more questions. Debate thus focuses not only on critique of research results but also on the contending priorities proposed in the research agenda—where should we direct our efforts in exploring world history?

The results of each individual study add incrementally to the broad picture of global patterns in human history. World history is so big that each new investigation, even of a big topic, can bring but a tiny change to the whole story. Yet each tiny change is significant: the accumulation of studies recently completed can already be seen to be filling in outlines of the past in some areas where it was blank and to be revising previous views of the past in other areas.

Chapter 10

Political and Economic History

S tudies in political and economic history have provided the backbone of world history since early times and continue as the largest sub-fields in world history. Political history addresses government, political conflict, and change, while economic history addresses the production of goods and services, the demand for goods and services, and their exchange through commerce, redistribution, or other processes. Politics operates through the structures of local political communities and the larger units known as states (including monarchies, empires, and nations); the institutions of economics have been households, mercantile firms, productive firms, and the states that have regulated economic life. The political questions of representation, administration, diplomacy, and war are in many ways distinct from the economic issues of land, labor, capital, money, and economic growth. Yet these two sub-fields of history overlap significantly because of the importance of government policies in determining economic outcomes, and because economic conditions often set the parameters for political communities.[1]

Research Agenda: Political and Economic History

The discussions of research agenda, in this and subsequent chapters of part III, review the most basic and most pressing questions about the past, in the minds of observers ranging from historians (both professional and amateur) to the general public.[2] In simple terms, one asks, what do we need to know about the global past? More precisely the research agenda, based on what is known of the past for a given field of history, consists of the questions of historians and the questions of the public about that field. Historical questions emerging out of public discourse

1. Of the four disciplinary areas I have discerned, politics and economics are old work along the historians' path, social history is new work along the historians' path; culture is old work along the scientific-cultural path; and environment-technology-health is new work along the scientific-cultural path.
2. The next several paragraphs define *research agenda* as the term is used in part III. In the more formal review of historical analysis in part IV, especially chapter 16, I return to discussion of *research agenda,* and contrast it with *research design.* By the latter I mean the analytical response to the questions of the research agenda—the scope and method of the analysis, the data to be investigated, and the work plan.

usually result from the dilemmas and problems posed by the current social situation. Thus the wave of democratization movements beginning in 1989 brought in its wake a smaller wave of studies of democracy in history.[3] Questions of historians, in contrast, are more likely to result from problems in the past: debates about historical processes or contradictions between research results are what cause historians to pose new questions for research. For instance, since it was realized that most silver mined in the seventeenth century was sold in China and elsewhere in Asia, was it implied that Asian economies were growing rather than stagnating in that era?

The questions that make up the research agenda, as I have suggested, depend on what is known about the past as well as what is not known. That is, a narrative of political and economic history provides the basis on which new questions are posed. The most prominent narrative in world history is its long-term political narrative—that is, the formation of early states, the development of empires, the periodic rise and fall of imperial systems in different areas of the world, and the succession of dominant powers up to the present. A related narrative recounts the evolving technology and social organization of warfare, with the result that advances in warfare facilitated the expansion of great states. A distinct political narrative traces the changing institutions of government and the social classes or interests that dominated government; of particular interest has been the expansion of representative government.

The leading global economic history narrative portrays the expansion of systems of long-distance trade, as they interacted with centers of production and of wealth. In this narrative, attention has focused on the major trade goods and on systems of money. But of nearly equal importance has been the study of different systems of production, focusing on the use of labor and particularly on the role of peasants, artisans, slaves, and wage workers in production. For recent centuries it has been possible to trace levels of aggregate output and wealth for national economic units and to observe the growing inequality between rich and poor nations.

The questions of historians can often be categorized, I find, into questions about the origins, the timing, the dynamics, and the legacy of past historical phenomena. (I find it helpful, in addition, to break the "dynamics" of the past into the functioning of historical systems, the connections among systems, and the transformations of systems.) For the origins of political and economic systems, historians' questions focus not only on the initial creation of states and markets but on creation of new stages in history.

The origins of capitalism and of industrial production are thus among the most hotly debated historical questions, as are the origins of nations. What were the origins of the commercial systems of the Mediterranean, of the Silk Road, the Indian Ocean, and (later) the Atlantic and Pacific? What were the factors promoting growth, decline, or transformation in the economic or political systems?

3. Larry Diamond, ed., *The Democratic Revolution: Struggles for Freedom and Pluralism in the Developing World* (New York, 1992).

For timing, one can ask when a world economic system came into being, and when a world political system came into being. In the functioning of political and economic institutions, historians have inquired into the workings of imperial governments and systems of long-distance trade. (Analyses of constitutional systems and trade diasporas may be seen as attempts to answer these questions.) How did trans-Saharan trading systems function for 1,500 years? For connections in political and economic history, historians have asked about the spread in types of government (such as Persian traditions of government) and about the way in which commerce in such stimulants as tea and coffee linked commercial systems around the world.

In posing questions about the transformation of political and economic systems, historians have asked about the rise and decline of modern slavery, the emergence of electric power, and the transformations of empires into nations or nations into empires. Sometimes the questions are about frameworks of analysis: should political and economic systems be explained in terms of the dominance of their central power or in terms of interconnections of constituent pieces? In politics, theses tend to focus on dominance and cataclysmic change in the history of states. Wars, for instance, come as a shock to the global system, as indicated in the aftermath of World War I, World War II, the Napoleonic wars, and the Mongol conquests. In economics the analysis is more likely to center on evolutionary change. What has been the balance of dominance and interconnection in the operation of global economic systems?

The legacies of Greek democracy and of Roman law for the modern world are often invoked, the first providing an inspiration and the second providing principles and procedures still in use. The legacy of the Mandate of Heaven provided successive dynasties with rationale for continued power and gave opponents a principle for overthrowing ineffective rulers. The legacy of the founding caliphs offered the Islamic world a model for piety and effective government. For economic affairs, the notion of legacy is less commonly employed, though one could think of African railway systems as a legacy of the colonial era.

In addition to these questions about the past arising from the historical record itself, the concerns of the present day provoke important questions about the political and economic history of the past. The state is seen today by some as the protector of social welfare for its citizens, but in other perspectives the state is seen as a source of oppression for some of its population: have states become more or less oppressive with time? For much of the twentieth century, the world experienced a competition between capitalist and socialist socioeconomic systems: what precedents were there for this struggle, and what legacy has been passed on from that struggle? Is corruption in public life becoming more serious? What precedent is there for democratic political institutions? What fates have previous states brought to small political communities, such as indigenous peoples? Will states bring employment and education to all citizens?

Having demonstrated that the questions are many but that they arise logically from existing knowledge and current problems in political and economic affairs, let us now consider the results of recent research in these two fields.

Political Dynamics: Rise and Fall

When did political history begin? Only when sizable political units emerged, according to the usual argument. These were monarchies that were themselves based on the localized institutions of government that preceded them by perhaps thousands of years. The analysis of this early stage of political history depends on the collaboration of historians and anthropologists.[4]

Political history focuses overwhelmingly but not entirely on the state. Christopher Chase-Dunn and Thomas Hall, in their macrohistorical studies of human societies, identify politics as one of several levels of large-scale organization.[5] As they present it, the scale of politics and government exceeds that of production and exchange of foodstuffs, but long-distance trade takes place at a scale beyond that of politics and government. Their analysis of expansion and contraction of political systems sets government in the context of a wider range of social processes.

Michael Mann proposes a history and theory of power relations, addressing power from early states to 1914 in the first two of a projected four volumes. He defines societies as "constituted of multiple overlapping and intersecting sociospatial networks of power," and focuses on ideological, economic, military, and political sources of power. His narrative of power, however, seems to respect well-known boundaries of societies.[6] Christopher Chase-Dunn and Thomas Hall set the same political systems in world-system perspective, emphasizing a similar but wider range of power relations through spatial and graphical portrayal.[7] Quite a different theory of power is that of Michel Foucault, who focuses on fields of power relations.[8] It may not be possible to unify or generalize theories of power, but the various approaches should be placed in contact with one another.

4. Timothy Earle, *How Chiefs Came to Power: The Political Economy in Prehistory* (Stanford, 1997). For Jan Vansina's interpretation of political culture in equatorial Africa before large states, see chapter 13.
5. Chase-Dunn and Hall 1997.
6. Michael Mann, *The Sources of Social Power*, 2 vols. (Cambridge, 1986–1993). Mann's definition of society and power seems closely related to that of Geertz: see page 238.
7. Chase-Dunn and Hall 1997; Chase-Dunn, *Global Formation: Structures of the World-Economy*, 2nd ed. (Oxford, 1998).
8. Foucault began with an emphasis on the complexity of social interrelations and the changes in systems of interrelations, working first with mental illness, then clinics, then prisons, and finally sexuality. He emphasized the ability of people to redefine meanings and relationships and thus to change aspects of the systems within which they lived. His emphasis on the interplay of power and knowledge at all levels of society effectively challenges the notion that power can be monopolized by political elites. Michel Foucault, *Mental Illness and Psychology*, trans. Alan Sheridan ([1954] New York, 1976); Foucault, *The Birth of the Clinic: An Archaeology of Medical Perception*, trans. A. M. Sheridan Smith ([1963] New York, 1973); Foucault, *The Order of Things: An Archaeology of the Human Sciences*, trans. anon. ([1966] London, 1970); Michel Foucault, *Power/Knowledge: Selected Interviews and Other Writings 1972–1977*, ed. Colin Gordon, trans. Colin Gordon et al. (New York, 1980).

The documentation of rise and fall in world political history can lead one not only into the chronicles of creating and eliminating great states but into the changing institutions and practices of governments over time. The anthropologist Bruce Trigger, in an innovative, cross-sectional approach to what he calls "early civilizations," has compared seven societies by type of government: territorial states (Egypt, North China under Shang and Western Chou dynasties, and Inca) and city-state systems (Mesopotamia, Mexico, Maya, and Yoruba).[9] He emphasizes that both models of political structure have been applied across the millennia and that each has characteristic benefits.

The notion of the monarchy has arisen so often that the idea of kingship may seem unproblematic, but the differences in types of kingship show monarchy to be worthy of detailed scrutiny. Is the monarch selected by heredity, individual accomplishment, or consensus among contending social institutions? What, further, is the history of political advisers? Informal advisors at one level are followed by formally appointed ministers. Still other institutions of government include courts, legislatures, laws, diplomats, armies, provincial governments, tribute, taxation, and public works. Writing a global history of developing political institutions is no easy matter. It has been attempted within civilizational context, as for Europe, for China, Japan, and the Islamic world.[10] But to expand the frame of analysis and trace the development and revision of governmental institutions across civilizational lines will be a complex task. One might compare this possibility with the work that has already been attempted in military history. Because of available documentation, military historians have tended to restrict themselves to analysis of the great powers. Within those limits, however, military historians have made courageous efforts to analyze military technology and institutions at a global level.[11]

World history, in politics, tends commonly to be explored at the regional level. One advantage to this approach is that it focuses attention on the common institutions of the various regional political systems. Regional networks of political systems may be identified on a large and small scale throughout the world, as successful innovations spread by conquest or imitation to neighbors. The Chinese system of politics is the largest and most durable, with its ideological rationalization in the Mandate of Heaven. The system of administration through scholar-gentry spread along with the frontiers of the empire, but other institutions and titles of Chinese inspiration spread to many parts of eastern Asia. Persian practices

Continued

For a concise and effective survey of Foucault's work, see Lois McNay, *Foucault, A Critical Introduction* (New York, 1994). For an approach linking Foucault's work to the natural sciences, see Pamela Major-Poetzl, *Michel Foucault's Archaeology of Western Culture: Toward a New Science of History* (Chapel Hill, 1983).

9. Bruce G. Trigger, *Early Civilizations: Ancient Egypt in Context* (Cairo, 1993).
10. J. R. S. Phillips, *The Medieval Expansion of Europe* (Oxford, 1988). For an effort at describing planetary rise and fall, see Knutsen 1999.
11. Martin Van Creveld, *Technology and War from 2000 B.C. to the Present* (New York, 1989). For a different approach to military history, based on quantification rather than institutions, see William Eckhardt, *Civilizations, Empires, and Wars: A Quantitative History of War* (Jefferson, N.C., 1992).

and symbols of government, developed especially in the Achaemenid period, maintained significance in the surrounding Caucasus, Central Asia, and North India well into the second millennium. Over the same time period, the political tradition of Mesoamerica underwent progressive development, though under a variety of regimes. The institutions of the Islamic Caliphate, assembled initially by founders of the regime in Medina and Damascus, provided governmental systems that lasted into the twentieth century and court systems that operate today.

In the language that has come to be used in recent centuries, all of these governmental systems had constitutions, in that they tended to work by certain regular political principles. The practice of writing and legislating constitutions began in the eighteenth century, and now most people are governed by states with written constitutions that can be traced directly to those of the eighteenth century.[12]

For the twentieth century, the issue of democracy—meaning political representation and participation by the majority of the population—has come to be an essential topic in politics. Consent of the governed appeared to be a less significant issue for large political units in earlier times, yet at local levels some sort of consensus was necessary for a governing authority to maintain power. For this reason the issue of democracy is of interest in human history over the long term: while one cannot simply project the current vision of democracy back into earlier times, one may seek to reconstruct the degree of participation and consensus of the governed in political affairs at local levels as well as in the state overall.[13]

Perhaps because the traditions of political history focus so firmly on nations and empires, one is hard pressed to find political studies reaching beyond individual states and their interactions. World historians have not relied much on the anthropological and political science literatures, with their emphasis on local government, representative institutions, and constitutional principles. It would be a step forward to have more world-historical studies addressing these issues, linking patterns of local, national, and imperial government and tracing long-term persistence and transformation of governmental institutions and principles.

States: Empire and Nation

The analysis of nations occupies a disproportionate space in the historical literature. This can hardly be surprising, since the field of history has built itself, in the last century, on the celebration and analysis of nationhood. The analysis of nationhood is one of the great arenas of struggle in which the field of world history has gained its identity in contrast to national history. Aside from those

12. Studies of constitutional history have so far been defined almost uniquely in national terms, though a rich array of evidence awaits scholars who wish to undertake study of the transnational spread and transformation of constitutional forms.
13. John Dunn, *Democracy: The Unfinished Journey, 508 B.C. to A.D. 1993* (Oxford, 1992); Mann 1986–1993; Craig N. Murphy, *International organization and industrial change. Global governance since 1850* (Cambridge, 1994); Donald Wallace White, *The American Century: The Rise and Decline of the United States as a World Power* (New Haven, 1996).

exceptionalists who studied one nation at a time, generally to set it above all others, there arose theorists of nationalism. Theorists of nationalism divided into those who wished to focus on European nations, concentrating on the rise and transformation of national ideas among them, and those who wished to explore nationhood in larger context.[14] The latter group addressed the Americas from the eighteenth century and Asia and Africa from the nineteenth century, constructing from them a global story of nationhood.[15] As a result, much of the identity of world history has emerged out of debates over nations and nationhood.

The nation, with its current meaning of an organized political community of shared identity and institutions (and almost always with a state), is a modern creation. The global analysis of nationhood and nationalism is a sophisticated and interactive area of study, in which local and global factors, culture and ideology, technology and economics, political tradition and social movements combine to create a pattern of accelerated creation and transformation of national political units.[16] Nations have been analyzed and theorized in detail. The political triumph of the nation led to a widespread belief that analysis of nations was sufficient for explaining global political dynamics in the twentieth and even nineteenth centuries.

Lacking, however, has been any equivalent global analysis of the politics of earlier times, particularly on the political history of empires. That is, global political analysis has neglected empires and the degree to which the political activity of great powers is that of empires and not just that of nations contained within empires. The recent rebirth of interest in imperial history, especially for the British empire, would seem to provide an opportunity to study the role of empires in politics. For the most part, however, the new studies in imperial history have focused empirically on a single empire at a time, rather than on the comparison and interplay of empires with each other and with nations.[17]

Empires are different from nations. The category of empire goes back thousands of years, and while empires have surely changed over time, the category

14. The most sophisticated founder of the Eurocentric study of nationalism was Hans Kohn, *The Idea of Nationalism: A Study in its Origins and Background* (New York, 1944); Kohn, *The Age of Nationalism: The First Era of Global History* (New York, 1962).

15. Emerson 1960; Elie Kedourie, *Nationalism* (Oxford, 1960); Ernest Gellner, *Nations and Nationalism* (Ithaca, 1983); B. Anderson 1983.

16. Linda Basch, Nina Glick Schiller, and Cristina Szanton Blanc. *Nations Unbound: Transnational Projects, Postcolonial Predicaments, and Deterritorialized Nation-states* (Amsterdam,1994); Buzan 1991; Peter Evans, Dietrich Rueshemeyer, and Theda Skocpol, *Bringing the State Back In* (Cambridge,1985); Hobsbawm 1975; Hobsbawm, *The Age of Empire, 1875–1914* (New York, 1987); Hobsbawm 1962; Hobsbawm 1990; Hobsbawm 1994; Charles S. Maier, *Dissolution: The Crisis of Communism and the End of East Germany* (Princeton, 1997); Maier, *The Unmasterable Past: History, Holocaust, and German National Identity* (Cambridge, Mass., 1988).

17. William Roger Louis, *Great Britain and Germany's Lost Colonies, 1914–1919* (Oxford, 1967); Louis, *Imperialism at Bay: The United States and the Decolonization of the British Empire, 1941–1945* (New York, 1978); Louis 1998–1999; Cain and Hopkins 1993; Bayly 1989.

seems likely to survive. To a remarkable degree, the study of individual empires has superceded any broad effort to explore the role of empires in world history, to explore the changing institutions of empires, or to investigate the patterns of relationships between empires and other political units.[18] Empires are distinguished by having a core area, outlying territories governed directly or indirectly under the aegis of the empire, and areas of informal influence, all under the hegemony of a governing elite centered in the core area. The relative sizes of the components and the institutions by which they were held together varied greatly, but with rare exceptions a military force, based primarily in the core area, guaranteed imperial power.

David Abernethy has provided one step toward a more general analysis of empires in his study of five centuries of European overseas empires. This work accounts for stages of expansion and decolonization through the interplay of public, private and religious sectors of the imperial order, but it restricts the analysis to European empires. A comparison of recent studies on the Mongol Empire, on Islamic states, Rome, and modern European empires might yield a clearer view of the general character of empires and their changes over time.[19]

Studies of global political order, while they might be developed for earlier times, are limited in practice to the nineteenth and especially the twentieth century. The field of international history developed out of multinational studies in diplomatic history and has provided interpretations of global politics especially through the work of Akira Iriye. Noam Chomsky's wide-ranging and eclectic studies, focusing on critique of U.S. policy in international affairs, are rich in historical comparisons and insights on military, political, and propaganda activities of governments and corporations. Political scientist Samuel Huntington, in summarizing a proposal for sustaining U.S. hegemony in international affairs, develops a long-term interpretation of global political order.[20]

At a more specific level in political affairs, a growing literature on frontiers in history is distinguishing with increasing clarity between frontiers as well-policed borders between adjoining powers and frontiers as zones at the edge of major powers, where contending influences overlap. Recent works have given particular attention to Eurasian frontiers.[21]

Two fascinating studies reveal the possibilities of exploring law in global historical context. Lauren Benton, addressing a wide sampling of colonial situations, emphasizes the place of cultural identities in what she labels as an example of

18. Ibn Khaldun [1377].
19. Abernethy 2000; Michael Adas, ed., *Islamic and European Expansion: The Forging of a Global Order* (Philadelphia,1993); Reuven Amitai-Press and David O. Morgan, eds., *The Mongol Empire and its Legacy* (Leiden, 1999); P. S. Barnwell, *Emperor, Prefects, and Kings: The Roman West, 395–565* (Chapel Hill, 1992); Armitage 2000; Pagden 1995.
20. Chomsky 1994; Samuel P. Huntington, *The Clash of Civilizations and the Remaking of World Order* (New York, 1996); Ranajit Guha, *Dominance without Hegemony: History and Power in Colonial India* (Cambridge, 1998); Iriye 1992; Iriye 1997.
21. Thomas J. Barfield, *The Perilous Frontier: Nomadic Empires and China, 221 B.C. to A.D. 1757* (Cambridge, Mass., 1989); Brower 1997; Daniel Power and Naomi Standen, eds., *Frontiers in Question: Eurasian Borderlands, 700–1700* (New York, 2000).

institutional world history. While her volume traces a long-term shift from multicentric law systems in early modern times to state-centered law by the twentieth century, she also shows that legal systems at any time have had to account for multiple perspectives. Jeremy Adelman achieves similar insights with a study of a single society, Argentina in the nineteenth century, where the struggles over law and political order centered on the place of Argentina in the Atlantic economic and legal system.[22]

Empires, while constructed on some notion of dominance, still existed as a balance of imperial and colonial influences. While there have been many studies of local resistance to colonial order, there are now some studies aggregating or generalizing these results into larger pictures of the relative place of metropolitan and peripheral peoples in the operation of imperial systems.[23] The operation of empires involved struggles in social life as well as political conflicts, and the social-historical dimension of empire has been explored most recently through various aspects of women and imperialism.[24]

While formal imperial control of territories has almost entirely lapsed, for the present, the exercise of hegemonic power at the global level, especially by the United States, may be reaching an unprecedented level. Michael Hardt and Antonio Negri claim that this represents a new set of institutions for political dominance, and they propose to call it *Empire*.[25] Their footnotes survey much of the literature in global political history, and debate over this book should help determine whether the current situation is sharply different from imperial balances of power in earlier times.

Microeconomies: Production, Commerce, and Money

Individual commodities, the money with which they are exchanged, and the commercial networks moving and exchanging goods provide small stories that are big stories in world history. Porcelain, silk, silver, iron, grains, people in chains, gold, coffee, tea, locomotives, and televisions are but a few of the key global commodities whose exchange tells tales of global connections.

Certain goods became commodities—that is, they had a value in long-distance trade, and to the degree that they were durable, the records of archaeologists show us their history.[26] Even in days before cities became significant, specific types of stone such as obsidian, turquoise, or emerald tended to be traded over long

22. Adelman 1999; Lauren Benton, *Law and Colonial Cultures: Legal Regimes in World History, 1400–1900* (New York, 2002). See also Jeffrey N. Wasserstrom, Lynn Hunt, and Marilyn B. Young, eds., *Human Rights and Revolutions* (Boston, 2000).

23. Adas 1979; Peter Wells, *The Barbarians Speak: How the Conquered Peoples Shaped Roman Europe* (Princeton, 1999).

24. N. Chaudhuri and Strobel 1992; De Pauw 1998. These and related studies are discussed in more detail in chapter 11.

25. Michael Hardt and Antonio Negri, *Empire* (Cambridge, Mass., 2000).

26. Pryor 1977.

distances. Copper, gold, and silver came to have such value, along with spices. One of few goods with high bulk to value to be traded over long distances was salt—highly desired in dry interior regions.[27]

The stages in development of commerce and markets present a major issue in world history. The earliest written records from Mesopotamia are those of commercial transactions, showing that commercial specialists had emerged. The Classical era of the Mediterranean, in the late first millennium B.C.E., was a time not only of active commerce but of great expansion of money to facilitate that commerce. Metallic coins, created under the authority of monarchs, circulated in their realms and beyond. The Romans were especially energetic in making coins, and their coins have been rediscovered in most parts of the eastern hemisphere.

Money consisted not only of coins stamped by governments. An alternative that may have been developed at a similar time was sea shells, and especially cowrie shells from the Maldive Islands off the western coast of India. These shells, resupplied by a regular trade, became money in Ceylon, Bengal, Yunnan, the Persian Gulf, and, with a remarkably long trek, in West Africa. Currency was less widespread in the Americas than in the eastern hemisphere, but cocoa seeds, for instance, served as a common currency in Mesoamerica, and cotton textiles were exchanged as a result.[28] In other areas and times, bricks of tea, squares of cloth, and other commodities were used as money. Of course many commercial transactions took place with little or no money exchanging hands if the parties were exchanging equal values of goods or if one of the parties provided credit to the other.

Dennis Flynn and Arturo Giráldez have done much to enunciate and spread the story of silver mining and trade from the sixteenth through the eighteenth century. As they recount it, the opening of major silver mines in Peru and Mexico in the mid-sixteenth century corresponded with a substantial increase in silver demand in China, occasioned especially by the Qing decision to collect taxes primarily in silver. As a result, from 1571 silver went around the world in two directions from Spanish America to China—across the Pacific on galleons to Manila, and across the Atlantic, through Europe, and by stages to India and China.[29] The unmistakably global connections of silver trade and the relentless demand for silver in China each had major implications: the existence of world-encompassing trade connections and the steady expansion of the Chinese economy.

A world-historical approach, giving emphasis to connection and therefore to commerce, will emphasize the pervasiveness of exchange and of monies. In very recent times, colonial rulers have claimed that their subjects did not know or were just learning about "the money economy." Such statements should generally be viewed with caution. In colonial Africa, for instance, European rulers commonly

27. Herbert 1984; Adshead 1992; Kurlansky 2002; Mark Kurlansky, *Cod: A Biography of the Fish that Changed the World* (Penguin, 1998); Mazumdar 1998; Michael. N. Pearson, ed., *Spices in the Indian Ocean World* (London, 1996).

28. Jan Hogendorn and Marion Johnson, *The Shell Money of the Slave Trade* (Cambridge, 1986).

29. Flynn and Giráldez 1995; Flynn and Giráldez 2002.

refused to recognize the local moneys, or meant by "the money economy" that their subjects should give up their own work and accept work in European enterprises at a unilaterally set wage. Thus, the story of world commerce is partly that of the spread of money, but it is far more the story of competition and domination in markets.

Among the topics in commercial history that have gained wide attention from world historians are the Silk Road and trade diasporas. The Silk Road connected East Asia to West Asia and the Mediterranean through the desiccated regions of Central Asia. Large-scale trading ventures on this route are known for at least two millennia, but archaeological records show that exchange of some sort has moved along this route far longer.[30]

Trade diasporas have gained wide attention through the work of Philip Curtin, who emphasized them to illustrate a pre-industrial commercial institution, working effectively at long distances through reliance on family and ethnic identity.[31] Claude Markovits demonstrated the place of South Asian merchants in global trade through a detailed study of merchants from two towns of the Indus Valley, who sustained long-distance trading networks for two centuries. Markovits contests Curtin's notion that trade diasporas were dependent on cultural difference, and challenges Curtin's idea that they were limited to pre-industrial times.[32]

While trade receives more attention in studies of world history, the production of food, handicrafts, and even luxuries has arguably been more fundamental to economic life, though it is not as easy to document as trade. The Danish economist Ester Boserup is one of the few scholars to find a way to focus on issues of economic production outside of industrial production. Her studies, based on field research in East Africa, proposed general statements on the relations among population density, agricultural innovation, and the activity of women in farming. Overall her approach has suggested ways of locating change as well as continuity in rural economic life.[33]

Macroeconomics: Growth and Transformation

For the twentieth century, macroeconomic statistics have been developed to permit the analysis of gross domestic product and its changes over time, and have greatly clarified the functioning of economies at aggregate levels. These figures on the total value of domestic output, when divided by population, give per capita output estimates that are often taken as a measure of relative economic welfare

---

30. Liu 1995b, 25–48; Liu 1995a; David Christian, "Silk Roads or Steppe Roads? The Silk Roads in World History," *Journal of World History* 11 (2000), 1–26.
31. Curtin 1984b.
32. Claude Markovits, *The Global World of Indian Merchants, 1750–1947: Traders of Sind from Bukhara to Panama* (Cambridge, 2000).
33. Ester Boserup, *The Conditions of Agricultural Growth* (Chicago, 1965); Boserup, *Woman's Role in Economic Development* (New York, 1970); Boserup, *Population and Technological Change: A Study of Long-term Trends* (Chicago, 1981).

from country to country.[34] Comparisons of GDP figures show the relatively steady growth of industrial economies, and the slower and more episodic growth of agricultural economies. The problem of wealth and poverty, debated since the late eighteenth century if not before, became a major focus of global economic analysis in the late twentieth century; with it emerged an analytical interest in economic inequality among (and within) nations, and an expanded interest in "development" policy as a means for minimizing inequality. Economic historians sustained the pattern of dividing the world of the twentieth century horizontally into the industrial powers and their colonies or semi-colonies, and dividing it horizontally into the spheres of influence of each of the great powers. Colonization and decolonization thus structured much of the economic history of the twentieth century.

For the nineteenth century, Kevin O'Rourke and Jeffrey Williamson have used microeconomic data to make macroeconomic arguments. They relied on trade statistics to project the fluctuations of growth and decline, openness and autarky, back through the nineteenth century and make an effective case for an era of globalization in the world economy from 1840 to 1914. The limit on their approach, however, is that it focuses on major industrial powers of the North Atlantic (plus Australia) and analyzes them in national units, because of the statistical difficulty of extending the argument to the world as a whole.[35] The vision of a single world economy for the nineteenth century is offered in this work, but it cannot be sustained until it is more directly connected to other regions of the world, perhaps following up on the type of global analysis of Asian and African economies conducted earlier by A. J. H. Latham, and on the two-stage interpretation of European and then global industrialization proposed by Peter Stearns.[36]

The world economy of the eighteenth century has received, in recent years, virtually as much attention as that of the nineteenth century, especially because of some skillful comparisons of China and Europe. Here again, scholars have been using microeconomic data to make macroeconomic arguments. Kenneth Pomeranz has received particular attention for his comparisons of life spans, textile output, fuel consumption, furniture holdings, sugar consumption, tobacco consumption, and more in the wealthy manufacturing regions of China and the manufacturing regions of England. He argues that the results show levels of output, income, growth, and ecological degradation that were remarkably similar for the two sets of regions. The "great divergence," therefore, was an affair of the nineteenth century: only then did British economic growth continue while that of China halted. The crucial advantages for Britain, Pomeranz argues, were the availability of coal near to factory sites and the availability of lightly populated lands and timber in the Americas. R. Bin Wong's comparison of Chinese and European government policy yields similar results. Using monetary flows to underscore their argument, Dennis Flynn and Arturo Giráldez argue in addition that there

---

34. Angus Maddison, *The World Economy: A Millennial Perspective* (Paris, 2001).
35. O'Rourke and Williamson 2000.
36. A. J. H. Latham, *The International Economy and the Underdeveloped World, 1850–1914* (London, 1979); Stearns 1993b.

was a single world economy for these centuries, linked especially by the trade in silver from the sixteenth century.[37]

Andre Gunder Frank proposed a global synthesis of these arguments, articulating the thesis that "there was a single global world economy with a worldwide division of labor and multilateral trade from 1500 onward." The centers of production and accumulation of wealth were East Asia and South Asia; other regions participated without dominating. He emphasized monetary flows of silver, gold, and credit as the lubricant for global exchange. He contested previous visions of Asian hoarding of precious metals and argued that inflows of money corresponded to settlement of new territories and expansion of production in Asia. The economic transformation from 1500 on, in this view, was neither the establishment of links between previously distinct regional economies nor the progressive incorporation of regions into a European-dominated economy or world-system. Instead, it reflected the fluctuations of regional control within an economy that was already global and in which European economies gained dominance only in the nineteenth century. Europeans achieved dominance because of their control of the resources (especially monetary) of the Americas, plus the profits gained from trade with the Americas and from growing participation in Asian "country trade." A cyclical pause in Asian economic growth in the late eighteenth century corresponded with rapid growth in Europe, and Europeans were able to turn this cyclical advantage into longer-term economic and political hegemony.[38]

At a level that is at once broader and more piecemeal, Angus Maddison has proposed a millennial macroeconomic synthesis of the world economy that leads in quite a different direction. This approach, based on studies of national income and economic growth in the twentieth century, develops estimates of gross domestic product and population for national units of recent times, and then projects them back into the past. The conclusions focus on Europe and East Asia but treat Europe as the one growing area of the world economy since about 1200, while other areas (notably China) are treated as having static or occasionally declining per capita GDP for the past millennium. Estimates for other areas are less well documented, but Maddison estimates that the populations of Europe, India, and China added up to 70 percent of the world total in both 1000 and 1700, so that changes in estimates for other regions would not change the global totals by much.[39]

---

37. Jack Goldstone has labeled the work of these scholars as "the California school," since virtually all are based in California. Goldstone 1991; Flynn and Giráldez 1995; Von Glahn 1996; Wong 1997; Pomeranz 2000.

38. Frank 1998. For an interpretation of global political economy over a longer time frame that has similarities to Frank's approach, see William R. Thompson, *The Emergence of the Global Political Economy* (London, 2000).

39. For the period from 1500 to 1820, Maddison estimated that Western European population grew by 0.26 percent per year, while per capita GDP grew by 0.15 percent per year; for China, the equivalent estimates were 0.41 percent growth per year for population and 0.00 percent for per capita GDP. Maddison 2001; Sweezy et al. 1976.

Wallerstein has argued for parallel worlds in the early modern period. He envisions the creation of a new world-economy based in sixteenth-century Europe, relying significantly on American colonies, and growing and transforming in various cycles thereafter. The growing empirical data on Asian economies have ultimately forced a revaluation, in which world-system analysts must decide whether there existed a separate Chinese world-system and whether there existed a world-encompassing global economy or a global economy made up of distinct regional systems.[40]

Contrasting views of Europe and Asia in the world economy are thus set up for debate in the years to come. The neoclassical economic historians such as Maddison work with comparisons of national units, the world-systems analysts focus on structural relations among core and outlying areas, and the California group conducts eclectic comparisons and linkages on a wider range of issues. Debate is already taking place in several fora, and the results are sure to be revealing both for the methods and the interpretations that are reaffirmed.[41] My own concern about the direction of this debate is that while it is clearly advancing the ability of historians to document and interpret economic history at the global level before the twentieth century, the arguments are too much about dominance and not enough about connection. Perhaps additional work from world-systems and area-studies perspectives will take us to the next level and develop a more comprehensive picture of transformations in an interconnected world economy of the early modern era.

The modes-of-production approach to global social and economic history, which flowered briefly in the 1970s and 1980s, has since dropped from sight.

---

40. Ho-fung Hung, "Imperial China and Capitalist Europe in the Eighteenth-Century Global Economy," *Review* 24 (2001), 473–513.

41. The electronic debate on the positions taken by David Landes and Andre Gunder Frank took place from 11 May to 9 June 1998, and was shared among four discussion lists: H-WORLD, H-Asia (Asian studies: www.h-net.msu.edu/~asia), EH.RES (Economic history research: www.eh.net/lists/archives/eh.res), and WSN (World systems network: www.csf.colorado.edu/wsystems/wsarch.html). Frank, 1998a; David S. Landes, *The Wealth and Poverty of Nations: Why Some are so Rich and Others so Poor* (New York, 1998). The debate of 2 December 1998 between Frank and Landes at Northeastern University is available in a transcript at www.worldhistorycenter.org/whc/seminar/pastyears/frank-landes/. It was broadcast, after a delay, on C-SPAN. In the first half of 1999, the World History Seminar at Northeastern University followed up the debate with presentations by Peter Gran, Kenneth Pomeranz, Bin Wong, Peter Perdue, and Sanjay Subrahmanyam. For an amplification and discussion of "California school" views, see the four elements of a forum: Patrick Manning, "Asia and Europe in the World Economy: Introduction." *American Historical Review* 107 (2002), 419–24; Kenneth Pomeranz, "Political Economy and Ecology on the Eve of Industrialization: Europe, China, and the Global Conjuncture," Ibid., 425–46; R. Bin Wong, "The Search for European Differences and Domination in the Early Modern World: A View from Asia," Ibid., 447–69; David Ludden, "Modern Inequality and Early Modernity: A Comment on the AHR Articles by R. Bin Wong and Kenneth Pomeranz," ibid., 470–80.

Part of the idea of this approach was to argue for local specificity in social and economic organization, yet also to emphasize connection ("articulation") among local systems. The analyses became too theoretically complex to sustain with empirical data, and the approach was largely dropped. It is my opinion, however, that future attempts to establish the place of local economic and social systems in global setting will lead to revising these approaches.[42]

No one has made an argument for an integrated world economy before the sixteenth century, but a number of cases have been made for broad and interconnected economic systems in earlier times. Janet Abu-Lughod stated this possibility forcefully in her interpretation of a fourteenth-century world-system stretching from the Mediterranean and East Africa to China. More vaguely, suggestions have been made of a global economic system associated with the Mongol conquests. In an argument based on development of institutions rather than the functioning of an economic system, R. J. Barendse has made the case for a relatively rapid and widespread development, especially in the tenth and eleventh centuries C.E., of decentralized, horse-based military systems reliant on subordination peasants to the warriors—in effect, "global feudalism." At a time still earlier, Liu Xinru has emphasized the Eurasian connections of the Kushana state.[43]

Several scholars have sought to generalize these patterns. Andre Gunder Frank and Barry Gills have made the argument for a continuous and gradually expanding world-system for the past five thousand years, and they gathered a group of analysts to debate this approach; David Wilkinson used the term "Central Civilization" in a similar interpretation. Christopher Chase-Dunn and Thomas Hall have gone on to propose a quantified version of this approach.[44]

The consideration of such long-term economic patterns raises the issue of the dynamics of change. E. L. Jones has emphasized incremental growth in most economies, with occasional periods of rapid growth, sometimes in response to ecological windfalls. Andre Gunder Frank has emphasized long-term economic and political cycles, with a period of some four hundred years. A different sort of dynamic is imposed by epidemic disease, expanding at times when human

42. Barry Hindess and Paul Q. Hirst, *Pre-Capitalist Modes of Production* (London, 1975); John G. Taylor, *From Modernization to Modes of Production: A Critique of the Sociologies of Development and Underdevelopment* (London, 1979); Roxborough 1979.
43. Abu-Lughod 1989; Bertold Spüler, *History of the Mongols, Based on Eastern and Western Accounts of the Thirteenth and Fourteenth centuries,* trans. Helga and Stuart Drummond ([1969] Berkeley, 1972); Liu 1995a; Liu 1995b; R. J. Barendse, "Chinese 'feudalism,'" H-WORLD (h-net.msu.edu/~world) 28 Oct. 1999; Stephen Morillo, "Global Feudalism," H-WORLD 5 Nov. 1999.
44. While in many ways these are new studies based on recently developed data, in other ways they restate ideas long explored. The first edition of the Encyclopaedia Britannica included an elaborate time line, picturing the rise and fall of individual civilizations and the temporally growing patterns of interaction. Frank and Gills 1993; Chase-Dunn 1997; Chase-Dunn and Hall, eds., *Core/Periphery Relations in Precapitalist Worlds* (Boulder, 1991); Christopher Chase-Dunn, E. Susan Manning, and Thomas D. Hall, "Rise and Fall, East-West Synchronicity and Indic Exceptionalism Revisited," *Social Science History* 24 (2000).

population and communication have expanded—the examples of plague in Eurasia in the sixth and thirteenth centuries stand out. Yet another dynamic is set in motion by the conquest of settled centers by mobile warriors, who bring destruction but also establishment of wider networks.[45]

In addition to these analyses of hemispheric and continental scope, a substantial number of studies appeared during the 1990s on the political economy of trade and empire. James D. Tracy edited two excellent volumes gathering individual contributions on merchant empires from the fourteenth to eighteenth centuries, and Sanjay Subrahmanyam synthesized his earlier work with an analysis of political economy in early modern South India. R. J. Barendse's analysis of *The Arabian Seas,* set at a wider scale, traces the integration of trade networks bordering the western Indian Ocean.[46] Anthony Reid's two volumes on Southeast Asia documented both commerce within the region and connections with East Asia, South Asia, and Europe in the early modern period.[47]

Much work remains to be done to develop details in each of these areas of study. But one can see the beginnings of a body of interpretation and debate on the world economy (or the collection of the world's main economic regions) over many centuries. These ultimately may provide a solid context for assessing recent economic change. Giovanni Arrighi has made an introductory foray into this type of thinking with his analysis of a long twentieth century in which he links contemporary political economy to patterns of development stemming from as early as the fourteenth century. The growing regional inequalities in levels of per capita income and output have been long known to economic historians, and Kenneth Pomeranz has given the phenomenon a new name—"the great divergence"—as well as documenting its timing in China. We may expect further studies addressing this divergence worldwide: in a broad and detailed analysis, Mike Davis has linked the acceleration of this divergence to a mix of ecological and political changes in *Late Victorian Holocausts.*[48]

Conclusion: Rethinking Old Categories

Political and economic history have provided the strength of world history. The recently completed work and the new questions it raises suggest that much more

45. E. L. Jones, *Growth Recurring. Economic Change in World History* (Oxford, 1988); Michael Rowlands, Mogens Larsen and Kristian Kristiansen, eds., *Centre and Periphery in the Ancient World* (Cambridge, 1987); Sanderson 1995; Gil J. Stein, *Rethinking World-Systems: Diasporas, Colonies, and Interaction in Uruk Mesopotamia* (Tucson, 1999); Frank 1993.

46. Tracy 1990; Tracy 1991; Subrahmanyam 1990; Von Glahn 1996; Barendse 2002; Barendse 2000; Vimala Begley and Richard Daniel de Pubma, eds., *Rome and India: The Ancient Sea Trade* (Madison, 1991); Fox-Genovese and Genovese 1983.

47. Reid 1988–1993. See also Jan De Vries, *The Economy of Europe in an Age of Crisis, 1600–1750* (Cambridge, 1976); and K. Chaudhuri 1990.

48. Arrighi 1994; Pomeranz 2000; Mike Davis, *Late Victorian Holocausts: El Niño Famines and the Making of the Third World* (London, 2001). For a comparative economic analysis with a shorter time frame, see Alice Amsden, *The Rise of "The Rest": Challenges to the West from Late-Industrializing Economies* (New York, 2001).

can be done. These studies correspond to work along the historians' path to world history in which new perspectives and new data are developing a much broader and more interconnected picture of global political and economic history. In economic history, scholars are showing considerable originality in developing data and comparative perspectives relevant to the issues; nonetheless, the difficulty of developing data on comparable regions for transnational comparisons remains a major hindrance. For politics, much new research has been framed at a broad level, and we await further analysis and debate that that will clarify larger-scale and longer-term patterns.

For recent times, I think it is a priority for historians to trace and interpret the rise of international organizations, as one may plausibly argue that they represent the most pervasive and dramatic change in political and economic orders within the past century. For earlier times, I think it is equally important to clarify the patterns and dynamics of empires, including their interactions with each other and with smaller political communities. For both economics and politics, we need to get beyond historians' habitual focus on the most apparently dominant elements, to study interactions of a wider range of elements.

Chapter 11

Social History

Studies of social history were at the cutting edge of historical research for most of the late twentieth century. Especially in studies of European and U.S. history, but also in ancient and medieval history and in the expanding area-studies fields, social history achieved a period of academic hegemony. The disciplinary and topical subdivisions within social history kept it from being a dogmatic or uniform movement and perhaps assisted in providing continuing excitement in the analysis known variously as "history from the bottom up" and studies of "everyday life."[1] World history underwent its own major expansion in the era when social history was booming, and world historians were generally ready to accept the notion that there is more to understanding the past than great men and great powers. Nevertheless, social history has not been at the forefront of global historical studies. Civilizations and nations, rather than communities and social classes, have been treated as the building blocks of world historical interpretations in textbooks and monographs. One obvious explanatory factor is that of scale: studies in social history are generally based on local-level data or sometimes on national statistics, and they do not lend themselves automatically to investigation at transregional or global levels.

The topics of social history begin, perhaps most basically, with human population: the size, age and sex structure, and growth of human populations. Migration, both for short and long distances, is a common pattern and an important source of historical change. Humans organize their lives into families and communities. Analyses of family account for issues of gender, age, residence, inheritance, and socialization. Communities include these issues as well, and in diverse fashions. Communities include those organized by language, ethnicity, occupation, class, and residence. Communities may be urban or rural. Social classes and castes vary greatly according to the social situation, including monarchs and other elites, and such commoners as peasants, artisans, and slaves. For all these treatments of the social history of groups, the biographies of individuals also comprise an important dimension of social history.

Can social history be analyzed and interpreted at the world-historical level? The answer is surely in the affirmative, and this chapter presents a summary of

---

1. Major journals include *Social History, Journal of Social History, Journal of Family History, Population Studies,* and *Le Mouvement Social.*

advances in recent research and proposes research that world historians can undertake. The methods of social history include methods of the adjoining disciplines on which the field draws: demography, sociology, social anthropology, psychiatry, and feminist theory. Most of these fields rely principally on microlevel, individual approaches, which do not automatically connect to the aggregate, macrolevel social phenomena in which world historians may be most interested. The analysis of social movements, however, is one area in which individual and aggregate analysis have been combined effectively, and suggests that other such linkages been micro- and macrolevel social history are feasible.

Research Agenda: Global Social History

The narratives of global social history have been slow to develop. We know that human population grew very slowly until recent times, though populations grew rapidly when new areas opened up to settlement. Migrations have been rapid, as in the Atlantic migrations of the nineteenth century, or slow, as in the Han or Bantu migrations of earlier millennia. Successive migrations have grown in magnitude, and new communities have formed through separation from home communities and interaction with new neighbors. Ethnic and language communities formed and reformed; new ones were created by migration, old ones were absorbed by contact. Cities began as soon as agriculture developed, and they expanded and declined in waves that are beginning to clarify. The occupations of farmers, merchants, artisans, herders, soldiers, transport workers, priests, and officials expanded and contracted in accord with the organization of the economy. The balance of social inequality shifted as elite and slave populations grew at some times, while social equality developed at others. Elites receive special attention in world history, partly because they are well documented, but also because the concept of "elite" provides one way to compare leaders of society across time and space. Elites clearly include such groups as royal families, military leaders, merchant communities, and religious leaders, but world historians will have to learn to define elites more precisely in order to make appropriate cross-cultural comparisons.

The connections among rural agricultural and herding peoples have varied sharply over time: the period of the Mongol Empire was one of several in which herding peoples had great power. More commonly, farming and herding peoples lived in close but uneasy interaction, while in recent times herding peoples have become marginalized. The history of family change is well known in the last two centuries: families became larger as infant mortality declined, but they became smaller as collateral relatives became less common. Still, the patterns of family life and gender relations in earlier times are open for study.

Historians have asked about the origins of various ethnic and religious communities, of social classes (especially peasants, slaves, and wage laborers), and of the nuclear family. Have family size and structure changed, and if so, is the change in response to migration, economic change, or health conditions? Social structures and institutions have been investigated for religious communities, occupational groups, gender roles, and family structures. Historians have asked about the timing of the development of classes of slaves and industrial workers and about patterns of childrearing.

Perhaps the most basic question in the dynamics of social change is whether human societies have developed on a localized basis or through interactions among localities and region. More specifically, historians have explored the dynamics of migratory movements, population growth and decline, and community growth and conflict.

These questions on social history, though broadly cast, are mainly about patterns at a regional rather than planetary level. In addition, social history is best documented for industrial regions in recent times, so there is a tendency to treat North Atlantic patterns as the norm. Even legacies tend to be thought of mainly at the societal level: the legacy of an earlier social history still influences the descendants of Irish migrants to Australia and the descendants of Maori migrants to New Zealand.

The questions coming out of current issues, however, tend to be as much global as regional. The social conflicts of life today raise questions about the past of identity by race and ethnicity, the nature and permanence of gender roles, whether class distinctions are necessarily passed from one generation to the next, and whether social inequality is growing or declining. The populations of both young people and old people seem to be expanding in different ways, raising questions, for instance, of whether there has ever been such a large proportion of old people in populations. The rise in divorce and revelations of conflicts between parents and children raise questions of whether families are now weaker than they were before. Similarly, patterns of migration, crime, and population boom or bust each suggest questions about the past.

Population and Migration

Estimates of human populations by region have been proposed for the last several centuries. With the development of archaeological and paleontological studies, such estimates have been pushed back millions of years, though they remain heavily speculative. For large parts of Europe, the Americas, and East Asia, there now exist population estimates going several centuries back, based on assemblage and estimation from detailed local records. For the Americas at the time of contact with the eastern hemisphere and for Africa in the era of slave trade, systematic but indirect estimates of population have been developed based on available records.[2] By continued application of these techniques, it will be possible in years to come to have improved estimates of the size and structure of human populations, especially for recent centuries and possibly for earlier times.

---

2. Lee and Feng 1999; Noble David Cook, *Demographic Collapse: Indian Peru, 1520–1620* (Cambridge, 1981); Manning 1990; Bruce Fetter, ed., *Demography from Scanty Evidence: Central Africa in the Colonial Era* (Boulder, 1990).

For times within the past 20,000 years, it is often possible to identify something of the path and character of migrations through the distribution of languages. For the Americas, the noted historical linguist Joseph Greenberg and his colleagues have shown that early migrations from Asia into the Americas included at least three major waves, represented by the Amerindian languages (the great majority of all Amerindian languages), the Na-Dene languages (located principally in Canada), and the Inuit languages (along the Arctic fringe). The range of migrations over time across the Eurasian steppe is reflected in the Altaic languages (from the Pacific to Turkey) and in Indo-European languages (with traces as far east as the fringes of China). The seaborne migrations of Austronesian speakers have been broken into their successive waves through linguistic analysis: from the south Chinese mainland to Taiwan, to the Philippines, to the southwest (Indonesia and Malaya), to the east (Melanesia), and then in later movements further to the west (Madagascar) and to the east (Polynesia).[3]

An interesting parallel of the Indo-European languages and the Niger-Congo languages appears on the map. Each of these large language groups has about seven major subgroups (with a putative homeland near the Black Sea for Indo-European and in the Niger Valley for Niger-Congo). In each case, part of a single subgroup spread east and south starting perhaps five thousand years ago, and left a large region of relatively similar languages: the Indo-Aryan languages (associated with the "Aryan migrations") and the Bantu languages. While the subsequent political and social histories of the Indo-Aryan speakers and the Bantu speakers were quite different, the process of expansion into vast new territories suggests important similarities between the two experiences.

An analogous deduction on early migration arises for the Chinese languages. Since the greatest variety in versions of Chinese language is south of the Yangzi Valley, it is possible that the similarity of language throughout North China is the result of migration from the south. Linking linguistic information with food crops, one may speculate that an early expansion of Chinese populations began in the south, relying on rice cultivation developed by nearby Austoasiatic speakers, and that the arrival of sorghum from the west enabled an expansion of agriculture northward. The establishment of major states in the north and political expansion to the south would then have been a later development.[4]

Other early migrations are attested by the records of literate peoples who experienced their invasions: the Huns and Germans as seen by the Romans, and the Yuezi-Kushan to the west of China.[5] In some cases the migrants were literate but still on the move: Turks, Mongols, Arabs, Berbers, and Vikings. From other records we know of migrations of Fulbe herders across the African savanna and Arawak mariners into the Caribbean.[6]

3. Greenberg 1987; Greenberg and Ruhlen 1992; Bellwood 1991.
4. Heine and Nurse 2000; Christopher Ehret, *The Civilizations of Africa: A History to 1800* (Charlottesville, 2002); Renfrew 1988. For an excellent source of information on languages, see the Ethnologue website at www.ethnologue.com.
5. Liu 2001; Stein 1999.
6. Cohen 1997; Thomas Sowell, *Migrations and Cultures: A Worldview* (New York, 1996); Robin Cohen and Steven Vertovec, eds., *Migration, Diasporas and Transnationalism*

Turning to modern times, much has been learned of the social history of the Atlantic slave trade. Recent work has replaced a previous narrative with a smaller total number of transatlantic migrants, yet showed that slave trade drew on commodities from around the world, reduced African populations, restructured African societies, and delivered slaves throughout the Americas. The more than 10 million forced African migrants to the Americas greatly exceeded the number of European migrants to the Americas up to 1840. Yet these studies, while they have succeeded in knitting together the social histories within several African and American regions, are still at the early stages in any linkage to the social history of all the regions around the Atlantic.[7]

Chinese overseas migration in the era since 1500, while not yet the subject of a systematic quantitative estimate, was nearly as great as that from Europe, with most migrants going to Southeast Asia. The patterns of that migration, focused far more on commerce than on conquest, have been clarified greatly in a spate of recent publications. Especially through the work of Adam McKeown, these studies are elaborating a new and more flexible conceptualization of migration and boundaries.[8]

Other migrations of the early modern era led across the Islamic world and across the Pacific.[9] The great European migration of the nineteenth and early twentieth century, mostly to the Americas, can now be studied in comparative perspective.[10] Twentieth-century migration, after a decline from 1920 to 1950, reached and exceeded its previous peak, in association with accelerating rates of population growth and urbanization that spread to every area of the world.[11]

Continued

(Cheltenham, U.K., 1999); Robin Cohen, ed., *The Cambridge Survey of World Migration* (Cambridge, 1995).

7. Curtin 1969; Philip D. Curtin, ed., *Migration and Mortality in Africa and the Atlantic World, 1700–1900* (London, 2000); Manning 1990.

8. Tim Wright, ed., *Migration and Ethnicity in Chinese History: Hakkas, Pengmin, and Their Neighbors* (Stanford, 1997); McKeown 2001; Lyn Pan, *Sons of the Yellow Emperor: A History of the Chinese Diaspora* (Boston, 1990); Elizabeth Sinn, ed., *The Last Half Century of the Chinese Overseas* (Hong Kong, 1998); Wang 1997.

9. Dale F. Eickelman and James Piscatori, eds., *Muslim Travellers: Pilgrimage, Migration, and the Religious Imagination* (Berkeley, 1990); Chappell 1997; W. George Lovell and Christopher H. Lutz, *Demography and Empire: A Guide to the Population History of Spanish Central America, 1500–1821* (Boulder, 1995); Ralph Shlomowitz and Lance Brennan, "Epidemiology and Indian Labor Migration at Home and Abroad," *Journal of World History* 5 (1994), 47–70.

10. John Bodnar, *The Transplanted: A History of Immigrants in Urban America* (Bloomington, 1985); Canny 1994; Hatton and Williamson 1994; Nugent 1992; Brinley Thomas, *Migration and economic growth* (Cambridge, 1954).

11. Elazar Barkan and Marie-Denise Shelton, *Borders, Exiles, Diasporas* (Stanford, 1998); Paul R. Ehrlich and A. H. Ehrlich, *The Population Explosion* (New York, 1990); Gabriel Sheffer, ed., *Modern Diasporas and International Politics* (New York, 1986). A substantial archive of data on domestic and international migration for Latin America in the twentieth century has been collected in the papers of Hector Enrique Melo at the World History Center, Northeastern University.

Community and Class

The notion of community is broad and contains many sub-sections. I will begin with the most problematic of these categories, race. The social history of the assumed physical distinctions of "race" applies principally to the last four or five centuries. Before then, while physical differences among people of different regions were known and rendered unmistakably by artists, transformation of those differences into categories of social discrimination had not yet taken place. Racial discrimination reached its peak in the nineteenth and twentieth centuries and was central to the history of social hierarchy in that era. Continuing research on the fifteenth through eighteenth centuries shows, however, that subtle yet tangible forms of racial discrimination developed among Europeans in that era. Of more practical and researchable interest to social historians are the maintenance and reproduction of racial differences over time, and especially the representation of race in recent centuries. The changes in racial ideology over the past two centuries are more than a simple rise and fall of racial discrimination. Of particular interest are the development of succeeding terminologies for race and a changing list of races identified.[12]

The notion of class has been widely debated in modern social thought. The affirmations and denials of class as a social structure have led to periodic swings in scholarly approach. E. P. Thompson led in posing class as a relationship to a social system, thus prefiguring a Marxist overlap with postmodernist thinking. More traditionally, social class is defined in terms of occupation yielding such groups as peasants, slaves, wage workers, petty bourgeoisie, bourgeoisie, and aristocrats. There has been little recent work focusing specifically on class in world history. While recognizing the problematic and situational meaning of "class," I would argue that the need for cross-regional comparisons requires that world historians conduct studies on the formation, rise, and decline of social classes by region and worldwide. One such study is the recent work on sailors in the Atlantic by Peter Linebaugh and Marcus Rediker, which extends the Thompsonian tradition of working-class social history beyond national boundaries (and also draws on subaltern studies) to give a lively portrait of work and social contestation at the turn of the nineteenth century.[13]

12. Cañizares-Esguerra 1999; Frederick Cooper, Thomas C. Holt, and Rebecca J. Scott, *Beyond Slavery: Explorations of Race, Labor, and Citizenship in Postemancipation Societies* (Chapel Hill, 2000); David Roediger, *Wages of Whiteness: Race and the Making of the American Working Class* (London, 1999). It remains of interest to deduce how and when physical distinctions arose among regional groupings of *Homo sapiens*. These must surely have developed no earlier than 100,000 years ago (probably much later), and they seem to have been well in place 10,000 years ago.
13. Linebaugh and Rediker 2000. See also Guha and Spivak 1988; Cooper et al. 1993; Doudou Diène, ed., *From Chains to Bonds: The Slave Trade Revisited* (Paris, 2001); Frantz Fanon, *The Wretched of the Earth*, trans. Constance Farrington ([1961] New York, 1968); Lovejoy and Hogendorn 1993; Moses I. Finley, *Ancient Slavery and Modern Ideology* (New York, 1980).

As with social class and elite, the category of ethnicity is problematic, having meanings and applications that can vary sharply from situation to situation. Anthropologists and area-studies historians have long worked on interpreting ethnicity, working to overcome imperial visions of "the tribe" as a timeless and essential identity and to show the negotiation of ethnic identity. More recently, historians with a global approach have completed several studies that begin to expand the groundwork for understanding the place of ethnic groups in world history.[14]

Communities are defined in many ways beyond race, class, and ethnicity. Among other bases for community are residence, legal status, religious affiliation, and combinations of these. In a small-scale but no less global study of residence, Ida Altman has traced the movement of sixty families from sixteenth-century Brihuega in Spain to Puebla, Mexico, and observed the transformations in social values and practices that took place with the migration, despite continuing ties between the two populations.[15] Occupational groups—including merchants, transport workers, soldiers, missionaries, and voluntary associations for any of these groups—have all benefited from investigations at transregional scale.[16]

14. According to Fredrik Barth, "Ethnic distinctions do not depend on absence of social interactions and acceptance, but are quite to the contrary often the very foundations on which embracing social systems are built." Barth, "Introduction," in Barth, ed., *Ethnic Groups and Boundaries: The Social Organization of Culture Difference* (Oslo, 1969), 10. See also Mark J. Hudson, *The Ruins of Identity: Ethnogenesis in the Japanese Islands* (Honolulu, 1999); Patrick Vinton Kirch, *The Lapita Peoples: Ancestors of the Oceanic World* (Cambridge, Mass., 1997); Jack D. Forbes, *Africans and Native Americans: The Language of Race and the Evolution of Red-Black Peoples* (Urbana, 1993); Hall 1992; Colin A. Palmer, "From Africa to the Americas: Ethnicity in the Early Black Communities of the Americas," *Journal of World History* 6 (1995), 223–36. Aside from occasional attention to Mongols and other peoples of Central Asia, there has been insufficient attention among world historians to comparative and interactive studies of nomadic peoples, especially in Africa and in Asia.

15. Altman 2000. For other community studies, see Blassingame 1972; Philip P. Boucher, *Cannibal Encounters: Europeans and Island Caribs, 1492–1763* (Baltimore, 1992); Julia Clancy-Smith, *Rebel and Saint: Muslim Notables, Populist Protest, Colonial Encounters (Algeria and Tunisia, 1800–1904)* (Berkeley, 1994); David Cohen and Jack P. Greene, eds., *Neither Slave Nor Free: The Freedmen of African Descent in the Slave Societies of the New World* (Baltimore, 1972); Thernstrom 1969.

16. G. Brooks 1993; Alden 1996; Dauril Alden, "Changing Jesuit Perceptions of the Brasis During the Sixteenth Century," *Journal of World History* 3 (1992), 205–18; Edmund Burke III, ed., *Global Crises and Social Movement: Artisans, Peasants, Populists, and the World Economy* (Boulder, 1988); Carney 2001; Allen Isaacman and Richard Roberts, *Cotton, Colonialism, and Social History in Sub-Saharan Africa* (Portsmouth, N.H., 1995); Pablo Pérez-Malláina, *Spain's Men of the Sea: Daily Life on the Indies Fleets in the Sixteenth Century*, trans. Carla Rahn Phillips (Baltimore, 1998); James Pope-Hennessy, *Sins of the Fathers: A Study of the Atlantic Slave Traders, 1441–1807* (New York, 1968); Mary Turner, ed., *From Chattel Slaves to Wage Slaves: The Dynamics of Labour Bargaining in the Americas* (Bloomington, Ind., 1995); Patricia Risso, "Cross-Cultural Perceptions of Piracy: Maritime Violence in the Western Indian Ocean and Persian Gulf Region during a Long Eighteenth Century," *Journal of World History* 12 (2001), 292–320.

One of the strengths of social historical studies is the cosmopolitan grouping of urban historians. While urban historians have not generally worked closely with world historians, they are an internationally connected group of scholars with deep experience in comparing and connecting the experiences of various urban centers. While the urban history literature retains its earlier focus on cities of Europe and North America, urban historians are now well aware of the magnitude of Asian cities in times before 1800 and the extension of urbanization to every region in recent times.[17]

Another section of the social history literature that should be of interest to world historians focuses on social movements. The term refers especially to the mobilization by class, gender, and race of subaltern groups to challenge restrictions on their social advance, but it also includes studies of conflicts among social movements. The analysis of social movements tends to focus on recent times (for instance, on instances of genocide in the twentieth century), but this approach is relevant to earlier times to the extent that relevant data are available.[18]

Gender

The distinctions between male and female are in place at all times. Certain biological differences are inescapable, but social differences are elaborated by choice and tradition. How does society respond to this difference? Are males and females kept together or separated? Are they set in hierarchy? Are the differences used to mark a division of labor? Are other differences in life explained using gendered metaphors?

World history, especially as a history of great states and long-distance trade, included little recognition of gender and little space for women. The first stage in introducing a gendered perspective to world history has been the effort to locate women in the past and include descriptions of their activities and contributions

---

17. Tertius Chandler, *Four Thousand Years of Urban Growth* (Lewiston, N.Y., 1987); Mike Davis, *City of Quartz* (London, 1990); Mumford 1938; Mumford 1961; Aidan Southall, *The City in Time and Space: From Birth to Apocalypse* (Cambridge, 1998). The H-Urban website includes a wide range of data on cities (www2.h-net.msu.edu/~urban/weblinks/index.htm).

18. Peter Gran argues that the methodology of social history allows the development of an approach to world history focusing on the countervailing pressures of hegemonic practices from elites and the social movements of other classes: Gran 1996. For studies relying on a social-movement approach, see Eugene D. Genovese, *From Rebellion to Revolution: Afro-American Slave Revolts in the Making of the Modern World* (Baton Rouge, 1979); René Lemarchand, *Burundi: Ethnocide as Discourse and Practice* (New York, 1994); Mark Levene, "Why is the Twentieth Century the Century of Genocide?" *Journal of World History* 11 (2000), 305–36; Mani, 1998; Justin McCarthy, *Death and Exile: The Ethnic Cleansing of Ottoman Muslims, 1821–1922* (Princeton, 1996); Rudé 1964; E. Thompson 1964; C. Tilly 1964.

to society. One result of this effort has been the development of several reference works on women in world history.[19]

A second approach has been to develop long-term interpretations of gender roles. The best-known such interpretation, by Gerda Lerner, locates the creation of patriarchy in the early agricultural age, followed by the reproduction thereafter of a social order subordinating women to men. In subsequent work, Lerner has sought to develop techniques for elaborating female roles in historical developments over shorter time frames.[20]

Another approach has been to explore the history of women in area-studies perspective. This approach extends the earlier work of restoring women to history in national perspective for the United States and European countries.[21]

One area of gendered history that has received wide attention from researchers is the place of women in colonized societies. For early modern times, studies have appeared on the lives of slave women in the Caribbean and on Native American women in North America.[22] A somewhat larger number of studies has appeared to address the place of women in the expanding European empires of the nineteenth and early twentieth centuries. Several studies have explored the roles of European women both as agents of the imperial order and as allies of colonized women facing pressures both from their own families and from the colonial system. Others focus on the women of the colonized societies as they seek to navigate a changing social order.[23]

19. Lynda G. Adamson, *Notable Women in World History: A Guide to Recommended Biographies and Autobiographies* (Westport, Conn., 1998); Gayle V. Fisher, *Journal of Women's History Guide to Periodical Literature* (Bloomington, Ind., 1992); Bella Vivante, *Women's Roles in Ancient Civilizations: A Reference Guide* (Westport, Conn., 1999).

20. Gerda Lerner, "Creation of Patriarchy," Vol. 1 of *Women and History,* 2 vols. (New York, 1986–1993). For other long-term presentations on women in world history, see De Pauw 1998; Fatima Mernissi, *The Forgotten Queens of Islam,* trans. Mary Jo Lakeland (Minneapolis, 1993). For an effort to locate ancestral matrilineal forms throughout Africa, see Murdock 1959; for new work on matrilineage in African history, see Ehret 2002.

21. Nikki R. Keddie, "The Past and Present of Women in the Muslim World," *Journal of World History* 1 (1990), 77–108; Barbara Ramusack and Sharon L. Sievers, *Women in Asia: Restoring Women to History* (Bloomington, Ind., 1999); Berger and White 1999; Marysa Navarro and Virginia Sánchez Korrol, with Kecia Ali, *Women in Latin America and the Caribbean* (Bloomington, Ind., 1999). For examples of similar approaches to women in European and U.S. history, see Renate Bridenthal and Claudia Koonz, eds., *Becoming Visible: Women in European History* (Boston, 1977); and L. Tilly and Scott 1978.

22. Barbara Bush, *Slave Women in Caribbean Society, 1650–1838* (Bloomington, Ind., 1990); Hilary McD. Beckles, *Natural Rebels: A Social History of Enslaved Black Women in Barbados* (London, 1989); Carol Devens, *Countering, Colonization: Native American Women and Great Lakes Missions, 1630–1900* (Berkeley, 1992); Carol Devens, " 'If We Get the Girls, We Get the Race': Missionary Education of Native American Girls," *Journal of World History* 3 (1992), 219–38.

23. Clancy-Smith and Gouda 1998; Lora Joyce Wildenthal, *Colonizers and Citizens: Bourgeois Women and the Woman Question in the German Colonial Movement,*

Further, and along with the elaboration of analytical frameworks for address-ing the full range of gendered issues in society, studies have begun to appear addressing gender in transregional perspective. Of particular note is Mrinalini Sinha's study of England and Bengal, in which she argues that a nineteenth-century debate about masculinity, with its stereotypes of the manly Englishman and the effeminate Bengali, transformed gender categories and social relations in both the colony and the metropole. Expanding the insights of a generally gendered approach, the team of Ruth Pearson and Nupur Chaudhuri has written a general statement on gender in the context of empire.[24]

For all the new works that one can cite, it remains striking that studies of women and gender roles in world history have developed so slowly and that their development has been restricted to a small number of themes. Why has gender been such a difficult issue to develop in world history?[25] What are the lines that have been taken and could be taken? The well established presumption that women's lives are acted out in the private sphere of the family rather than the pub-lic spheres of economy and politics has suggested that women's history is family history, and family history plays no great part in world history. By this logic, the expansion of studies of women and colonialism (or gender and colonialism) has come about precisely because, in colonial situations, the state interferes in the working of families and in social values generally. As the private becomes public—as in the debate about *sati* in nineteenth-century India—the lives of women enter at last into government documents and into the purview of historians.[26]

Let me conclude this section by briefly introducing a framework that may be of assistance in broadening the analysis of women in world-historical studies. Focusing on women and on labor, I argue that women's work can be divided into reproductive, domestic, and social labor. Reproductive life is the bearing of children and bringing them through infancy, mostly completed between ages twenty and thirty-five. Domestic labor is that of maintaining one's own house-hold, the household of one's parents, or the household of one's children or of

Continued

   1886–1914 (Ann Arbor, 1994); Antoinette Burton, *Burdens of History: British Feminists, Indian Women and Imperial Culture, 1865–1915* (Chapel Hill, 1994); N. Chaudhuri and Strobel 1992; Marnia Lazreg, *The Eloquence of Silence: Algerian Women in Question* (London, 1994); Strobel 1991.

24. Sinha 1995; Joan Wallach Scott, ed., *Feminism and History* (New York, 1996); Robert Shoemaker and Mary Vincent, eds., *Gender and History in Western Europe* (London, 1998); B. Smith 1998; Ruth Roach Pierson and Nupur Chaudhuri, eds., *Nation, Empire, Colony: Historicizing Gender and Race* (Bloomington, Ind., 1998).

25. Joan W. Scott continues to offer leadership in developing gendered approaches to the past, and Peter Stearns has recently focused his comprehensive vision of social history on gender. Scott, ed., *Feminism and History* (New York, 1996); Stearns, *Gender in World History* (New York, 2000).

26. Mani 1998. Ann Laura Stoler has also addressed directly the private sphere under colonial rule and its encounter with hierarchies of race and state power. Stoler, *Race and the Education of Desire: Foucault's History of Sexuality and the Colonial Order of Things* (Durham, 1995).

SOCIAL HISTORY 211

other families; it often takes place for all of one's years. Social labor is the production of economic goods; it may include agricultural work, marketing, artisanal work; it might be paid or unpaid. This work is done especially by young adult women and by women no longer having children, but in some cases it is done for all of their lives. The lives of most women involve all three types of work, but not usually at one time. For women as individuals and for the societies relying on women's work, the timing of moving from one type of work to another and the social value accorded to each type of work provides statements of the destiny of individual women and the structure of society overall. Attention to the patterns of these types of work and to changes in them may offer a way to include the lives of women in history, to link them to gender roles generally, and to show the place of family history in society more broadly.[27]

Family

The questions on family in world history are large and intriguing, but evidence on family life is scattered and partial at best. Even where data exist, there remains the problem of how to tie the history of families in localities to global issues. The origins of human families, the role of families in early migrations, the development of lineage structures, patterns of inheritance and of childrearing—these are potential elements of an interpretation of transformations in human family life which may be pieced together in years to come.[28]

For the existing literature, European studies of family history contain much of the best and most innovative work in family history. In this field new techniques were developed for analysis of individual population registers (especially religious parish records, but also civil records), analysis of censuses at local and regional levels, reconstructing families over multiple generations from these records, and classifying families according to structural types. From these studies, some broad syntheses of family history have developed, giving an image of European family history over several centuries. With few exceptions, however, these studies minimize attention to migration.[29] As a result, this approach to family history restricts

27. This approach is implemented in a web-based instructional module authored by Shelley Stephenson, in the web course "Navigating World History," produced and distributed by iLRN, for which I was the creative consultant and general editor.

28. Kuper 1988; Snooks 1996.

29. Seccombe 1992; Michael Anderson, *Approaches to the History of the Western Family, 1500–1914* (London, 1980); Gary S. Becker, *A Treatise on the Family* (Cambridge, Mass., 1991); John Bongaarts, Thomas K. Burch, and Kenneth W. Wachter, *Family Demography: Methods and their Application* (Oxford, 1987); Gene H. Brody, and Irving E. Siegel, eds., *Methods of Family Research*, 2 vols. (Hillsdale, N.J., 1990); André Burgière, Christine Klapisch-Zuber, Martine Segalen, and Françoise Zonabend, eds., *A History of the Family*, trans. Sarah Hanbury Tenison, Rosemary Morris and Andrew Wilson, 2 vols. ([1986] Cambridge, Mass., 1966); M. Gordon 1973; Laslett 1966; Seccombe 1993; J. Smith, Wallerstein, and Evers 1984.

itself to the study of localized units rather than explore the links of families across distances and across social, geographical, and cultural borders.

A few studies have attempted to pull together individual articles or individual insights focusing on families as they are spread across regions, focusing either on European colonial expansion or on long-term patterns in family life.[30] A more systematic attempt to document and analyze the formation of families in various social situations, accounting especially for the influence of families, may yield insights into the global history of family change. Put in terms of models, we need a model to explain how it is that family formation has changed over the past several centuries in many parts of the world from a pattern in which young people required parental approval to marry and settled on lands of their parents, to a pattern in which young people select their own mates and move to new residences.

The distinctions in age have ramifications throughout society, but especially in the operation of families. While infants are usually cared for by their mother, thereafter arrangements can vary widely. Who cares for children? When do they begin to work? Is their instruction handled by the family, or by other institutions? When and how are they initiated into adulthood? Do they marry formally to start their families, and if so, how and when? For those who have reached adulthood, is there another stage to indicate that they have become seniors or elders? The treatment of old people varies, from society to society, ranging from veneration and obedience to neglect.

Biography

World history, while it addresses global processes and long-term change, must also account for individuals in history. Few world historians are likely to subscribe to the "great man" approach to history, since they are concerned with the range of political, social, and environmental factors that set the context for the life of each individual. At the same time, world historians are definitely interested in individuals: in individual responses to global processes, and in the impact that individuals can and do have on the passage of events. A number of biographic studies in world history have already become influential in the teaching of world history.

One pathbreaking study is Ross Dunn's biography of Ibn Battuta, the fourteenth-century jurist and traveler, who visited every major region in Asia and Africa in which Islam was prominent in trade or government. Through one man's view of so many regions, the reader is able to gain a sense of the unity and variety in the Islamic world of that time, and also the development of Ibn Battuta's sense of the world. A very different biography appeared at much the same time, as William H. McNeill published his study of the life of Arnold J. Toynbee. For one

30. Maria Beatriz Nizza da Silva, ed., *Families in the Expansion of Europe* (Aldershot, U.K., 1998); G. Robina Quale, *A History of Marriage Systems* (Westport, Conn., 1988); G. Robina Quale, *Families in Context: A World History of Population* (New York, 1992).

great world historian to review the life of another provides the reader with a set of personal reflections on the task of conceptualizing world history.[31]

In more recent biographic studies, Frances Karttunen combined a set of biographical chapters about several interpreters and guides who facilitated the connections among peoples in the great cultural encounters of the early modern world. Roxann Prazniak reconstructed what she called "dialogues across civilizations," summarizing individual Chinese and European views of each other's society and of world history. Sally Wriggins's biography of Xuanzang traces the life of the Chinese pilgrim of the seventh century C.E. who traveled to India to obtain sacred Buddhist texts for study and dissemination in China.[32]

In biography as in social history generally, the small beginnings summarized in this chapter have not yet become a large part of teaching and research in world history, and their global approach has not yet become a large part of studies in social history. But these are promising beginnings. If, as I think is likely, the study of social patterns in world history continues to expand, the results could ultimately provide a solid picture of the past interactions in social history, which can be compared and linked more usefully to the understanding of the social interactions that are so obviously taking place in the world of today.

Conclusion: Viewing Society Globally

There is hope for social history at the global level. It has been curious to see the expansion of world history focused mainly on areas other than social history in the era when social history generally has undergone its greatest development. But a continuing effort to link the two may yet yield interesting and important results. In particular, if each camp concentrates at once on its own characteristic strength and on that of the other camp, the paths of linkage may widen. World historians need to find ways to address in more detail the topics that have been developed so effectively by social historians at national and local levels. And social history needs to find more ways to explore transregional interactions—for instance, by including rather than neglecting migration in studies of local communities. The other main hope for the advance of global studies in social history lies in exploration of the anthropological literature. This extensive theoretical and empirical literature includes debates, perspectives, and data of great importance at all levels, from long-term social change at the civilizational level to patterns of family interaction. World historians should get beyond the stage of testing the waters of anthropology and dive in.

31. Ross E. Dunn, *The Adventures of Ibn Battuta: A Muslim Traveler of the Fourteenth Century* (Berkeley, 1989); W. McNeill 1989.
32. Frances Karttunen, *Between Worlds: Interpreters, Guides, and Survivors* (New Brunswick, 1994); Prazniak 1996; Wriggins 1996. For additional biographic studies linked to global issues, see Emily Ruete (Salme Said), *Memoirs of an Arabian Princess from Zanzibar* (New York, 1989); Jason Thompson, "Osman Effendi: A Scottish Convert to Islam in Early Nineteenth-Century Egypt," *Journal of World History* 5 (1994), 99–124; and Helen Wheatley, "From Traveler to Notable: Lady Duff Gordon in Upper Egypt, 1862–1869," *Journal of World History* 3 (1992), 81–104.

Chapter 12

Ecology, Technology, and Health

The themes of ecology, technology, and health are relatively new to historical studies, and they are much of what has made world history into a distinctive field. This is work along the external path to world history, the scientific-cultural path; many of the authors were trained principally in fields outside history. They have begun to write on the subject of history as they have discovered patterns of significance at the global level and have developed a desire to show the links of their discoveries to other aspects of history and present them to a wider audience. The wider audience has grown, particularly out of concern for the environment of the planet Earth. Many humans understand, at last, that we are not only exploiting the planet in pursuing our lives, but we are transforming it as well.

This chapter addresses research on a very wide range of topics. In considering the reproduction and degradation of the ecology, it addresses land, climate, fresh water and the oceans, plants, animals, and microbes. For human technology, the chapter addresses issues in food, shelter, tools, metallurgy, transportation, communication, construction, and industry. For issues of human health, the chapter addresses nutrition, disease, and a range of medical and healing practices.

Research Agenda: Ecology, Technology, Health

The separate narratives of ecology, technology, and health in human history may not combine seamlessly. But it makes sense to begin with stories of the changing ecology of Earth, because the planet has had a busy history even before it had humans to bring more changes. The later stages of continental drift conditioned the development and distribution of plant and animal species. Long-term changes in climate brought successive ice ages and caused ocean levels to rise and fall, influencing migration by animals and humans and changing the vegetation in each region. A warming trend for the past ten thousand years, accelerated in the very recent past, has raised ocean levels and caused formerly fertile areas, such as the Sahara region, to become desert. The short-term fluctuations of the El Niño oceanic cycle have brought climatic fluctuations that have been unpredictable so far. Human technology has changed in a way that is more predictable only in that the innovations have accelerated with time. The making of tools is now known through archaeology to have gone on for some two million years.

The development of microlithic technologies associated with *Homo sapiens* has now been traced back, in some African sites, over one hundred thousand years. Human migration to Australia and New Guinea, beginning about sixty thousand years ago, makes clear that techniques for water transport developed early on. Later innovations in transportation brought advances in transport by horses, oxen, and camels, then long-distance sailing ships, followed by the transportation advances of the industrial age. In human health, a characteristically broad diet has distinguished humans from the first, and it was supplemented by use of herbs and the development of healing specialists. As human populations grew and entered new environments, new diseases developed, sometimes with disastrous consequences, but also resulting in new immunities. In very recent times, expanding knowledge of public health and medicine have greatly extended the expectation of life by reducing the number of early deaths, though not yet by extending the maximum of human life.

Historians' questions about the past of ecology, technology, and health may be classified, as I suggested in chapter 10, into questions about origins, timing, dynamics, and legacy. The origins of technology present very pertinent questions, as one seeks to know the place and the social circumstances of the development of agriculture, gunpowder, or Velcro. For ecological history, in contrast, questions of origins are not so obviously relevant, as changes in geology, in species development and in climate come in patterns that are more clearly cyclical than unique.

Questions of timing are complex but significant. Did agriculture arise once, perhaps in the Fertile Crescent, with subsequent and secondary inventions in other areas? Or did agriculture arise at much the same time in six to eight areas of the world, perhaps all conditioned by changing ecological conditions? Did the nineteenth-century decline in death rates in so many parts of the world come after the development of new levels of understanding of public health and medicine? Or did the death rates decline before human knowledge could have affected them, perhaps because of higher immunological resistance that resulted simply from the passage of time after contact among regional populations? How rapidly did the American crops—especially maize, manioc, and potatoes—become significant contributors to nutrition in the Old World? Has forest clearance been a phenomenon restricted to the nineteenth and twentieth centuries, or has the clearing of forests been an ecologically significant factor for a longer period?

The dynamics addressed in these sub-fields of history involve the interplay of humans, animals, plants, and land. The history of ecology or health draws the analyst into numerous and overlapping processes. What models or metaphors does one adopt to provide a simplified but appropriate summary of such processes as the adoption of new crops or the development of healing practices? To focus just on health as an example, the nature and impact of endemic and epidemic disease are very different in general, and epidemic diseases themselves spread according to very different dynamics: by water for cholera, by personal contact for smallpox, by mosquitoes for yellow fever. Within the same field of health, the dynamics of curing and caring differ greatly, and historians who live in an era of curative medicine must adjust their minds to recapture a past in which the main effort in health focused, necessarily, on caring.

What has been the legacy of earlier patterns in ecology, technology, and health? Jared Diamond has offered an unusually strong thesis on historical legacy: that the early complex of agriculture and domestic animals formed under human direction in Eurasia, because of the range of biota that it included (but especially because of the disease vectors developed within this multispecies community), was destined to dominate every continent in subsequent migrations, right up to the present day.[1] Other legacy arguments include the argument that the Black Plague in Europe set the logic of social relations for the time since, and that the scientific method, once enunciated by Francis Bacon and others, set into action a chain of investigation and innovation that transformed the world. Quite a different sort of argument is that indigenous healing systems—focusing at once on a wider range of pharmacological, psychological, and spiritual principles than scientifically based medicine—have preserved alternative approaches to healing and caring that much to add to today's systems of health care.

Contemporary concerns have been important in highlighting questions about the world history of ecology, technology, and health. Concerns have grown about global warming, the limits on fossil fuels, the disappearance of forest cover, and the current loss of plant and animal species. The development of new technology appears in some ways to be transforming human existence so rapidly that it is cutting us off from human life in earlier times. The emergence of new medical technology is extending human life, but it may also be creating new moral dilemmas. Have the transformation of society and the aggravation of human inequality in the past two centuries created the potential for new crises or new diseases? Those who raise such questions may not go first to historians for answers, but each of these current issues has a substantial historical dimension, and historians can already throw light on many of them.

Ecology: Reproduction and Degradation

Perhaps more than any other field of study, ecological history requires the reader to think at once of long-term transformations and short-range changes. For surveys that make this point in various ways, Stephen Jay Gould conveys the importance of learning to accommodate to geological time, and William Durham proposes some challenging theses on the interplay of social structure and biological change in human communities.[2]

Several broad syntheses in ecological history have clarified the linked processes of reproduction and degradation in ecological systems. Clive Ponting's popular survey gives especially effective portrayals of the human exhaustion of natural resources through examples of excesses in grazing, whaling, and fishing.[3]

1. J. Diamond 1997.
2. Stephen Jay Gould, *Time's Arrow, Time's Cycle: Myth and Metaphor in the Discovery of Geological Time* (Cambridge, 1987); Durham 1991.
3. Ponting 1991; Vasey 1992; Simmons and Simmons 1996; Worster 1993; O'Connor, 1998.

The Braudelian approach to regional analysis, in which natural and social processes are described in intimate interconnection, has led to effective presentations for the Mediterranean and the Indian Ocean, but also for the impact of commerce and iron production in West Africa and for the hunting of elephants in Northeast Africa.[4] Alfred Crosby's approach to ecological history, in contrast, has given particular attention to various sorts of migration: people, plants, animals, and diseases.[5]

For the impact of natural processes, William Atwell's survey of volcanism from the thirteenth through seventeenth century—based on data drawn from ice cores, tree rings, and compilations of textual evidence—suggests strong correlation between episodes of volcanism and short-term climatic changes, mostly with negative effects for agriculture. Brian Fagan's survey of the El Niño phenomenon links the downwelling of salt in the North Atlantic to changing patterns of wind and temperatures in the South China Sea, and then to sharp climatic change in recent times and in earlier times.[6] David Keys and Mike Davis, in times separated by fourteen centuries, provide memorable tales of the occasionally sudden change as human society encounters natural processes, the first from volcanism and the second from El Niño. In addition, each highlights the aggressive opportunism of humans seeking to benefit from disaster.[7]

Following the patterns laid down earlier by E. L. Jones and William Cronon, a significant number of studies have explored the interactions of economy and environment in the second millennium. Both Robert Marks and Kenneth Pomeranz have dug deeply into the ample documentation on the southern and central coasts of China to argue that the regional ecology, though pressed hard by the dense population and intensive exploitation, nonetheless was managed with considerable skill and did not conform to the previous scholarly impression that Chinese history was a long ecological disaster. David Watts, exploring another intensively exploited environment, the West Indies, came up with a less optimistic interpretation, based in part on his assessment of colonial economic policies.[8] James McCann's survey of environmental history in Africa during the past two centuries reveals, through detailed comparisons of a few regions, at once the

4. Braudel 1995; J. McNeill 1992; K. Chaudhuri 1990; Brooks 1993; Burstein 1995. Burstein proposes a link between trade activity and the size of elephant populations in Northeast Africa.
5. Crosby 1986; Crosby 1994.
6. William S. Atwell,, "Volcanism and Short-Term Climatic Change in East Asian and World History, c. 1200-1699," *Journal of World History* 12 (2001), 29–8; Fagan 2000a; Brian Fagan, *The Little Ice Age: How Climate Made History 1300–1850* (New York, 2000); H. H. Lamb, *Climate, History and the Modern World* (London, 1982).
7. David Keys, *Catastrophe: An Investigation into the Origins of the Modern World* (New York, 1999); Davis 2001.
8. Marks 1998; Pomeranz 2000; David Watts, *The West Indies: Patterns of Development, Culture and Environmental Change since 1492* (New York, 1989); Lynne Withey, *Voyages of Discovery: Captain Cook and the Exploration of the Pacific* (Berkeley, 1989); Curtin 1990b; J. McNeill 1994; Jones 1981; Cronon 1983.

processes of environmental degradation and skilled human exploitation of the land in precolonial, colonial, and postcolonial eras.[9]

In the most comprehensive ecological history to date, John R. McNeill has offered an interpretation of ecological issues in the twentieth-century world that makes unmistakably clear the depth of transformations of the century in land use, agricultural production, life in the oceans, and mineral resources.[10] His conclusion balances optimism and pessimism in assessing the prospects for transformation, degradation, and reproduction of the planet's ecology, yet makes clear that the political, social, and cultural spheres of human activity will have to take account of the ecological transformations he has traced.

Technology: Invention and Diffusion

Three remarkable volumes summarize much of what is known of the history of human technology and at the same time leave us with enough questions about the topic to keep historians busy for many years. Joel Mokyr, a leading economic historian of industrial technology, has written a sweeping study of advanced technology from Classical times to the present, emphasizing the contributions of technological change to economic growth. While his focus on creativity at the societal level tends to rely on essentialized models of civilizations, his detailed breakdown of technology and technological change yields an impressive analytical structure. With it he analyzes technologies for classical Greece and Rome, Medieval Europe (including a bit on the Islamic world), China (drawing especially on the work of Joseph Needham), and Europe in early modern and industrial eras. His analytical paradigm addresses the causes and restraints of technological innovation. The approach, developed out of his strength in industrial-era technology, focuses on advanced technology and on the process of developing higher productivity. Arnold Pacey, exploring a somewhat shorter period of time but a wider geographic and social range, considers the interplay of advanced, intermediate, and basic technology in many areas of the world. While his approach gives little insight into the process of innovation, it provides substantial information on the way that different levels of technology can reinforce each other in a productive system. Johan Goudsblom's exploration of fire in history focuses not only on the origin of human control over fire, but on the succession of technological advances associated with combustion, and especially on the social changes associated with each new stage in the use of fire.[11]

Historical studies of technology may be divided into various subgroups, the largest of which studies advanced technology in very recent times. There the

9. J. Donald Hughes, ed., *The Face of the Earth: Environment and World History* (Armonk, N.Y., 1999); Ian Tyrrell, *True Gardens of the Gods: Californian–Australian Environmental Reform, 1860–1930* (Berkeley, 1999); McCann 1999.

10. J. McNeill 2000.

11. Mokyr 1990a; Pacey 1990; Johan Goudsblom, "The Civilizing Process and the Domestication of Fire," *Journal of World History* 3 (1992), 1–12.

relationship between science and technology, with the apparently orderly creation of new knowledge and new technology, can be treated as a whole new stage of human history cut off from earlier times. A second approach, widely represented in textbooks and in history of science, is to begin with the seventeenth-century scientific revolution in Europe and trace technology and science forward in time. For both these approaches, the topic is science and technology, and the implicit assumption is that scientific knowledge creates advances in technology. Studies of the history of technology focus overwhelmingly on recent times. This is understandable, given the dramatic transformation of industrialization, and it is even more understandable in the current wave of innovation in biological science and in communications technology. Meanwhile, a third and much more scattered category of writings on history of technology deals with basic and intermediate levels of technology and regions outside the centers of wealth and power.

The possibility before us is that of linking these approaches and exploring at once the process of innovation, the varying levels of technology that play off one another in any productive system, and the social impact and social sources of technological change. The framework of world history presents a marvelous possibility for integrating short-term and long-term change in basic and advanced technology. The material base of most technology makes it among the most readily available and the most directly comparable of historical artifacts. The assemblage of a comprehensive interpretation of human technology would not only provide a clear picture of one of the most important arenas of human achievement, but would also be perhaps the first of any arena of human activity for which such an interpretation would be developed. Thanks to the records of archaeology, linguistics, and the written record, we have information of technology for most times, regions of the world, and levels of society. While there exists a strong tendency to treat technological issues and innovations in isolation, there is also a countervailing tendency to view technology as a web of interacting practices and ideas. From the perspective of this technological web, even the high-tech dimensions of today's economy rely on supplies from and consumption by intermediate and basic levels of technology. Technology changes, but it provides the most consistently documented area of human history—the one that best enables us to see what is similar and different from society to society.[12]

In recent times, science and technology have come to be linked very closely to each other, with the benefits flowing in each direction. Computers, for instance,

12. History of technology has moved far beyond its earlier celebration of inventions. Historians of technology now address not just invention but the social pressures encouraging or restricting innovation; not just diffusion but the transformation of innovations as they are adopted; not just the economic impact of technology but also the social consequences; not just the technology itself but the appropriation and representation of technology; and not just progress through technical innovation, but new problems brought by each change in technology. I am indebted to William Storey for my introduction to these perspectives, which make history of technology a most fascinating field of study, for which the practitioners in the United States are gathered into the Society for the History of Technology.

resulted from scientific advances, and the development of computer technology led to many further advances in science. In earlier times, technological changes came about through the experience and insight of the artisan, not through contributions by specialized scientists. Scholars debate over whether the contribution of science became crucial to technological change as early as the seventeenth century or as late as the twentieth century.[13]

I will discuss technology in this section and address science separately afterward. What follows is a roughly chronological review of technology in world history, noting some of the issues that are under discussion or ought to be. In this review, I will emphasize the wide range of technologies for which historical information is being collected.

Research on the domestication of plants and animals, summarized in a number of excellent reviews, is doing much to clarify patterns in world history while relying heavily on the work of area-studies scholars.[14] Despite all the recent advances, however, results are not definitive for several important topics. In particular, clarifying the location and timing for the domestication of food crops in Southeast Asia is of great importance for the understanding the expansion of peoples speaking Austronesian, Austroasiatic, and Sino-Tibetan languages. For later times, more research is needed to clarify the timing of the spread of crops to new areas, such as the spread of sorghum from Africa to China before the common era, the spread of new varieties through the Asian rice-growing regions, and the adoption of large-scale maize production in several areas of the eastern hemisphere.

The construction of individual homes and communal buildings provides an area in which we may hope to reconstruct patterns for much of humanity. In an exceptional beginning for this sort of world history, H. Parker James has traced the stilt house, an easily constructed but sturdy and convenient form of housing, from its origins on the mainland of what is now South China, to the limits of Southeast Asia, and from there to most tropical regions.[15]

13. David S. Landes, *The Unbound Prometheus: Technological Change and Industrial Development in Western Europe from 1750 to the Present* (Cambridge, 1969); Mokyr 1990a.

14. Roger A. Caras, *A Perfect Harmony: The Intertwining Lives of Animals and Humans throughout History* (New York, 1996); M. Crawford and D. Marsh, *The Driving Force: Food in Evolution and the Future* (London, 1989); Kiple and Ornelas 2000; Carl O. Sauer, *Agricultural Origins and Dispersals: The Domestication of Animals and Foodstuffs* (Cambridge, Mass., 1969); Vasey 1992; Carney 2001; Herman Viola and Carolyn Margolis, eds., *Seeds of Change* (Washington, D.C., 1990); Zohary and Hopf 1993; Larry Zuckerman, *The Potato: How the Humble Spud Rescued the Western World* (Boston, 1998); Salaman 1985.

15. The principal data of the study are descriptions, images, and archaeological remains of stilt houses. In addition, a substantial section on historical linguistics traces the early movement of stilt houses throughout the Austric zone (the Austroasiatic and Austronesian languages) of Southeast Asia. The later chapters of the study trace the place of Europeans, especially English, in moving stilt houses to other tropical areas, and they focus ultimately on the Caribbean, where this imported architectural tradition competed and combined with an independent regional tradition of stilt houses.

The development and spread of metallurgy has been an important area of technological change, and the work of specialists has been too infrequently linked to other patterns in history. Archaeologists have traced many of the sites of early development in the working of copper, tin, silver, gold, and iron, but the established categories of Bronze Age and Iron Age are no longer sufficient to articulate what is known. The debates on diffusion versus independent invention continue as recent research gets closer to verifying an independent African development of iron technology. In addition to the initial development of metallurgy, the history of mining, improved manufacturing techniques, and trade in finished goods could be traced in far more detail.

The history of writing provides an opportunity for tracing important early links and later developments. This is work for specialists in each language more than for generalists, but Albertine Gaur, in an accessible summary, argues that most of the many scripts used around the world can be traced to a very few independent inventions of writing.[16] The development of additional techniques for representing each language and the spread of later innovations maintain significance alongside the original invention.

A number of important studies have described changing systems of transportation. While the development of horse transport took place at a time when records would be scarce, camels were domesticated later, and the spread of camel transport took place in literate societies. Richard Bulliet's detailed analysis of the development of successive saddles and systems of caravans shows why it was that camel transport displaced wheeled transport in the arid regions where camels thrived. Ian Campbell has summarized the development and spread of the lateen sail, and Clark Reynolds has surveyed the history of navies from the Classical era.[17] In earlier times, maritime technology developed in several great seas: the Mediterranean, the Indian Ocean, the South China Sea, and the South Pacific. Melanesian and Polynesian mariners may have been the first to develop sophisticated techniques for navigating the high seas; elsewhere, shipping developed primarily for commerce and focused on linking centers of population.[18] Comparative study of early maritime technology would surely be enlightening.

Documentation improves as one reaches more recent times, so that numerous studies of technology in the early modern world have appeared. These studies,

Continued

H. Parker James, "Up on Stilts: The Stilt House in World History" (Ph.D. dissertation, Tufts University, 2001).

16. Albertine Gaur, *A History of Writing* (New York, 1985). See also David Ewing Duncan, *Calendar: Humanity's Epic Struggle to Determine a True and Accurate Year* (New York, 1998).

17. Richard Bulliet, *The Camel and the Wheel* (New York, 1990); I. C. Campbell, "The Lateen Sail in World History," *Journal of World History* 6 (1995), 1–24; Clark G. Reynolds, *Navies in History* (Annapolis, 1998).

18. Peter Bellwood, *The Prehistory of the Indo-Malaysian Archipelago*, revised ed. (Honolulu, 1997); Peter Bellwood, Darrell Tyron, and James J. Fox, eds., *The Austronesians* (Canberra, 1995).

which give significant attention to military technology, have focused first on Europe and then on China, but studies on technology in textiles are now appearing for India and Africa.[19] For the technology of empire in the nineteenth and twentieth centuries, Daniel Headrick has been a one-man industry, producing four volumes that demonstrate with remarkable specificity how technologies from steamships to telegraphy to chemistry facilitated European imperial expansion.[20] Michael Adas, in a complementary approach, has traced the developing ideology of technological mastery in the minds of Europeans: the growing prestige of science and of European leadership in world affairs reinforced each other for more than a century.[21]

Health: Disease and Healing

The study of human health is difficult to conduct for early times because of the scarcity and imprecision of data. Nevertheless, archaeological investigations sometimes give indications on health and sickness of persons whose remains are studied, and textual evidence can be helpful if rarely definitive in the study of epidemics. For the latter, David Keys has compiled substantial evidence on the plague epidemic of the sixth century, arguing that the plague spread from East Africa to the Mediterranean in 540, and then eastward as far as China. More generally, Jared Diamond has provided an excellent description of the processes by which viruses and other microbes transform themselves to remain effective agents of infection.[22]

In a well-known story of long-term change in population health, it has been shown that the gene for sickle-cell anemia, a blood disorder, became especially prominent in populations exposed systematically to malaria—most notably the people of West and Central Africa. The recessive sickle-cell gene provided protection against malarial flukes for all who carried it and brought anemia only to those who inherited the gene from both parents.[23] The expansion of knowledge about the human genome makes it likely that other such stories of the interplay of disease and genetics will be discovered.

19. Landes 1969; Roger C. Smith, *Vanguard of Empire: Ships of Exploration in the Age of Columbus* (New York, 1993); Mazumdar 1998; David B. Ralston, *Importing the European Army: The Introduction of European Military Techniques and Institutions into the Extra-European World, 1600–1914* (Chicago, 1990); William K. Storey, ed., *Scientific Aspects of European Expansion* (Aldershot, U.K., 1996); Robert M. Adams, *Paths of Fire: An Anthropologist's Inquiry into Western Technology* (Princeton, 1996); Parthasarathi 1998; Mintz 1985.
20. Headrick 1981; Headrick 1991; Daniel R. Headrick, "Botany, Chemistry, and Tropical Development," *Journal of World History* 7 (1996); Headrick 1988.
21. Michael Adas, *Machines as the Measure of Men: Science, Technology, and Ideologies of Western Dominance* (Ithaca, 1989).
22. Keys 1999; J. Diamond 1997: 195–214.
23. The sickle-cell gene presents no advantage, of course, once exposure to malaria is ended. Kiple and King, 1981; Durham 1991.

The impact of disease in history has received a new and comprehensive review in the work of Sheldon Watts. His study explores the disease patterns, the human cost, and the ultimate recession of plague, leprosy, smallpox, syphilis, cholera, yellow fever, and malaria in the past millennium. Each epidemic has its own patterns, and it has been hard to know how to prepare for the next.[24]

The demographic literature on historical changes in death rates has expanded greatly in recent decades, but Philip Curtin has been most prominent among those linking such research to issues in world history. Curtin found first that European military commands gained improved control over public health for troops posted throughout the world in the nineteenth century. But in a subsequent study he found that European officers in the conquest of Africa were inconsistent at best in using the available technology in medicine and public health, so that their soldiers' death rates were, especially in the 1895 French campaign in Madagascar, unexpectedly high.[25] Other studies drawing on the medical and administrative records of the British Empire have given particular attention to the health conditions of India in the late nineteenth and early twentieth centuries.[26]

The range of relevant issues in the world history of health is far greater than the studies that have so far been undertaken. Studies in communicable diseases continue, with many important topics awaiting further exploration: influenza, infantile paralysis (polio), tuberculosis, AIDS, and malaria are clear examples. Cancer, while generally not communicable, has expanded greatly in incidence because of new environmental hazards and because of the decline in other causes of death. The field of nutrition is an active field of investigation that has not yet been linked closely to world history.[27] Global historians of health will not lack for topics of interest.

Humanity and Nature

Mankind has studied and interfered with nature from the beginning. With various levels of tools, humans have seized control incrementally of new segments of nature's province. But the interaction has more dimensions. Humanity has changed nature, as with the burning of great swaths of Australian vegetation since early times. Nature has changed mankind, as in developing and spreading new diseases. At each stage, some humans conclude that they do indeed have control over their destiny; at each stage, most of them receive humbling lessons to the contrary. Chinghiz Khan dominated all of Eurasia, but his dream of

24. Watts 1999; W. McNeill 1976; Crosby 1972.
25. Kiple and King 1981; Crosby 1991; Curtin 1989; Curtin 1998.
26. Catanach, I. J., "The 'Globalization' of Disease? India and the Plague," *Journal of World History* 12 (2001); Shlomowitz, and Brennan 1994; Sheldon Watts, "From Rapid Change to Stasis: Official Responses to Cholera in British-Ruled India and Egypt, 1860 to c. 1921," *Journal of World History* 12 (2001), 321–74.
27. Matthew Gandy and Alimuddin Zumla, eds., *Return of the White Plague: Global Poverty and the New Tuberculosis* (London, 2001). The journal *Food and Foodways* addresses current work in the history of nutrition.

avoiding death led nowhere. He would have done better to plan more carefully for his death.

Still, in technology, health, and control of our environment, humans continue to make steady and even accelerating advance. As Wolf Schäfer has recently argued, it is perhaps time for us to look beyond the opposition of man and nature, in that humanity has become one of the principal agents for change in the natural world: he distinguishes a "first nature" of geological and biological forces, which he links rhetorically to the time of Pangaea, from a "second nature" in which humanity joins in transforming the world. The latter phase he identifies as "Pangaea II," arguing that while the continents are still dispersed, human society has now converged to a level of unity and coherence that echoes and supersedes the prior commingling of the continents.[28]

In the attempt to understand the changing balance of man and nature, many of the clearest voices have come from specialists in natural science and in consciousness, rather than the traditional line of historians who focus on social affairs. James Burke and Robert Ornstein, in *The Axemaker's Gift*, pursue the logic of invention and its influence from earliest and recent times. Burke, a biologist, and Ornstein, a psychologist, begin their story with the creation of the axe, a double-edged tool. They emphasize not only the physical consequences of the axe for food supplies and later for forests but the long-term consequences of what they call "sequential thinking," in which each problem is broken down into smaller problems, with the result that small problems are solved but create unnoticed large problems.[29]

For Burke and Ornstein, as for Schäfer, mankind is at a turning point, and old habits are changing or must change to avoid disaster for our species and our planet. Their arguments certainly deserve serious reflection. At the same time, ours is not the first generation to express the belief that mankind is at a turning point or to look to the future with a mixture of eager anticipation and dread.

Graeme Snooks, in another approach to the logic behind long-term human development, sets his story in the context of the periodic (and accelerating) rise and fall of life forms on earth. For humans he identifies several basic dynamics, all of them in operation across all of history, but each of them dominating for a certain time: the dynamic of family expansion for the Paleolithic, recurrent cycles of conquest for the Neolithic (from the beginnings of agriculture to the eighteenth century), and linear waves of technical change for the Industrial Revolution. Snooks concludes, too, that the basic patterns of human interaction with one other and with nature must change in the near future.[30]

---

28. Wolf Schäfer, "The New Global History: Toward a Narrative for Pangaea Two," *Erwägen Wissen Ethik* 13 (2002), 1–14. Despite the evidence of forces unifying humanity, there may still be significant centrifugal forces bringing about new sorts of differentiation.

29. J. Burke and Ornstein 1995. In addition to the original axe, the analysis addresses agriculture, writing, religion, science, technology, class, and medicine. The stages are reminiscent of those identified two centuries earlier by Condorcet, though Burke and Ornstein emphasize the limitations of this sort of progress.

30. Snooks 1996.

To generalize these points, the historical record is now broad and comprehensive enough to permit empirical as well as conceptual exploration of the changing relationship of humanity and nature. Philosophers and, more recently, scientists have sought to conceptualize various aspects of the relationship. It might now be appropriate to define a sub-field of world history to document the interrelationship of man and nature as it has changed over time. Historians who have chosen to work in this area have chosen mostly to focus on science, the formal attempt to explain the natural world. While this is an important portion of the terrain, it might be wise for historians to direct some additional energy to study of the much larger changes in the relationship of mankind and nature through the anonymous and artisanal efforts to develop improved technology and, thereby, greater human control over nature.

Joseph Needham's long-term project (supported by many other scholars) to document and assess scientific study and technical development in China shows strength on both the scientific and technical sides of the equation. This work has conveyed the depth of achievement of the Song era and has also successfully challenged the notion of Chinese stagnation in the time since the Mongols replaced the Song dynasty.[31] Similarly, work on scientific study in the medieval Islamic world has reaffirmed the continuing development of observation and analysis, and the flow of treatises from one region of the eastern hemisphere to another.[32]

As a result, it is becoming increasingly clear that European scientists of the early modern era gained their conceptual breakthroughs not only from their improved techniques of study and their connection with works of the Classical era, but also from their thoroughness in locating, translating, and studying the works of scientists in regions distant from their own.[33] Similarly, in the eighteenth and nineteenth centuries, the scientific advances that were summarized in Europe often came from observations in other parts of the world, and in collection and systematizing of knowledge of local scholars. The great botanical gardens of Kew and Paris were built up through cooperation across global networks. The itinerant German scholar, Alexander von Humboldt, collected his wealth of information in five years of travel and study in Latin America in large part through the support of local scholars who shared their findings with him.[34] The study of science at a global level, encompassing national or civilizational

31. Robert Finlay, "China, the West, and World History in Joseph Needham's Science and Civilisation in China," *Journal of World History* 11 (2000); Needham et al. 1994; Derek Bodde, *Chinese Thought, Society, and Science: The Intellectual and Social Background of Science and Technology in Pre-Modern China* (Honolulu, 1991).
32. Hugh R. Clark, "Muslims and Hindus in the Culture and Morphology of Quanzhou from the Tenth to the Thirteenth Century," *Journal of World History* 6 (1995), 49–74.
33. Chaplin 2001; John M. Headley, "The Sixteenth-Century Venetian Celebration of the Earth's Total Habitability: The Issue of the Fully Habitable World for Renaissance Europe," *Journal of World History* 8 (1997), 1–28; Toby E. Huff, *The Rise of Early Modern Science: Islam, China, and the West* (Cambridge, 1993).
34. Drayton 2000.

levels, is beginning to show some additional patterns in the collection and exchange of knowledge.[35]

Conclusion: Learning in New Categories

The fields of ecology, technology, and health have been very much in the public eye, because of the recent changes, discoveries, and innovations in these areas. In that public discussion, opinion swings from optimism to pessimism, from hope to fear, about the future in each arena. And when the future is discussed, there are always some who seek to prepare for it by studying the past.

The results of historians' work in these new area are uneven but promising. It is probably the history of disease that has done most to demonstrate that historians must take more than politics, economics, and religion into consideration when explaining the past: accounts of the Black Plague and the Columbian Exchange are sufficient to make that point. The new data on short-term climatic change are perhaps the next most powerful indicators that historians must consider additional factors, as the history of volcanic eruptions and El Niño episodes and their impact is now being chronicled. The history of technology is not likely to lead to such spectacular results, yet it seems to document the constancy of change. Study in each of these areas alerts the historians to a mix of long-term and short-term dynamics, and encourages them to recognize the differing levels and types of human agency that are to be found in the past.

The issues of ecology, technology, and health are now part of historical studies. History will be more complex for their inclusion, but the attempt to encompass the interaction of a wider range of factors, if carried out with sufficient care, can make history a more satisfying field of study.

35. James R. Bartholomew, "Modern Science in Japan: Comparative Perspectives," *Journal of World History* 4, 1 (1993), 101–16; Kuhn 1962.

Chapter 13

Cultural History

A focus on cultural history gives greater attention to individual human agency than do studies of political, economic, or environmental history.[1] Although people have been limited in their ability to control the ways of the world, they are usually able to express their response to the events and processes in which they participate. Many cultural expressions of past times were ephemeral and are lost forever, but the remains of others survive in ceramics, burials, words, and habits of dress. In broadest terms, these cultural remains fit into a pattern beginning with many millennia of gradual differentiation in cultural patterns, followed by more recent millennia of cultural convergence. Among early humans, localized groups adapted to different ecologies and developed further distinctions in ideas and institutions, while in the last few millennia connections among populations have grown, and technology has been able to overcome ecological distinctions.

The elements of cultural history have been studied for a long time, but not—with the exception of literature, religion, and intellectual history—in much detail by historians. The recent development of much new evidence for cultural history, especially the development of new theories and new perspectives in cultural studies, now enables the field of history to be opened to the full breadth of cultural issues. The dramatic cultural changes of the present day, most insistently in that vast array of media known as popular culture, add significant pressure for a more thorough analysis of cultural issues in society both contemporary and historical. It is entirely possible that a steady advance in the clarity of conceptualization in cultural studies will lead to a flowering in cultural history in the time to come.

The historical profession, which turned in the mid-twentieth century from a uniform focus on political history to an exploration of the possibilities of social

---

1. This chapter relies substantially on portions of my article, "Cultural History: Paths in Academic Forests," in Robert W. Harms, Joseph C. Miller, David S. Newbury, and Michele D. Wagner, eds., *Paths Toward the Past: African Historical Essays in Honor of Jan Vansina* (Madison, 1994), 439–54. I am grateful to David Newbury and Joseph Miller for their commentary on draft versions of that article.

history, shifted again in the late twentieth century to take up interest in cultural history. The expanded interest in cultural history centered especially on recent centuries, but it extended as well to earlier times. World historians, however, have been relatively slow to take up a substantial focus on cultural issues (as they were slow before to take up social history). The potential for global cultural studies remains nonetheless exciting.

Alfred Kroeber and Clyde Kluckhohn catalogued the anthropological definitions of culture in a 1952 analysis that retains great value.[2] There exist many categories of culture under any definition: cuisine within material culture, dance within expressive culture, ideology within reflective culture, and the macroculture of civilization. A further complexity in cultural analysis is that most cultural artifacts are doing double duty. They are representations of human experience and at the same time are utilitarian items. Writing is a technology of communication, but it is also a vehicle for cultural expression. Religion serves at once as expression of spirituality, as means of social control, as statement of philosophy, and as an arena for aesthetic and representational creativity. The days of attempting to analyze music and painting on purely aesthetic criteria are passing rapidly, and the new cultural theories, while seeking to sustain aesthetic analysis, forthrightly link it to social and economic patterns.

The analyst, in addressing these overlapping categories, faces choices on when and how to do cultural analysis. Analysts explore culture through such lenses as the contrasting viewpoints of the creator, the producer, and those who experience the work; culture as artifact, as individual identity, and as synonym for society; culture to emphasize unity or diversity; and global or localized ways of looking at culture.

Research Agenda: Cultural History

The narrative of cultural history—the record of changing patterns in human representation and understanding of their world—is not easily summarized. If there is an overall narrative, it might begin with a review of early burial practices and pictorial representations of humans and animals. Early pottery gives a sense of the effort to beautify utilitarian goods, and early sculptures are especially powerful in the visions of social values that they convey. In archaeological remains, cities and

---

2. A. L. Kroeber and Clyde Kluckhohn, *Culture: A Critical Review of Concepts and Definitions* (New York, n.d. [1952]). This volume includes scores of anthropological definitions of culture in seven categories, followed by six categories of statements about culture. Adam Kuper, in his review of the Kroeber-Kluckhohn analysis, argues that the authors develop a genealogy of definitions of culture, in which "the scientific conception of culture emerges in opposition to humanist conceptions," and thereby follows the leadership of sociologist Talcott Parsons, "who created the need for a modern, social scientific conception of culture." The more basic lesson of the Kroeber-Kluckhohn volume for historians, however, is that anthropologists have debated the meaning of culture and have seen it as problematic rather than as a known quantity. Adam Kuper, *Culture: The Anthropologists' Account* (Cambridge, Mass., 1999), 59, 68.

stone construction have been most easily located and best preserved. Working from these relics of past moments of cultural flowering, archaeologists are moving steadily to fill in a picture of the fabric of early human society, so that the term *prehistory* is beginning to fall into disuse.

New media have been created from time to time, providing new venues for cultural expression. Dress, the most inescapable of cultural forms, changed whenever new materials became available, and also whenever communal fashion or social hierarchy dictated a new look. Music provided a way for each society to reaffirm its heritage, but music changed with the development of new instruments or the production of new sounds from existing instruments. Writing did not create poetic, dramatic, or narrative expression, but it preserved the record of these earlier forms before transforming them; histories and novels supplemented and ultimately displaced epics. With the advent of printing, the market for the written word could expand dramatically, and some languages became standardized. As photography arose in the mid-nineteenth century, the number of images expanded rapidly, but the new images were all in black and white for most of a century until additional technology arose and color returned. The twentieth century has developed such a panoply of new media—and such a range of new relations among artists, audiences, and intermediaries—that it becomes hard to keep up.

Reflective culture works through the social devices for preserving past knowledge, passing it on, and expanding it. For the individual, knowledge expands as one passes through the life cycle; for the society, passing knowledge to the next generation is imperative, and occasionally one is able to add to the fund of knowledge. For millennia, education, philosophy, and religion were all issues of responsibility for the family, until other social institutions became strong enough to remove part of each from the family. Religion remained alive and well in the family, but large-scale, organized religion took over in many parts of the world. Scientific investigation was virtually never done at the level of the family, and required state or other institutional support. Education for daily life remained in the family; villages or ethnic organizations took on education for certain social values, though specialists in priesthood and government received specialized education. The spread of near-universal primary education for literacy during the past two centuries represents a massive reorganization for both family and society.

Historians are used to asking questions on cultural issues that are focused on local or civilizational levels, and they can readily expand their efforts to exploring local aspects of other (anthropological) aspects of culture. For world historians of culture, the issue is to explore interconnections and to find an appropriate balance of local, societal, regional, and global levels of cultural expression. Has human culture developed primarily *in situ*, autonomously emerging among each local grouping? Has it developed mainly through interaction and exchange among groups?

Historians of culture will ask about the origins of each art form—music, dance, divination—and about the narrative of change and connection in each, as well as the values that each art form reinforces. What sort of periodization and chronology will cultural historians adopt? Often they have tied art forms and philosophies to the history of states or to religious institutions. Do the patterns of literature have a chronology distinct from that of political regimes? That is, the description of cultural artifacts and the chronology of their changes leads to questions about the dynamics of cultural production and cultural contact: What sustains traditions over time, and what brings about major innovations?

The products and processes of human creativity, interestingly enough, have value not only for their own time but also for later times. It is in cultural affairs that the term *legacy* has the greatest resonance. The legacy of the prison colonies structures much of Australian life even today, as the legacy of the Inca state conditions life in the Andes highlands. And every baseball season in Boston reminds Red Sox fans of the curse of Babe Ruth.

The cultural patterns and choices of the contemporary world do much to condition our view of the past. The recent development of a powerful and influential set of popular cultural media raises the question of whether popular culture is brand new or has significant antecedents. The current wave of globalization makes it appear that we face cultural convergence, with differences disappearing. Are there forces creating cultural differences as well as similarities? The spread of new media seems to be intensifying cultural experiences. But who exercises most power over the art now being created: is it creators, audiences, disseminators, or funders? Our effort to understand these and other current issues in culture will surely lead to innovative studies of the past.

Scope: Microcultural and Macrocultural Analysis

The word *culture,* in English, has meanings at many levels. It refers to immense social aggregates, such as "Asian culture" or "Islamic culture," but also to very specific and localized elements of culture, such as a work of sculpture or a village marriage pattern. Scholars have not used such terms as *macrocultural* or *microcultural* studies to distinguish analyses of culture at these poles, but I suggest that it would be wise to do so. While there is no clear boundary between macrocultural and microcultural arenas, the logic and the language used at the aggregate end are quite different than at the specific end of what may be a continuum of cultural analysis. In this section I develop the argument for this distinction, focusing on examples of macrocultural and microcultural analysis in the fields of linguistics and cultural anthropology

At the aggregate pole, one tends to use the term *culture,* in the noun form. It refers to a subject or an object at the societal level: inevitably, "French culture." Culture in this sense serves as identity. In macro studies of culture, one emphasizes distinctions and differences, as between nations and civilizations. At the broad level, these national or civilizational "cultures" are seen as objects to be classified and identified. Culture as identity, as object, becomes a synonym for society or even for ideology.

In cultural analysis focusing on specifics, the term *culture* tends to be used in the adjective form, as a modifier rather than as a subject. The emphasis tends instead to fall on the action of creation or of experiencing a creation, especially at the individual level. One focuses on the activity of cultural production rather than on the object produced as a result. In micro studies of culture, one emphasizes linkages and connections, as with links in a creative chain. In the terms I will develop later, "macrocultural" approaches use the "old" definition of culture, while "microcultural" approaches use the "new" definition. As we will see, both microcultural and macrocultural analysis can be carried out at the global level.

To begin with microcultural analysis, the terms *material culture, expressive culture,* and *reflective culture* each address microcultural issues. They deal with the creation, commissioning, production, dissemination, and experience of culture. For expressive culture, the art forms include music, dance, literature, and visual art. For material culture the art is supplemental to other social needs, as in architecture, dress, cuisine, and many areas of technology. Reflective culture, including philosophy, religion, education, and science, carries importance at all social levels but is most easily studied at elite levels. The specifics of microcultural analysis—of cultural production and dissemination—lead the observer into unending details, including shapes, sounds, colors, textures, and tastes; these specifics also include the logic of rhythms, the logic of colors, and other logics.

Macrocultural analysis begins with broad social distinctions rather than with individual acts of creativity. To begin with, there are societal categorizations: identifying and generalizing about cultural patterns for regions, civilizations, ethnic groups, nations, or religious communities. Within these groupings, macrocultural distinctions have been made among high culture, elite culture, popular culture, mass culture, subculture, and counterculture.[3] In the comparison of national or civilizational cultures, analysis seek to synthesize the many cultural activities, drawing out of them an aesthetic or philosophical essence to sum up the culture in question.

Of course the microapproaches and macroapproaches to culture are artificial distinctions in the minds of analysts. The actual cultural patterns of society cross the boundaries between the individual and the society, and analysts must find conventions to permit their studies to mirror this reality. One approach, popular among some anthropologists in the early twentieth century, was to define cultural "traits" as the atoms of cultural aggregates. It was hoped that each trait could be identified and traced in its contribution to the overall culture of the society in which it was found. Material traits (food crops or housing styles) were to be added up along with expressive traits (patterns of lineage structure) and reflective traits (beliefs in ancestor spirits) to give an overall picture of regional culture. This approach was repudiated by scholars who noted that no trait could exist in isolation, that each trait existed and was defined only in association with other aspects of culture, and that the society itself, rather than existing as a distinct unit, also depended on overlaps and interactions with other societies. If the notion of atomistic traits was too mechanical, there remained the need for a solution to the problem of how to link microlevel culture to macrolevel culture. In practice there are various attempts to make this link. Archaeologists, for instance, do their actual field work in tiny and widely separated excavations, yet generalize from their

---

3. Gans 1974; Theodore Roszak, *The Making of a Counter Culture: Reflections on the Technocratic Society and its Youthful Opposition* (Garden City, N.Y., 1969); Roland Marchand, *Advertising the American Dream: Making Way for Modernity, 1920–1940* (Berkeley, 1985).

results to provide interpretations of whole regional cultures. Similarly, architectural historians working with the size and shape of buildings and art historians working with paintings and murals have been willing to offer generalizations on the spirit of an age for all of Rome, all of Europe, or all of East Asia.

The old definition treated "a culture" as an identifiable entity, a "complex whole" of beliefs, of institutions and artifacts.[4] A culture (or an ethnic group or a society) served as the unit of analysis in cultural historical study. Within the framework of this definition, a range of approaches struggled for dominance. That range included, at one limit, analysts treating culture as coherent, bounded, and internally homogeneous (whom we may label as "lumpers"). At another limit, it included others (whom we may label as "splitters") treating culture as a shifting collection of attributes, without sharp boundaries, and containing competing influences, though still susceptible to holistic analysis.[5]

The new definition of culture focuses on the activities of cultural production and transformation. It centers on the struggles and ideas of individuals and groups of peoples and on the interaction of their contradictory ideas. In these terms, culture is "the semantic space, the field of signs and practices, in which human beings construct and represent themselves and others, and hence their societies and history."[6] The new framework is more explicitly historical than the old.[7] The unit of analysis in this framework is not generally agreed upon by its practitioners, but I would label it as the *debate*: analysis centers on a debate of some social import, and the people and events analyzed are parties and events drawn into the resolution of that debate.[8]

The contrast of old and new frameworks in studies of cultural history reflects the philosophical shift from modernism to postmodernism that has pervaded academic

---

4. According to Edward Tylor (1871), culture is "that complex whole...which includes knowledge, belief, art, morals, custom and any other capabilities and habits acquired by man as a member of society" (quoted in Kroeber and Kluckhohn 1952: 81). This anthropological idea of culture remained in tension with an alternative view, dominant in humanist circles and especially in Europe, which addressed the outlook and creativity of individuals and groups rather than patterns of society as a whole. See Kuper 1999: 23–46.

5. The terms "lumpers" and "splitters" have been used in European historiography by J. H. Hexter, and in American historiography, to distinguish those making overarching generalizations from those emphasizing particularities. Hexter, *The History Primer* (New York, 1971).

6. J. and J. Comaroff 1992: 27.

7. In anthropology through the mid-twentieth century, the American school and Kroeber in particular analyzed culture in historical terms; British and French anthropologists were less systematically historical. In the new framework, all claim the centrality of history to cultural change.

8. As the Comaroffs note, "We are the first to acknowledge that it is not easy to forge units of analysis in unbounded social fields." J. and J. Comaroff 1992: 32. The identification of a historically situated debate or discourse, however, may be a good start at establishing the unit of analysis appropriate to this framework. A debate can, for instance, be studied at levels from the familial to the global, so that analysts can achieve their desire to be freed from the old analytical units of kin, ethnicity and nation.

debates since about the 1960s.[9] The old definition is positivistic: within its framework, one may seek to delineate the elements of culture, the impact of various factors on culture, or the determinants of cultural change. The new definition is postmodern: it focuses on relationships and discourse, not on objects; it emphasizes indeterminacy, not cause-and-effect; it emphasizes change as the rule rather than the exception. Whereas the old framework centers on locating causality, the new framework focuses on identifying contingency. The old framework relies on the social sciences; the new framework displays a revived influence of the humanities.

To adopt the convenient terminology of Thomas Kuhn, a paradigmatic shift has occurred. Pressures grew within the old framework, as thinking about culture evolved and developed contradictions. Then innovators such as Clifford Geertz proposed a new paradigm to encompass the field as the old framework burst its limits.[10] More recently, the work of John and Jean Comaroff has gained recognition as a statement of the new outlook.[11]

The rethinking of cultural history forms but a single facet of the past generation's metamorphosis in scholarship. The broad changes, as described in part II, include more theory, new philosophies, greater analytical rigor, and interdisciplinary expertise set atop disciplinary specialization. For cultural history, the equivalent scholarly interaction and transformation is unusually complex. Cultural history draws at once on studies in cultural anthropology, archaeology, linguistics, literature, art, architecture, music, religion, and philosophy. With such a range of disciplines, it is hardly surprising that cultural studies should differ in both empirical content and theoretical orientation from continent to continent, from country to country.

The methods of linguistics, for instance, loom relatively large in cultural history as practiced for Africa, but these methods are diverse and convey a complex heritage. Lexical studies range from simple word lists to the elaborate technique of glottochronology; structural studies, less numerous, focus on comparative grammar. Various classifications of languages (drawing sometimes on lexical and sometimes on structural data) were central to debates on race, cultural groupings and migrations. Joseph Greenberg's breakthrough in classifying the genealogy of African languages—so central to our understanding of the Bantu migrations and much more—relied on a balance of lexical and structural elements and on "mass comparison" of data. More than anything, however, Greenberg relied on lexical comparisons of words and things.[12]

9. For an interpretation of this transition focused in the areas of economic and architectural thought, see Harvey 1989.
10. Kuhn 1970; Geertz 1973. Most of Geertz's essays were first published during the 1960s.
11. Comaroff 1985; J. and J. Comaroff 1992; J. and J. Comaroff 1993.
12. "There are three fundamentals of method underlying the present classification. The first of these is the sole relevance in comparison of resemblances involving both sound and meaning in specific forms.... The second principle is that of mass comparison as against isolated comparisons between pairs of languages. The third is the principle that only linguistic evidence is relevant in drawing conclusions about classification," Greenberg 1966: 1.

Yet another impact of linguistics has been indirect. Early in the twentieth century the Swiss linguist Saussure developed the notions of "the signifier" and "the signified," thus introducing formally the consciousness of the speaker as well as the word and the thing. His work became influential beyond the field of linguistics only much later and mainly through the work of French scholars including Lacan, Foucault, and Derrida. But its influence was nonetheless profound, as it sustained new departures in literary theory, cultural anthropology and, generally, the development of postmodernist philosophy.[13]

World historians would benefit from a comprehensive review of the various fields of cultural studies and their interactions, in both past and present. Such a review would assist historians in drawing critically on the evidence and analyses of linguists, anthropologists, art historians, and the many other breeds of cultural analysts. Assembling this review will doubtless require years of research and discussion; in the meantime, we can begin by exploring the evolution of individual fields of cultural study (including, of course, history). At present, the anthropologists have done best in reconsidering their own field, and Adam Kuper has been outstanding among the anthropologists, as he has traced the development of the two main threads of anthropological inquiry. In *The Invention of Primitive Society* (1988) he explores kinship theory from the establishment of its hegemony within anthropology to its virtual collapse in the late twentieth century. In *Culture: The Anthropological Account* (1999) he analyzes the rise of cultural anthropology in the United States, the debates within that tradition, and the interactions with other visions of culture. These volumes trace the preoccupations and transformations in anthropology with a clarity that provides excellent guidance for world historians seeking to draw on the vast anthropological literature.[14]

Kuper found, in studies of culture as within kinship theory, a broad continuity in anthropological analysis from the founding of the discipline to the 1970s. The work of George Peter Murdock provides a prism that displays many facets of that continuity. Murdock's 1959 survey of the "peoples and culture history" of Africa exemplifies a positivistic modeling of the results of cultural processes in a form characteristic of the "splitter" tendency within the old approach.[15] Murdock developed interpretations of long-term cultural change based on maps and tallies of such results of cultural production as kinship systems, crops, and political institutions.[16] Murdock's unit of analysis shifted between the tribe and the culture

13. Jameson 1981.
14. Kuper 1988; Kuper 1999. For a review of the history of *mentalités* centering on Europe, see Jacques Le Goff, "Les mentalités: une histoire ambiguë," in J. Le Goff and P. Nora, eds., *Faire de l'histoire* (Paris, 1974), 76–94. I know of no review of linguistics sufficiently broad for present purposes.
15. Murdock 1959. For a "lumper," see C. G. Seligman, *The Races of Africa* (New York, 1930).
16. Murdock, generalizing this method, created the Human Relations Area Files. These files assembled a great amount of data, but the data were coded on the assumption that they reflected independent observations and independent cases. For an example a study conducted through regression analysis of these data, see Pryor 1977.

province (the latter consisting of a collection of tribes), but the effect of his work was to reify ethnic groups. Although much of his interpretation was focused at the level of the culture province, and his text noted the overlaps and internal distinctions in ethnic groups, the organization of his argument and his widely reprinted ethnic map of Africa emphasized discrete and bounded ethnic territories.

Murdock's models, applied through his maps and tallies, centered on diffusion and differentiation of language groups, crops, kinship and political systems. He thus focused on the results rather than the processes of cultural production: he labeled his data as "traits." Still, he criticized "trait-chasing," by which he meant supporting hypotheses through searching out the traits under study in isolation from other data. His own main hypothesis was that an ancestral system of matrilineal descent had dominated the continent and that it had evolved by stages toward patrilineal descent, at rates varying with location. The hypothesis itself reveals the causality in his reasoning, and his investigation led at best to dubious results. With more success, he supported the thesis of independent invention of agriculture along the West African desert fringe. In this and some other cases, Murdock skillfully confirmed his analysis of innovations and population movements by collecting observations on several types of evidence.[17]

In the forty-plus years since the publication of Murdock's study of Africa, cultural studies have changed greatly. As Adam Kuper argues, "Mainstream cultural and social anthropology today has abandoned primitive society and, with it, society itself. Instead it is embracing the second tradition of anthropology, the anthropology of Tylor and Frazer rather than Morgan and Rivers, the anthropology of culture."[18]

17. In another successful line of inquiry, Murdock developed his thesis on the existence of "megalithic Cushites," overlaying distributions of several types of data to show that Cushitic-speaking peoples had earlier occupied zones of Tanzania now occupied by Bantu- and Nilotic-speakers.

Murdock's analysis of crop distribution showed the importance of the Western Sudan and Ethiopia as centers of agricultural innovation. His reliance on incomplete and faulty agronomic data led him to conclude that yams and rice were first imported to West Africa rather than domesticated there. The error over yams, while empirical rather than methodological, was especially costly, as it forced Murdock to postpone his date for the start of Bantu migrations until the arrival of Asian yams. In contrast, his hypothesis on megalithic Cushites has fared rather better in the light of subsequent work. Murdock 1959: 199, 222–25.

18. "Meanwhile, on the margins, there is the third tradition of anthropology, which has at its heart the theory of biological evolution." Kuper 1988: 243. As evidence for Kuper's point on the third tradition of anthropology, one may note William Durham's study of the intersection of genetic and cultural inheritance, which has gained popular as well as academic attention. Durham subscribes to "ideational theories" of culture and treats his definition of culture as similar to that of Geertz. (Durham defines culture as "systems of symbolically encoded conceptual phenomena that are socially and historically transmitted within and between populations.") Durham's purpose, however, is to seek out the links of biology and culture through a complex but positivistic analysis, and in doing so he is happy to utilize Murdock's Human Relations Files. Durham 1991: 8–9; see also 3, 193–96.

Clifford Geertz became the most prominent prophet of the turn to the anthropology of culture. His definition of culture as "a set of control mechanisms" is oft-quoted. More influential in practice were his expository technique of "thick description," intended to convey a multiplicity of viewpoints on any set of events, and the particular case of a Balinese cockfight along with the responses of community members when it was broken up by police.[19] The emphasis on thick description may be seen as an attempt to avoid modeling and thereby to sustain consideration of more variables.

The critique of anthropology associated with decolonization resulted, as Kuper has argued, in rejection of the idea of primitive society. In an important contribution to the critique, Johannes Fabian argued that the ethnographic present was not simply an erroneous assumption of social stasis, but an "allochronism," a device for placing the "Other" (the subjects of anthropological study) into a different time so as not to have to share the world with them.[20] The results of such critique devastated social anthropology, and left cultural anthropology as the main surviving branch of the field.

Yet the paradigm for cultural analysis developed into a rather different form from that proposed by Geertz. For not only had decolonization brought a change to the focus and outlook of anthropology, but new analytic devices had come forth, notably the Saussurean linguistics of the signifier and the signified and its more recent variants in philosophy and literary theory.[21] Thus, for the Comaroffs, culture is "a historically situated, historically unfolding ensemble of signifiers-in-action, signifiers at once material and symbolic, social and aesthetic." The term *culture* in the noun form virtually disappears from the lexicon of those utilizing this new definition, and the adjective form *cultural* takes its place.[22]

19. Geertz's definition is as follows: "culture is best seen not as complexes of concrete behavior patterns—customs, usages, traditions, habit clusters—as has, by and large, been the case up to now, but as a set of control mechanisms—plans, recipes, rules instructions (what computer engineers call 'programs')—for the governing of behavior." In his essay on the Balinese cockfight, Geertz ranges from categorical statements of national character ("...the Balinese, for whom nothing is more pleasurable than an affront obliquely delivered or more painful than one obliquely received..."), to nuanced statements of postmodernist philosophy ("The culture of a people is an ensemble of texts, themselves ensembles, which the anthropologist strains to read over the shoulders of those to whom they properly belong."). In sum, Geertz enunciated a transitional doctrine rather than setting forth a fully developed new framework. Geertz 1973: 44, 433, 452.
   For an insightful reconsideration of the colonial era in Bali, which structured the society that Geertz described, see A. G. Hopkins, "Globalization with and without Empires: From Bali to Labrador," in Hopkins, ed., 2002: 220–42.
20. Fabian 1983.
21. Kuper emphasizes the importance of kinship terminology as a complex analytical game; one may wonder whether the notions of signifier and signified provide a new and equally engrossing game. Kuper 1988.
22. J. and J. Comaroff 1992: 27.

Johannes Fabian applies the new framework by focusing significantly on language itself: the rise and elaboration of Shaba Swahili.[23] His multivalent vision of that process emphasizes that the language "emerged" as a range of speech patterns rather than as an authorized version, and he challenges the notion that Swahili "diffused" to Shaba from some point in East or Central Africa. Still, the language did not simply arise as a folk practice: Fabian emphasizes the importance of European grammarians in structuring and codifying the language, and the importance of the Belgian colonial regime in ensuring its spread at the end of World War I. The locus of Fabian's analysis of Shaba Swahili is the debate on the question of what was to be the vehicular language of Shaba. The participants in the debate were dominated by those in the colonial order who had influence over language policy, but they included all the speakers of the various dialects and tendencies in the language.

Fabian's analysis of language change addresses words, things, structures, and consciousness. To underscore the historic diversity within Shaba Swahili, he develops concepts consistent with that diversity.[24] His tracing of the emergence of Shaba Swahili, as compared with other dialects of Swahili, focuses on the developing structure of the language. While many of his sources are labeled "vocabulaire," his use of lexical data is subordinated to the exploration of putative pidgin and creole stages of the language. Fabian has shown how even words and grammars, so seemingly arbitrary in their symbolism, are ideologically charged.[25]

Fabian lambastes positivistic models of language change. Thus, "Centreperiphery thinking applied to language development shares with similar views in politics and economics an opprobrious logic of tautological definitions of correctness and deviance (the center is correct, the periphery deviates)."[26] As he argues, such notions of deviance serve to marginalize evidence on the "peripheries," to reinforce assumptions that the deviations will soon be overcome, and to draw the observer (and subsequently the analyst) away from focusing on processes reproducing cultural specificity in each "periphery."

Whereas work within the old definition focused on central tendency, work within the new definition privileges local variation. Murdock, reflecting the "splitter" tendency in the old approach, was sensitive to local variation in cultural makeup, but he treated such variation as subsidiary to his main concern, which was to locate centers of innovation and broad patterns of cultural change.

---

23. Shaba, a large province in the southeast of Congo-Kinshasa, lies at the headwaters of the Congo, Kasai, and Zambezi valleys. Swahili-speaking merchants from the East African coast arrived in numbers in the nineteenth century; Swahili became and remains a vehicular language for the region. Johannes Fabian, *Language and Colonial Power: The Appropriation of Swahili in the Former Belgian Congo 1880–1938* (Cambridge, 1986).

24. "Instead of seeking spots (centers of diffusion) I shall attempt to identify *spheres* or *fields of interaction* in which not 'Swahili' but varieties of Swahili, became one medium of communication among others." Fabian 1986: 9.

25. Ibid.

26. Ibid., 12.

For Fabian, local variation is more than incidental: it is the focus of his stories of cultural change. The differences among these studies provide reminders of the substantial changes in the language, theory, and empirical focus of studies in cultural history over the past generation.

Politics presents a likely field on which to test the differences between the two definitions of culture: the old definition centers on structure, the new definition focuses on process. Supporters of the new definition argue that the old definition helped sustain the colonial regime and reified hierarchy within that regime. The new definition, it is argued, leads to critique of colonialism and of hierarchy in general.

A new paradigm has thus emerged to challenge the old. The old paradigm, however, has not yet withdrawn from the field of discourse, either for scholars or among the general public. The difference between microcultural and macrocultural issues helps explain why both paradigms continue in use: the new approach fits better with microcultural analysis, and the old approach fits better with macrocultural analysis.

A Cultural Case: Equatorial Africa's last 5,000 Years

Jan Vansina's *Paths in the Rainforests* provides a useful example of a cultural study set at a large scale. I will rely on it to expand on the distinctions in the previous section, and prepare for the review of recent contributions to global cultural history in the sections to follow. I selected it as the case study because it provides a cultural history of politics, because it addresses a region usually left outside discussion of world history (particularly for early times), and especially because Vansina is so explicit about his methods.[27]

The book reviews the migration of Bantu-speaking people through Africa's equatorial forest for the past five thousand years and focuses on the development of political tradition in the region for the past three thousand years. Relying primarily on evidence from historical linguistics, Vansina traces the development of a common tradition based on families, villages, and districts and then explores, for region after region, the rise of lineages, states, associations, and brotherhoods all before the impact of the Atlantic world in the sixteenth century. *Paths in the Rainforests* reaffirms and extends Vansina's leadership in historical methodology in at least two ways.

First, it demonstrates the power of historical linguistics as an analytical tool. It shows the time depth and topical breadth of issues that can be explored through intensive study of historical linguistics—in this case, through an exploration of political tradition over a period of several millennia. The use of historical linguistics, which has become as important in African history today as it was in the development of Indo-European linguistic studies a century ago, is deserving of far more attention by historians of other areas of the

---

27. Vansina 1990. For some of Vansina's main contributions: on oral sources, Vansina [1961]; on written sources, Vansina 1966; on art history, Vansina 1984. See also his autobiography, Vansina 1994.

world.[28] Vansina portrays some five millennia of creation and transformation in the political tradition of Africa's equatorial forest through a lively narrative replete with empirical, methodological, and philosophical jewels. The result delivers a powerful blow to any remaining notions of the impossibility of recovering the outlines of the distant past or of the unchanging nature of African society.

Second, the book is a tour de force in cultural history, and this is the issue on which I wish to focus here. While Vansina gives prime emphasis to the history of political tradition in the equatorial forest, he explores many areas of material culture, expressive culture, and ideology in constructing that narrative. He draws on anthropology, broadly defined, in his study of history. His sophisticated use and critique of theory and method in anthropology and history make *Paths in the Rainforests* an exemplary study in cultural history. The notion of "cultural history," however, is a moving target. During the course of Vansina's career, the dominant scholarly definition of "culture" has changed from a focus on the *results* of cultural production (the "old" definition), to a focus on the *process* of cultural production (the "new" definition).

Vansina, in his study of political tradition over a wide area and a long time, has summarized a generation's research on the Western Bantu languages and their speakers.[29] He portrays a culture unfolding over time. Using linguistic data to recreate the spread and development of Western Bantu society, he chronicles first the elaboration of an ancestral tradition and then its modification through various perturbations to the equilibrium of the system. His topics are numerous, but his focus always returns to politics.

Vansina's presentation is, first of all, a narrative. The narrative begins with the ancient and common tradition. Here he centers on big men; on the three institutions of house, village and district; on such issues in economic life as farming, finding food, industries, and exchange; on beliefs in heroes, spirits, medicine men, witchcraft, and charms; and on the common preference for low population density. Expansion of the tradition accelerated with the adoption of metals and the banana. The "historical watershed" and the impulse to institutional change came with the occasional rise of population density to higher levels.[30] In recounting the

---

28. Bantu languages have been the subject of two generations of intensive linguistic analysis. One can speculate about the results that might emerge from the linkage of available historical data with similarly thorough linguistic analysis for Amerindians, Inner Asia, Southeast Asia, or the Mediterranean. For the early work, see Malcolm Guthrie, *Comparative Bantu*, 4 vols. (Farnham, U.K., 1969–70); 29.

29. Christopher Ehret contests Vansina's assertion that the Bantu languages may be divided into Western and Eastern subgroups, and argues instead that the languages of Eastern and Southern Africa, though encompassing a large area and population, stem genealogically from a subgroup within the languages that Vansina labels as Western. The implication for historical reconstruction, if Ehret is upheld, is that Vansina would have to modify his linguistic analysis of earlier but not later times. Ehret, "Bantu Expansions: Re-envisioning a Central Problem of Early African History," *International Journal of African Historical Studies* 34 (2001), 5–41; see also Vansina's response, ibid., 52–54.

30. Vansina 1990: 6, 99–100, 259–60.

subsequent development of descent systems of patriliny and matriliny in various regions, he emphasizes both the contingency of historical development and the degrees of cultural interaction and differentiation within the equatorial forest. But he also addresses interactions with other traditions—with Africans in adjoining regions and with the European influences of the Atlantic slave trade and colonial rule.

Second, for all its complexity, *Paths* relies heavily on lexical analysis and on discourse. Most of Vansina's method relies on historical linguistics and on a technique he labels as "words and things," or "the combination of linguistic and ethnographic data."[31] Vocabulary studies are "the most rewarding to historians because of the special properties of words as joiners of *form* and *meaning*.... The history of the *form* tells us something about the history of the *meaning*: the institution, belief, value, or object to which the form pertains." The analysis is thus based overwhelmingly on lexical studies and on the inheritance, borrowing, and innovation in words and in the associated things.[32]

Third, in his interpretation of cultural change, Vansina exhibits faith in the perfectibility of the old framework. In describing his research design, Vansina asserts that "the scholar working with 'words and things' is like a mosaicist or a pointillist painter." Ethnic units "must be abandoned as unanalyzed units for study," and the analyst must seek instead to pinpoint the geographical location of each observation.[33] The work of the mosaicist takes place, however, within a very capacious frame: the unit of analysis in this study is the "common tradition" shared by the people of equatorial Africa. Vansina traces the notion of the "tradition" back to the work of Alfred Kroeber.[34] Within this large unit, Vansina centers his attention on "major lineaments of the original tradition: the economic, social and political institutions...and key elements of worldviews and ideologies." Overall, then, Vansina's definition of his task—with his focus on words, things, the "common tradition" and institutions—reflects an approach grounded in the old definition of culture, yet open to the new.[35]

31. For "words and things" Vansina cites the linguist R. Anttila, but I think it is likely that he was also influenced in this choice of terms by Foucault's *Les mots et les choses* (translated as *The Order of Things*). R. Antilla, *An Introduction to Comparative and Historical Linguistics* (New York, 1972), 291–92; Foucault [1966].
32. Vansina 1990: 11.
33. Ibid., 20, 31.
34. Ibid., 6–7.
35. In constructing his analysis, he began with an ethnographic baseline just before the European conquest, then utilized a technique of "upstreaming" to determine earlier institutional changes. From the other side of his time frame, Vansina elicited the process of settlement from the genealogical model for Western Bantu languages. "First the petrified face of continuity, then the mobile face of change." Despite the broad strokes of this research design, in his handling of empirical data Vansina is more of a "splitter" than Murdock. His focus on "words and things" echoes, at some level, the emphasis of Foucault on "les mots et les choses." Ibid. 11–16, 31–33; Foucault [1966].

Fourth, Vansina draws another arrow from his interpretive quiver: the difference between the consciousness of historical actors and the world they perceived. This distinction makes clear his reliance on the new framework as well as the old. Before launching his narrative, Vansina presents this conceptual tool in a section entitled "reality and reality."

> I will use the expression "physical reality," in the sense that all observers, whatever their cultural background, agree on the action, situation, or object, not that it "really" exists in an absolute philosophical sense. Most records, however, are cultural interpretations shared by the members of a community. They are "collective representations" and refer to a different reality. I will use "cognitive reality" to designate it.... Among the sources, vocabulary testifies to cognitive and physical reality separately. An item meaning "sun" deals with physical reality, an item meaning "family" with cognitive reality.[36]

Thus, for example, districts were a physical reality, in that all early observers noted them; they also shared a single ancestral term. The notions of "maximal lineages" and "subtribes," in contrast, were cognitive expressions imposed later by outsiders.[37] Then in his conclusion, Vansina utilized "reality and reality" to restate the definition of "tradition:" they "consist of a changing, inherited, collective body of cognitive and physical representations shared by their members."[38] In short, this set of distinctions represents Vansina's handling of the issues of textuality and subjectivity, which are central to the new definition of culture.

Fifth, Vansina's models make each of his major interpretive and methodological points. He emphasizes the complexity of his interpretation by proposing multiple models for a given phenomenon. For instance, in addressing the initial Bantu expansion, Vansina argues that "It is unwise to rely on a single model of expansion"—he prefers to hypothesize an alternation between slow movement into unfamiliar habitat, and a rapid dash ahead in familiar habitat. In one instance he condemns "one-way models" of the relationship between ecology and community.[39]

In methodology, his lexical analysis centers on the starred (or putatively ancestral) form of each word and on the "tree model" of linguistic change, though he also notes that the "wave model" shows how change ripples through language, and he asserts that "both models work together."[40] In another instance of methodological modeling, he notes that a uniform mental scheme or model underlay the ethnographic questionnaires utilized by colonial-era ethnographers and officials.

36. Vansina 1990: 72.
37. For this example, Vansina's meaning of physical reality is as he introduces it. In other cases, he moves to a more shorthand approach of assuming that physical reality is really real. Ibid., 81.
38. Ibid., 259.
39. Ibid., 55, 255.
40. A "tree model" emphasizes language change through descent from an ancestral language through localized innovation, while a "wave model" emphasizes change through interaction among languages, as by borrowing of words or imitating syntax.

As he argues, the advantage of that single mindset is that we are left with parallel data from many regions; the disadvantage is that the system of selection incorporated prejudice and excluded much information of importance.[41]

Vansina assembles the elements of his interpretation by alternating among interpretive models. Thus we may emphasize, on the one hand, Vansina's metaphoric generalizations drawing on a positivistic intellectual heritage. Utilizing a mechanical image, Vansina once labels the equatorial tradition as "a gyroscope in the voyage through time."[42] The central metaphor of *Paths*, however, is that of a social system in equilibrium: "The system was in a stable dynamic equilibrium, because all the Houses and districts were *similar*, equal in manpower, and hence in military strength. But the system was potentially chaotic: one small change could trigger…a chain reaction…which would stop only when it reached the limits of the system." This metaphor, repeated at several points in the book, underscores the main line of Vansina's argument: "As soon as that balance was broken by an innovation, diffusion of the innovation or a counterinnovation followed in a continuing attempt to restore stability."[43]

On the other hand, Vansina relies on a metaphor that is postmodern in its contingency. He adopts from the equatorial African tradition the metaphor of the leopard and its spoils for analysis of the political system: "the disposition of the spoils of the leopard…is the best indicator of the political structure." Since the forest peoples recognized the leopard as the symbol of power, the successful leopard-hunter stood in the position of reaffirming or challenging the political order by delivering portions of his kill either to a series of local authorities, to a chain of ascending powers, or directly to a central authority.[44]

Vansina thus alternates among new and old approaches, yet relies most heavily on the old. In a methodological defense of his overall interpretation, Vansina characterizes it as a "complex hypothesis" whose very complexity provides support for its validity. The propositions in the book, he argues,

> achieve a high order of validity because they are interconnected (the relation to the ancestral tradition) and because they claim to account for many data. The quality of hypothesis varied with the density of the interconnections between its parts and with the number of the elements it attempts to explain.… This is why the dominant hypotheses in physics are so convincing: they address a multitude of disparate data by a single integrated hypothesis.… The main hypothesis laid out in this book is complex enough to induce confidence, even though the interconnections of its

41. Ibid., 11, 27–28.
42. The gyroscope metaphor is presented as follows: "…the debate between materialists and idealists as to the priority of physical or conceptual reality makes little sense, because change is perpetual, forever yoked to continuity. During this process of innovation all the principles and the fundamental options inherited from the ancestral tradition remained a gyroscope in the voyage through time: they determined what was perceivable and imaginable as change." Ibid., 195.
43. Ibid., 100, 193.
44. Ibid., 104.

various component propositions often remains loose. An alternate overall hypothesis to account for all the data is possible but unlikely.[45]

This notion of hypothesis-testing relies far more on the causality of positivistic thinking than on the contingency of postmodern thinking.

Vansina's concluding chapter argues for broad historical reconstruction, again within the old framework. He develops explicitly his notion of the tradition. "Traditions, as fundamental continuities which shape the futures of those who hold them, are not just in the minds of observers. They are 'out there.' They are phenomena with their own characteristics...." In a logical next step, Vansina lays out, ardently and skillfully, a call for "comparative anthropology." He rejects much earlier work ("the usual methods of comparative anthropology are flawed, if not bankrupt") but argues that a modified approach can lead to valid results. "The study of cultural tradition can change this situation and make the dream of controlled comparison come true... by following the historical course of a single tradition."[46] The accompanying argument lays out a macrolevel approach that has more in common with the goals (if not the techniques) of Murdock and Kroeber than with the more local level of studies of scholars working in the new tradition.

Vansina's sense of the perfectibility of the old framework appears to be strong enough that he declines to treat the new framework as representing any major breakthrough. In a review of the Comaroffs' *Ethnography and the Historical Imagination*, Vansina asks of the methodological propositions offered there, "How does this actually differ from the work of the garden variety of historians?" He thus argues that recent changes in the framework and methodology of cultural studies are more incremental than fundamental.[47]

On the other hand, for all the positivistic underpinnings of his *analysis*, Vansina's *conclusions* do exhibit a clearly postmodern dimension. They emphasize the uniqueness and contingency of the historical patterns, the mutability and interpenetration of historical processes in equatorial Africa. He stresses the numerous exceptions to evolutionist schemas of political development from local

45. Ibid., 250. This reasoning parallels, in the more complex realm of cultural history, Joseph Greenberg's methodological axiom of mass comparison, affirmed in his classification of African languages.

   "The importance of mass comparisons as opposed to isolated comparisons between pairs of languages has become clear to me as a result of certain questions of a general nature raised by a number of critics.... When resemblances can be assembled which are recurrent in many languages, which extend over vast and widely separated geographical areas and which encompass elements with morphological functions, pronouns, and the most stable part of the vocabulary... then common origin is the only adequate explanatory hypothesis.

   "The importance of resemblances recurrent in a large number of languages as plausible outcomes of some single ancestral form as elicited by mass comparison is of very great evidential power in excluding either chance or borrowing as explanations. Considerations derives from the elementary theory of probability helps to make this explicit." Greenberg 1966: 2–3.

46. Vansina 1990: 251–57.

47. "At heart anthropology is the comparative study of culture." Ibid., 260, 263.

community to state, arguing that the observed political patterns of the forest zone resulted from the coexistence of ideologies extolling the success of big men and stressing the ideal equality of all. He argues that environmental factors, while significant, did not determine the development of institutions: low population density and decentralized political systems were a choice and not a necessity.[48]

Cultural history, as this case indicates, is not easily divided into discrete subtopics and discrete methods. Vansina has put more effort into explicit discussion of his methods than most historians will want to. But in doing so he conveys at once the exhilaration and the complexity associated with attempts to interpret cultural change in the past. His techniques and dilemmas will reappear in the following discussion of the wider range of global studies in cultural history.

Disciplinary Themes

In scrutinizing the specifics of cultural production, one cannot help but be struck by the immense ranges of types of cultural production and by the wide (but not equally wide) range of disciplines developed to study these fields of activity. The fields include literary studies, linguistics, religious studies, visual arts and art history, theater and cinema studies, music studies and ethnomusicology, studies of dance, anthropology (social and cultural), popular culture, and more. There can be no single methodology for analyzing culture when there exist so many disciplines of practice and analysis within cultural studies. One might set the task of developing an academic map of cultural studies, defining the various disciplines, along with their attendant methodologies and theories, that apply to various subregions of the vast terrain that we call "culture."

Christopher Ehret's reconstruction of the history of Eastern and Southern Africa, relying on linguistic and archaeological records, gives substantial emphasis to such issues in material culture as food crops, domestic animals, and the structure of houses. He also notes the connections of these regions to the commercial systems of the Mediterranean and the Indian Ocean. Liu Xinru documents the place of material culture not only in trade but in religious exchanges between China and India. Lynda Shaffer, in her well-known article on "Southernization," relies on cottons, sails, silks, and spices to document her case for a wave of innovations in South and Southeast Asia that launched an expansion of cultural and commercial ties for the Southern Seas and beyond. Parker James, in an exemplary world-historical study at the dissertation level, has focused on a single architectural form, the stilt house of Southeast Asian origin, and traced it from its earliest archaeological records to its spread throughout the tropics.[49]

48. "Much of what the Comaroffs discover is like reinventing a partial wheel." Vansina, review of J. and J. Comaroff 1992, in *International Journal of African Historical Studies* 26 (1993), 417–20.

49. Ehret 1998; Liu 1995a; Lynda Shaffer, "Southernization," *Journal of World History* 5 (1994), 1–22 (reprinted in Dunn 2000: 175–91; Robert Finlay, "The Pilgrim Art: The Culture of Porcelain in World History," *Journal of World History* 9 (1998), 141–88; James 2001.

The most fully developed study in material culture at the global level is Adshead's comparison and linkage of material culture in Europe and China from 1400 to 1800. It provides a model that might be followed for other pairs of regions. Zeynep Çelik has used the devices of world fairs as a means for exploring the architecture of the Islamic world. In other cases, national or local studies of material culture may be compared and combined to develop a sense of broader patterns in cuisine, architecture, or dress.[50]

Studies in language and linguistics have great potential for revealing evidence in many aspects of cultural history. Such studies have been carried out in the greatest depth for Indo-European languages in Eurasia and for Niger-Congo, Nilo-Saharan, and Afroasiatic languages in Africa. Equivalent study for other language groups—such as Sino-Tibetan, Austronesian and Austroasiatic, Altaic and Amerindian, for instance, could add greatly to our understanding of early cultural history.[51] Even for cases in which substantial written texts remain, as in Egyptian and Greek, application of the techniques of historical linguistics may reveal substantial new information on social and cultural history.[52] Recent studies in the connections among languages have led to much clearer definitions of the categories of "pidgin" and "creole" languages and the processes of migration and social change that create and transform such languages.[53]

Recent studies on literature in world-historical context have emphasized European views of other regions through travel writers and views of Europe by visitors from other regions.[54] Where documents permit, tracing individual consciousness of global and local changes, and their interaction, can provide substantial evidence on the changing perceptions and changing realities of the world from the point of view of individuals.[55] Studies in visual art, music, and dance are

50. Adshead 1997; Zeynep Çelik, *Displaying the Orient: Architecture of Islam at Nineteenth-Century World's Fairs* (Berkeley, 1992); Katarzyna Cwiertka, *The Making of Modern Culinary Tradition in Japan* (Leiden, 1999); Harvey Levenstein, *Revolution at the Table: The Transformation of the American Diet* (New York, 1988).
51. Ehret 2002; Mallory 1989; Renfrew 1988; Heine and Nurse 2000.
52. Gaur 1985.
53. David Buisseret and Steven G. Reinhardt, eds., *Creolization in the Americas* (College Station, TX, 2000). For a thought-provoking application of the notion of "creolization" to visual arts, see Ben-Amos 1977. On languages, see John A. Holm, *Pidgins and Creoles*, 2 vols. (Cambridge, 1988–1989).
54. Mary B. Campbell, *The Witness and the Other World: Exotic European Travel Writing, 400–1600* (Ithaca, 1988); Teresa Hubel, *Whose India? The Independence Struggle in British and Indian Fiction and History* (Durham, 1996); Jameson 1981; Susan Gilson Miller, ed. and trans., *Disorienting Encounters: Travels of a Moroccan Scholar in France in 1845–1846* (Berkeley, 1992); Mary Louise Pratt, *Imperial Eyes: Studies in Travel Writing and Transculturation* (New York, 1992); Merle C. Ricklefs, *The Seen and Unseen Worlds in Java, 1726–1749: History, Literature and Islam in the court of Pakubuwana II* (Honolulu, 1998).
55. Sarah Swedberg wrote a dissertation that fits easily into the framework of early national U.S. history but at the same time draws substantially on world-historical thinking. This exploration of the Cranch family, a group writing letters from Hingham, Massachusetts from 1770 to 1810, wrote letters to correspondents in the

beginning to expand from their European base.[56] The issues of sport and physical education, which have earlier been considered in global historical perspective, are ripe for reconsideration.[57]

The study of religion in world-historical perspective is focusing on times before the past five hundred years. Studies of Zoroastrian religion by Mary Boyce, and of links between religious communities by Richard Foltz and Liu Xinru, suggest that it will be possible, with time, to develop broad statements on the development and interconnection of religious faiths from roughly 500 B.C.E. to 1500 C.E.[58] It is remarkable, in contrast, how little discussion of religion appears in the world history literature for recent centuries. I think this absence reflects the confluence of two trends. First, with the rise of secularism in the West, historians analyzing broad connections have tended to neglect studies of the spiritual dimensions of religion, especially from the eighteenth century forward. Christian missionaries and Islamic revivalists, in this context, tend to be treated as part of political history for the purposes of world history. Second, the numerous studies of religion that have been conducted for many areas of the world tend to be treated as local histories rather than elements of an interconnected global history of religion. I think it is certain that religion will reappear in world-historical studies of recent times, partly because of the dramatic changes in religious communities (e.g., in Buddhism, Pentecostalism, Islam, and the religious traditions of the African diaspora) and partly because of the difficult questions in spirituality and understanding of the supernatural that are raised by the changes in human life everywhere.[59]

Continued

United States and Britain, and on subjects ranging around the whole Atlantic. Drawing on literary theory, Swedberg traces the evolving consciousness of family members in an era of revolutionary change. Sarah Swedberg, "The Cranch Family, Communication, and Identity Formation in the Early Republic" (Ph.D. dissertation, Northeastern University, 1999).

56. W. Johnson 1988; Vansina 1984; Suzanne Preston Blier, "Truth and Seeing: Magic, Custom, and Fetish in Art History," in Bates, Mudimbe, and O'Barr 1993: 139–66; Blum, Bohlman, and Neuman 1991; W. McNeill 1995.

57. For a world history of physical education that is unconnected to other historical studies yet shows remarkable geographic and temporal breadth, see Deobold B. Van Dalen, Elmer D. Mitchell, and Bruce L. Bennett, *A World History of Physical Education* (New York, 1953). See also Kenneth J. Carpenter, *Protein and Energy: A Study of Changing Ideas in Nutrition* (Cambridge, 1994); Kiple and Ornelas 2000.

58. Mircea Eliade and Ioan P. Couliano, *The Eliade Guide to World Religions* (New York, 1991); Julian Baldick, *Mystical Islam: An Introduction to Sufism* (New York, 1989); Boyce 1992; Eickelman and Piscatori 1990; Brian Fagan, *From Black Land to Fifth Sun: The Science of Sacred Sites* (Reading, Mass, 1998); Valerie I. J. Flint, *The Rise of Magic in Early Medieval Europe* (Princeton, 1991); Foltz 1999; Alexander Knysh, *Islamic Mysticism: A Short History* (Leiden, 2000); Liu 1995b; Phillip Sigal, *Judaism, the Evolution of a Faith* (Grand Rapids, 1988); Huston Smith, *The Religions of Man* (New York, 1958).

59. Recent migrations have brought the varying traditions of Mahayana and Theravada Buddhism into contact with one another, stimulating both debate and confluence.

Philosophy and ideology, in contrast to religion, turn out to be areas of reflective culture that gain substantial attention from world historians. Perhaps the largest part of this literature has been reflections on how Europeans have looked at the world, spurred in part by Edward Said's 1978 critique of Orientalism.[60] Among the most thoughtful of works in this tradition are reviews of historical thinking in areas outside Europe, as well as studies of the historical work of specific groups within the Western tradition, such as African American writers of the nineteenth and twentieth centuries and writers based in Mexico in the eighteenth century.[61] Science has grown substantially as a field of interest, so that the history of science, which long existed as a field very distinct from other fields of history, is now being connected to other realms of world history. The heritage of Joseph Needham's years of study on science in China is one direct cause, but an additional trend linking science to empire has also resulted in several substantial recent books.[62]

The developments in various specific disciplines are surely the source of much of the recent excitement in cultural studies. Examples to which one may allude include literary theory, the rise of ethnomusicology, the flowering of studies in pidgin and creole languages, and the great debates in social and cultural anthropology. But for all the study and discussion on cultural specifics, the analysis of culture never seems to abandon a substantial focus on the macro vision of culture as identity, object, and destiny. Debate at the macrolevel of culture identity often derails or swallows up microlevel discussions on creativity and experience. Without judging the merits of this interplay, we may observe empirically a tension between specifics and aggregates in cultural studies.

Interpretive Themes

Macrolevel studies address culture at the ethnic, national, or civilizational level and explore cultural conflicts and connections at an aggregate rather than individual level. Such studies are of interpretive importance and are in wide demand by audiences seeking clear statements on cultural comparisons. The topics of these studies include identity, comparison of cultural types, cultural evolution, and interactions of aggregated cultures, as well as global studies of cultural specifics.

60. Mary Helms, *Ulysses' Sail: An Ethnographic Odyssey of Power, Knowledge, and Geographical Distance* (Princeton, 1988); Armitage 2000; Adas 1989; Headley 1997; James Muldoon, "Solórzano's De indiarum iure: Applying a Medieval Theory of World Order in the Seventeenth Century," *Journal of World History* 2 (1991), 29–46; Pagden 1995.
61. Dorothea A. L. Martin, *The Making of a Sino-Marxist World View: Perceptions and Interpretations of World History in the People's Republic of China* (Armonk, N.Y., 1991); Keita 2000; Cañizares-Esguerra 2001.
62. Bodde 1991; Bartholomew 1993; Chaplin 2001; Drayton 2000; Huff 1993; Deepak Kumar, *Science and the Raj, 1857–1905* (New York, 1995); Roy MacLeod, "Passages in Imperial Science: From Empire to Commonwealth," *Journal of World History* 4 (1993), 117–50; Philip F. Rehbock, "Globalizing the History of Science," *Journal of World History* 12 (2001), 183–92; Storey 1996.

One of the hazards of interpretive analysis at the macrocultural level is the quantity of general pronouncements and critiques. A great deal of effort has gone recently into rather general critiques of the analysis of culture in history.[63] These are critiques and analyses of such concepts as objects, identities, concepts, dominance, agency, debate, hegemony, administration, discourse, knowledge, and empowerment. Although I can see no precise fashion in which to summarize the main lines of these arguments, I think that in general they call for more specific definitions of culture and its attributes. One can only applaud this approach, and join in the critique of those who claim that "culture matters!" without being able to specify what is meant by culture.[64] At the same time, there is a tendency for these manifestos to address each other rather than to focus theory on the specifics of historical analysis.

The empirical analyses of topics relevant to cultural history at the macrolevel, in my opinion, are most simply classified into those that trace cultural development within a single society (or region) and those that address contact or interchange among societies. For the former, the logic of the analysis must emphasize the internal patterns of contradiction and evolution. For the latter, the logic must emphasize varying models of interaction.

Studies emphasizing cultural evolution, though bounded by the limits of the region under study, may nevertheless address very large topics, and may lend themselves to comparative analysis. Robert Bartlett's study of four centuries of cultural evolution in medieval Europe provides a coherent summary, ripe for comparison to other regions. John Gillis's edited collection on national commemorations shows how a common practice of invented tradition held many nations together. Arthur Marwick's analysis of cultural revolution in four North Atlantic countries during the 1960s emphasized the evolution of a common tradition, and Adam Zamoyski's study extends a similar sort of reasoning to the romantic and revolutionary activists of the North Atlantic in the century ending 1871.[65] In two critiques of criticism, Regina Bendix analyzes of the creation of folklore studies and Deborah Root scrutinizes changing values in the market for

63. Appadurai 1996; J. and J. Comaroff 1993; J. and J. Comaroff 1992; King 1991; Kuper 1988; M. Lewis and Wigen 1997; Robertson 1992; Deborah Root, *Cannibal Culture: Art Appropriation, and the Commodification of Difference* (Boulder, 1996); Said 1993; Jonathan Friedman, *Cultural Identity and Global Process* (New York, 1994); Immanuel Wallerstein, "Culture as the Ideological Battleground of the Modern World-System," *Theory, Culture and Society* 7 (1990), 31–55.

For earlier studies providing critiques of cultural analysis, see Marshall Sahlins, *Historical Metaphors and Mythical Realities: Structure in the Early History of the Sandwich Islands Kingdom* (Ann Arbor, 1981); Sahlins, *Islands of History* (Chicago, 1985); Said 1978; Fabian 1983; Geertz 1973; H. White 1973.

64. Landes 1998.

65. Robert Bartlett, *The Making of Europe: Conquest, Colonization, and Cultural Change, 950–1350* (Princeton, 1993); John R. Gillis, ed., *Commemorations: The Politics of National Identity* (Princeton, 1994); Hobsbawm and Ranger 1983; Iriye 1997; Arthur Marwick, *The Sixties: Cultural Revolution in Britain, France, Italy and the United States, c. 1958–c. 1974* (Oxford, 1998); Emiko Ohnuki-Tierney, *Rice as Self: Japanese Identities Through Time* (Princeton, 1993); William Preston, Edward S. Herman, and Herbert I. Schiller,

visual art.[66] Studies of the Islamic world, its history and current cultural directions, run parallel to these studies of the West.[67]

Studies of cultural contact have become extremely numerous. I found it surprising to see how many such studies have appeared in recent years. While they address a wide range of social and cultural connections, they have not yet been generalized or linked into a larger narrative that would attract attention to these studies as a group. I think it is reasonable to expect in the next few years several efforts to make general statements on cultural contact in world history.

It may be that earlier studies helped to initiate this wave of analyses in cultural contact. Martin Bernal's *Black Athena* attracted substantial attention not only to the place of Egypt in Greek civilization but to cultural contact and borrowing in general. Jerry Bentley's *Old World Encounters* brought a formal statement on connections to the world-history literature and may have encouraged some of the subsequent studies.[68]

The sub-categories of cultural contact turn out to be numerous indeed. Religious connections have been analyzed in several fashions. In a long-term analysis focusing on current political anxieties, Peter Partner analyzed holy wars over the past two millennia, especially those of Christianity and Islam. Other studies of religious conversions in early times and recent times present tales of changing cultural values.[69] Ethnogenesis, the creation of new ethnic groups as a result of cultural contact, was the topic of a special issue of the *Journal of World History*.[70] Perceptions of the "other," as recorded in literary and other

Continued

    Hope & Folly, the United States and UNESCO 1945–1985 (Minneapolis, 1989); Simmons and Simmons 1996; Stephen Vlastos, *Mirror of Modernity: Invented Traditions of Modern Japan* (Berkeley, 1997); Adam Zamoyski, *Holy Madness: Romantics, Patriots, and Revolutionaries, 1776–1871* (New York, 2001); R. H. Tawney, *Religion and the Rise of Capitalism* ([1926] Gloucester, Mass., 1962).

66. Regina Bendix, *In Search of Authenticity: The Formation of Folklore Studies* (Madison, 1997); Mary Helms, *Ulysses' Sail: An Ethnographic Odyssey of Power, Knowledge, and Geographical Distance* (Princeton, 1988); David Hillman and Carla Mazzio, *The Body in Parts: Fantasies of Corporeality in Early-Modern Europe* (New York, 1997); Root 1996. See also the influences of cartography on development of national identity, as portrayed in Thongchai Winichakul, *Siam Mapped: A History of the Geo-Body of a Nation* (Honolulu, 1994).

67. Bernard Lewis, *What Went Wrong? Western Impact and Middle Eastern Response* (New York, 2002); Albert Hourani, *A History of the Arab Peoples* (Cambridge, Mass., 1991).

68. Bentley 1993; Bernal 1987–1991; Von Laue 1987.

69. Richard Fletcher, *The Barbarian Conversion: From Paganism to Christianity* (New York, 1998); Partner 1998; H. Clark 1995; Jessica A. Coope, "Religious and Cultural Conversion to Islam in Ninth-Century Umayyad Córdoba," *Journal of World History* 4 (1993), 47–68; James C. Russell, *The Germanization of Early Medieval Christianity: A Sociohistorical Approach to Religious Transformation* (New York, 1994).

70. David A. Chappell, "Ethnogenesis and Frontiers," *Journal of World History* 4 (1993), 267–76; Chappell, "Frontier Ethnogenesis: The Case of New Caledonia," *Journal of World History* 4 (1993), 307–24.

texts, can provide detailed commentary on the process and results of cultural contact.[71]

Even the term *contact* benefits from being broken down. On one hand, one can use it to refer to the first impressions of peoples newly in contact, during a brief but formative time: a number of studies provide details on the dynamics of these first impressions, especially for nineteenth-century encounters.[72] But contact can also continue for decades or generations. A greater number of studies address long-term contacts of peoples and the resulting social transformations. Almost all of these are colonial situations.[73] These studies explore a wide range of issues indeed, but it may be that they are linked by patterns in the historical data or patterns in the minds of the analysts. For the latter reason, any prospective reviewer of this literature should ask about the metaphors and models of cultural contact used in these analyses.[74]

Macrocultural analysis has also been developed for other sorts of aggregates— popular culture, elite culture, counterculture, cosmopolitan culture—set within a larger culture or overlapping cultures. These studies address such issues as cultural production, performance, ritual, and innovation, but they also address the manipulation of culture by powerful interests.[75] Paul Gilroy's interpretation of the

71. Dathorne 1996; Tapan Raychaudhuri, *Europe Reconsidered: Perceptions of the West in Nineteenth-Century Bengal* (New York, 1988); David C. Gordon, *Images of the West* (New York, 1989); Obeyesekere 1992; Withey 1989; Samuel M. Wilson, *The Emperor's Giraffe, and Other Stories of Cultures in Contact* (Boulder, 1999).

72. Patricia Seed, *Ceremonies of Possession in Europe's Conquest of the New World, 1492–1640* (New York, 1995); Alex Calder, Jonathan Lamb, and Bridget Orr, eds., *Voyages and Beaches: Pacific Encounters, 1769–1840* (Honolulu, 1999); Edward L. Schieffelin and Robert Crittenden, *Like People You See in a Dream: First Contact in Six Papuan Societies* (Stanford, 1991); Peter R. Schmidt and Roderick J. McIntosh, eds., *Plundering Africa's Past* (Bloomington, Ind., 1996); Znamenski 1999.

73. Liu 2001; Colin G. Calloway, *New Worlds for All: Indians, Europeans, and the Remaking of Early America* (Baltimore, 1997); I. C. Campbell, "Culture Contact and Polynesian Identity in the European Age," *Journal of World History* 8 (1997), 29–56; Cooper and Stoler 1997; David Northrup, *Africa's Discovery of Europe, 1450 to 1850* (New York, 2002); Mark Francis, "The 'Civilizing' of Indigenous People in Nineteenth-Century Canada," *Journal of World History* 9 (1998), 51–88; Guha 1998; Hall 1992; Frances Karttunen and Alfred W. Crosby, "Language Death, Language Genesis, and World History," *Journal of World History* 6 (1995), 157–74; Frances Karttunen, "After the Conquest: The Survival of Indigenous Patterns of Life and Belief," *Journal of World History* 3 (1992), 239–56; Lipman 1997; Terence Ranger, "Europeans in Black Africa," *Journal of World History* 9 (1998), 255–68; Seed 2001; William K. Storey, "Big Cats and Imperialism: Lion and Tiger Hunting in Kenya and Northern India, 1898–1930," *Journal of World History* 2 (1991), 135–74; J. Comaroff 1985.

74. For instance, whether they rely on concepts of progress, stages, diffusion, divergence, or convergence. See chapter 16 for further discussion of this point.

75. Paul Gilroy, *The Black Atlantic: Modernity and Double Consciousness* (Cambridge, Mass., 1993); Gans 1974; John Mackenzie, *Propaganda and Empire: The Manipulation of British Public Opinion 1880–1960* (Manchester, 1986); Mackenzie 1992; Sabean 1984. For an interpretation of cosmopolitan Francophone culture in twentieth-century Africa, see Patrick Manning, *Francophone Sub-Saharan Africa 1880–1985*, (Cambridge, 1988).

Black Atlantic, for instance, is a study in a cosmopolitan, modernist culture within a larger Western culture.[76]

Conclusion: Representing the World

Every work of art is unique. To label a work as "derivative" or "a copy" is usually to disparage it. To the uniqueness of each work, cultural historians have commonly asserted the uniqueness of the society in which it was created. Cultural history, interpreted in this fashion, is a mosaic of locally distinct and locally focused creations. Cultural historians have focused within real or artificial cultural boundaries, studying "our heritage" or "other cultures" for their specific values and qualities and even for their specific patterns of evolution.

The world historian considering cultural issues may wish to take a slightly different approach. Every cultural creation is indeed about the specifics of the space in which it is created. But the space in which it is created may be the world, as seen by the creator, and not just a locality. There is room, therefore, for world historians to balance the localized studies of culture with greater emphasis on connection and commonality, and to focus on global patterns in representation. Scholars have not been at the cutting edge in developing global approaches to cultural issues: in teaching, for instance, the many compendia on world literature present students with an eclectic sample of literary works from a wide range of societies and time periods. It is time for historians (and literary scholars) to study culture in this way alongside the students.

While some studies in material culture and expressive culture have begun to show the possibilities of explorations at the global level, it is the studies in reflective culture and especially in intellectual and religious history where the most work has been done. Even here, we need more work on connections and not just on comparisons.

How should one select strategies for global cultural analysis? The most pressing task, I think, is to clarify the multiple meanings of culture and the multiple fields and methods for its study. That is why I found Vansina's *Paths in the Rainforests* to be such a helpful study: it is at once broad and specific, and the author is quite explicit about his concepts and his methods. Vansina offers cultural historians a methodological bridge linking what I have called old and new approaches to cultural analysis.[77]

The bridge dramatizes the problem of how to link results from incommensurate frameworks.[78] The two ends of this bridge may be labeled "culture as object"

76. Analysis of "popular culture" is applied often to such fields of expressive culture as music and print culture, but much less frequently to material culture.

77. A suspension bridge spanning a forest creek, as sketched by an American artist for an 1890 book, graces the cover of *Paths in the Rainforests*: Vansina may have chosen it as a metaphorical bridge over time and space, linking the reader to the long tradition of the forest peoples. I propose to extend the image of the bridge one step further, to the methodology of cultural history.

78. This problem arises in each area of historical studies, but the most complex and interesting version of it arises in the study of cultural history.

and "culture as practice." Vansina's approach has been to conduct his analysis in part from each of these ends and also from the bridge surmounting them. Postmodernist analysts of "culture as practice" argue that treating culture as entity—using positivistic cause-and-effect reasoning, focusing on central tendency, using mechanical and organic metaphors—tends to privilege hierarchy and justify the hegemony of the powerful over the weak. Thus the definition of culture as practice leads, according to its proponents, to critique of colonialism and of hierarchy in general. The macroeconomic view of culture, though vulnerable to this critique, cannot be made to go away. Societies and individuals have too much invested in reverence for cultural icons and in generalizations about cultural identity to cease analyzing at this level.

From either end of the analytical bridge, the historian may use multiple lenses through which to view the subject matter: the numerous artisanal and analytical disciplines in culture, and the differences among local, civilizational, and global perspectives. The complexity of the view from the bridge stems from the many choices involved in manipulating whole frameworks of analysis, and not just the elements of a single frame.

It is too late to turn back: historians have already committed themselves to applying multiple disciplines to their study of the past, and they cannot escape having to address the conflicting philosophies within each discipline. The choice of strategy in addressing alternative analytical paradigms will therefore preoccupy historians in years to come. One approach is to fix oneself within a given paradigm—as Murdock and Fabian attempted to do. Another approach is to attempt the blending or synthesis of paradigms: this tactic, I think, usually yields inconsistent and overly eclectic results. A third approach is alternation among paradigms: I attempted to show how Vansina's models, metaphors, and overviews correspond to such an alternation among paradigms. Perhaps there exists a fourth strategic approach, in which various disciplines and paradigms may be linked together in some logical and encompassing frame. But such a broad exercise in theory exceeds our abilities at present—certainly in the field of history and especially in cultural history.[79] We are left with the alternation of frameworks as the broadest practical strategy for interpreting the past.

Global studies in cultural history should help clarify the changing ways in which people have represented the world. There can be no revelation of a single great pattern—the variety in human expression is too great to be neatly categorized. The many historical studies of cultural contact, which have quietly been moving ahead, may be the best hope for substantial advance in the interpretation of cultural history. These studies require attention to the perspective of each party under study, along with the perspective of the author. In their empirical results and their developing analytical methods, these studies may be laying the groundwork for a distinct field of global cultural analysis.

---

79. Marxian philosophy sought for over a century to provide such an interdisciplinary framework, but the postmodernist critique has shown that it failed to be comprehensive.

Chapter 14

Debating World History

H istorians are a contentious lot. While their arguments are usually conducted in polite language, the disputations are conducted on a number of fronts at once, and the frame of mind of the disputants ranges from a sporting pleasure with making point after point to a savage determination to win the day. Contending narratives, analyses, evidence, and assumptions come into play at once, so that the debates among historians threaten to become as complex as the history they seek to interpret.

World history, as a new field of scholarship, is relatively innocent of debate. For much of the twentieth century, the arguments were over whether there could be a history of the world or whether it was too expansive an area of inquiry for historians to address. Those active in world history found that their work involved much more synthesis than debate: assembling a great narrative took precedence over analyzing the global past and developing contending interpretations. At the end of the century, debates within world history began to emerge, and they developed connections to earlier debates in other fields of study. The purpose of this chapter is to provide a rapid survey of major debates in world history and to anticipate the shores that will experience rising tides of debate.

The potential number of topics of debate in world history is huge, and the potential number of positions far more so. In the following rapidly shifting discussion, I begin first with issues in long-term human development, and then alternate among disciplines in an order paralleling that of chapters 10 through 13. In each disciplinary area I give attention first to current debates among world historians—those featured in world-history journal articles, conference papers, and electronic postings. Second, I address recent debates on global issues in adjoining fields (for instance, economic historians have been debating global economic inequality in recent years, but world historians have yet to join that discussion). Third, I identify some issues for which there ought to be debates among world historians.[1]

---

1. In addition, one could list debates inherited from earlier times as well as those that have emerged recently.

Long-term Human Development

Domestication of plants and animals—the pattern and impact of this great set of transformations, especially about ten thousand years ago—has become an area of significant debate among world historians. The development of agriculture, once presented as a unique development in the Fertile Crescent that gradually spread to all the Old World, is now beginning to be presented as a largely coincidental development, about 10,000 years ago, of agricultural techniques in as many as ten areas of the world. Jared Diamond's summary argues that the world today remains deeply structured by consequences of that transformation, and that the people of Eurasia have been favored fundamentally by the consequences of geography.[2] The responses to this thesis are leading to debates about both early and recent times.

The debate on the origin of *Homo sapiens,* while it has not involved world historians actively, is nonetheless of great importance for world history. While the debate on ultimate human origins has declined, the disputations on the origins of our own species have become more active. Archaeological work of the late twentieth century developed a firm consensus that the earliest hominids developed in East Africa. The differences develop on the emergence of *Homo sapiens* in the past 100,000 to 200,000 years. Those working from analyses of mitochondrial DNA emphasize the African origins of the latest evolutionary change, and the displacement of previous hominid populations in Africa and Eurasia by these new creatures. Those working from archaeological remains in Eurasia argue, in contrast, that *Homo sapiens* developed in parallel at various points of the Old World, or (alternatively) that the new populations and the old intermingled.[3]

Another debate extending far into the hominid past is that on gender and aggression in human development. This discussion, carried on by human biologists and physical and social anthropologists, arises as a result of expanded gender studies, and in turn should have significant impact on gender studies generally. Earlier notions of man the hunter are being replaced by visions of early hominid communities relying heavily on gathering and on cooperative activities, in which women played a leading rather than subordinate role. If such a picture of pre-agricultural humanity could be confirmed, then it would require the identification of a process of transition to the male-dominant societies of earliest recorded history.

The debate over the relative changeability of human nature, though it has declined in recent years, needs to be revisited as long-term studies in human history gain strength. The ancient works of literature still retain their interest for readers of today because they convey the impression that, while customs have changed, the underlying passions, desires, and frustrations of humans have changed little over time. In contrast, analysts such as Sigmund Freud and Norbert Elias have argued that the development of civilization has taught humans to repress more and more of their elemental urges in order to gain the benefits of a more cooperatively governed society.[4]

2. J. Diamond 1997.
3. McBrearty and Brooks 2000: 533–34.
4. Freud 1930; Elias [1939].

These latter interpretations provide an account of progress and perhaps of its inevitability in world history. Those arguing within a time perspective of fifty to one hundred years can often make the case that patterns of progress in history are often illusory, and that things as often get worse as better in the lives of most. But to argue the same over the course of thousands of years is more difficult, as humans have obviously become more numerous and more effective in their control over nature. Yet arguments against the vision of steady progress in human history can be made effectively, and from many viewpoints.[5] For instance, while agriculture expanded productivity, it was also very laborious and cut into the leisure time of those who lived by farming.

In these issues and in others, the patterns and dynamics of human history are themselves a topic of debate. Evolutionary interpretations tend to assume that change was gradual and incremental. Cyclical interpretations assume times of rapid change or expansion, followed by eras of decline or stagnation.[6]

Economy and Politics

Perhaps the most fully developed debate in world history today is the one that can be called "Europe and Asia in the world economy." From the 1970s, economic historians have written of European miracles—of them, E. L. Jones has been most effective, and David Landes has articulated the most recent general statement.[7] Then came the countervailing work of historians of Asia and of the sea lanes. Dennis Flynn and Arturo Giráldez have led historians of the sea lanes with their analysis of the early modern silver trade, which proposes a China-centered world economy. They have become associated with the "California school" of Asia-based world historians, and with R. J. Barendse, Sanjay Subrahmanyam, and Andre Gunder Frank to synthesize the approach.[8] One formulation of the question is, "When did Europeans gain dominance in the world economy?" Another is, "When did European traders gain dominance in Asian markets?" Yet another is, "When did European productivity and output exceed that of Asia?" David Landes would argue that Europeans created the world economy in the sixteenth century and dominated it from the moment of creation. Frank would argue that Asians dominated the world economy until the end of the eighteenth century, when they were eclipsed by Europeans. In the latter view, how far back can one trace an Asian-dominated world economy?

The current debate on Europe and Asia in the world economy is of course an extension of the long debate on the rise of the West. That debate can be traced to eighteenth-century writers, and each generation has brought restatements of the

5. Bury 1932; Nisbet 1980; Headrick 1988.
6. Frank and Gills 1993; Johan Goudsblom, *Fire and Civilization* (London, 1992); Jones 1988.
7. North and Thomas 1973; Rosenberg and Birdzell, 1986; E. Jones 1981; Landes 1998.
8. Flynn and Giráldez 1995; Wong 1997; Pomeranz 2000; Subrahmanyam 1990; Barendse 2002; Frank 1998; Goldstone 1991. Jack Goldstone, a Europeanist, coined the phrase and joins with the California school.

problem and new visions of the answer. The announcement of the thesis on the modern world-system by Immanuel Wallerstein in his 1974 book launched the previous phase of this debate. While historians participated actively in early stages of that debate, with time they have tended to leave discussion of the modern world-system to historical sociologists.

A second debate, which may be labeled "global feudalism," has begun recently and is restricted to a few world historians, but is none the less fascinating and promising. This is a debate that developed around H-WORLD postings by R. J. Barendse of Leiden University, who argued that the tenth century C.E. brought a series of simultaneous developments across the Eastern Hemisphere, in which peasants came to be settled down, and horse-based elites developed political power in localized rather than centralized political units. Stephen Morillo responded by arguing that such transitions did indeed take place but gradually over a period of several centuries rather than within the short range of time identified by Barendse.

Historians have focused much energy on interpreting nations, but they have tended to neglect any general interpretation of empires. The term *empire* is used to refer to polities from the time of the Achaemenids to the contemporary United States. Empires have been studied comparatively within certain time frames, as in the early modern era or the nineteenth and twentieth centuries. But empires have not undergone the sort of systematic analysis to which nations have been subjected. One possible task of world historians is a more systematic analysis of empires, their constituents, their political dynamics, and how they have changed over time.

Society

In social history, it is probably the debate over modern slavery that has done most to advance the interpretation of world history. National and global historians of slavery have combined to show the connections and interpenetrations of slave systems in every area of the Atlantic world and beyond, and in so doing have shown ties with industrialization, the emergence of citizenship, and patterns of cultural exchange.

A second major area of debate in social history is that of revolution in the modern world. The issue of revolution is prominent for much of the world in the nineteenth and twentieth centuries. One debate is whether revolutions are limited to cases of class upheaval within a society, or whether the term *revolution* may also be applied to successful independence movements, as for the Americas in the eighteenth and nineteenth centuries, and for Asia and Africa in the twentieth century.

More discussion ought to be given to the issue of social and economic inequality at a global level. One of the most striking contrasts in all of world history is the development, in the past two centuries, of a widespread ideology of social equality for citizens within nations and equality among nations, at a time when the economic inequality within nations and between nations has grown to unprecedented extremes. Social historians have addressed this issue with studies of slavery and emancipation; economic historians have addressed it with studies of wage levels.

Meanwhile, the categories of class, community, family and ethnicity are each deserving of more thorough analysis at the world-historical level.

Debates on Man and Nature

Environmental degradation and renewal are the main themes in current work in ecological history. Human populations have brought significant environmental change in every era, and humans have had to adjust to environmental changes beyond their control. In assessing ecological history, analysts must determine whether humans have done more to work their will on nature or to bring about environmental degradation. The forces of ecological renewal are not to be underestimated, however.

The choice between emphasis on diffusion or independent invention continues to divide historians of technology. The contrasting emphases of analysts seeking out the sources of innovations lead some to trace all to a single primordial invention and others to find multiple developments of each new idea.

In recent times—once a scientific community became organized and productive, rather than primarily speculative—there has been a debate on the degree to which science determines the pace of technological change. For times well into the twentieth century, some scholars argue that technological change comes more from practical, hands-on improvements than from theory-based laboratory work. For those arguing the other side, that scientific research does govern the pace and direction of technical change, we must ask whether this means that we are now in a fundamentally new age.

Culture

Defining the scope and content of global cultural studies is a debate waiting to take place. It should be a revealing and productive debate, as we learn more about identifying global cultural connections. At present, religion is the easiest cultural topic to discuss at the global level. Most successfully, world-historical studies of religion have drawn heavily on comparative religion—that is, on comparisons among the major organized religions. But world historians have not been very effective at including more localized or less formalized religious traditions within their purview, or at exploring the new religious trends of the twentieth century. For the former issue, one might explore the religious practices of shamanism, widespread in East and Central Asia and also in the Americas. On the latter issue, one might explore either the spread of Pentecostalism to most areas of the world, or the growing pressure on all religious communities to adopt ideologies of toleration for other religions, which may in turn lead to fundamental theological change.

There ought to be more world-historical analysis of reflective culture. Is there a global history of ideas? For the twentieth century there can be no doubt of the worldwide patterns of ideas in nationalism, socialism, and religion. In earlier centuries, the greater divisions among political and cultural communities made it less likely that ideas would develop unmistakably global impact. There were certain broad communities in early times—the Islamic world, the empires of China and their surroundings, and the world of Christendom—in which the ideas of elites circulated far and wide and in which ideas among the common people could also circulate widely. Determining the key issues, patterns, and dynamics in the global history of ideas is, however, an arena in which only the most elementary work has been done.

The existing studies of material culture and expressive culture have the potential to be extended to the global level, as do studies of social values. As an example of the latter, one might seek to trace the forces leading at once to expansion and restriction of brutal punishment and torture, especially over the last three centuries. In region after region, dominant social values have changed to rejection of torture, mutilation, and public executions. At the same time, the development of new technology and new social conflicts has brought new forms of brutality and torture.

Conclusion: How to Conduct Debates in World History

In one sense, debates in world history are simply more debates about history. These are discussions among historians who share training, styles of discourse, and problems in documenting the past. In another sense, debates in world history should have a special quality, related to the focus on interconnections and broad patterns. Debate among world historians examines connections in history and in historiography. In history, world historians seek to locate connections and broad patterns in the past. In historiography, world historians seek to establish connections among localized historiographical debates—for instance, connections among the debates identified in this chapter.

Debate in world history should emphasize what may be called transparency. The notion of transparency in debate is intended to address the fact that world historians are a group only in their common interpretive interests, while the specific materials in which they are expert vary sharply from person to person. Debates in world history must certainly privilege those who know the evidence well for the issue under debate. At the same time, the debates must address the framework of analysis and the comparative basis of assertions, to enable others to participate in debate even if they do not know the evidence well. The point is to create a world-historical discourse in which each argument is linked to others rather than a series of airtight specialist monographs.

Both for the transparency of a historical interpretation and for its inherent strength, world historians need to be clear in summarizing their evidence. For evidence, the researcher is often torn between the access to secondary sources that provide breadth of coverage and primary sources that offer depth at the cost of slower and more difficult research. The distinction between primary and secondary sources, however, is situational rather than absolute, so the researcher should indicate how each is defined for any study.

Another key contribution to transparency in world-historical debate is for authors and critics to identify the perspectives with which they are working.

Debates among historians are partly about the best perspective. While the multiplicity of perspectives involved in any study of world history makes it improbable that one perspective can be identified as "best," the comparison of perspectives of various historians and their historical subjects is an important part of sorting out interpretations of the past.

Replication of studies will be important in advancing world-historical discourse. Because world history is a fledgling field holding many exciting topics in relation to the number of scholars, it is tempting for each researcher to seek out a new topic of study in hopes of filling in a bit more of the immense map. For this reason I think it is important to emphasize the importance of replication and competition in studies of world history. Only when more than one scholar reviews the evidence and interpretations on a given topic can we hope to know the range of possible interpretations of that topic. It is unwise to allow scholars to feel proprietary about "their" topic or documents. The exciting debates and the real advances in knowledge come when multiple scholars are working on related topics, testing their assumptions, data, and interpretations against each other's.

All of these recommendations for the conduct of world-historical debates entail a hope for progress in the interpretation of world history. If the historian must remain fundamentally skeptical about whether the global past can be told as a tale of progress, that same historian, working toward artisanal and interpretive advance, measures the work of the field in terms of its progress.

Part IV

Logic of Analysis in World History

The events and patterns of the past seem sometimes to follow a clear logic, but for other times they seem chaotic. The earthquakes that destroyed Port Royal, Jamaica in 1692 and Lisbon in 1755 had no human logic. Yet the historian's presentation of the past must be logical in order to retain the reader's attention. Given the breadth of scope in world history, the importance of logical handling in analysis and presentation is underscored.

I return to method in part IV, giving a closer scrutiny to the logic of analysis in studies of world history. Part II introduced the range of changing theories and disciplines on which world historians may call, and part III reviewed the initial results of world historians' applications of them to the global past. The chapters of part IV put the methods of disciplines, area studies, and global studies to work in a different way: I propose guidelines for the scope, framework, strategy, and execution of world-historical analyses.

These chapters distinguish among the various elements of framework and strategy in analyzing history. How to combine these elements is dependent on the topic, discipline, and question under study. No single element of these frameworks or strategies is brand new, nor is it even new to combine them. But the systematic effort to compare and contrast, to link, and to explore historical systems, to develop coherent strategies for analysis and for presenting the results—this adds up to a revolutionary step forward. With these techniques, historians are able to articulate and document interpretations at a historical breadth that was rejected before as sheer speculation but now can be advanced as serious historical hypotheses.

The broadest outlines of a historical study are defined by its limits in space, time, and topical coverage. World history necessarily involves analysis at several levels, and the exterior scale of the historian's framework needs to be linked to the interior scale and dynamics of space, time, and topic. In proposing the appropriate logic of analysis at each scale, I have expressed it in terms of research agenda and research design and have broken down research design into frameworks for analysis, strategies of analysis, and modeling the dynamics of historical change. Then I turn to the important matters of verifying the validity of interpretations in world history, and presenting the analysis effectively to the readers.

Overall, the chapters of this section emphasize the design, execution, validation, and presentation of world-historical research, with attention to assuring that studies in world history account for a range of analytical perspectives and a range of social perspectives.

Chapter 15

Scale in History: Time and Space

World history is big. Such current terms as *world history* and *global history*, as with the earlier *universal history*, obviously reflect the intent to encompass a large story. But how big is the world, and what is encompassed in the global? The variety of meanings we give to these terms conveys the range of potential scales of "the world." I remember being struck by this point when I was a graduate student in Madison in the 1960s: my interludes in front of the television during study breaks led me to encounter, all too frequently, the ads from World of Dinettes, inviting me to choose from the widest range of kitchen furniture. The days have gone past when I could illustrate the variety of meanings of "world" to undergraduates with reference to "Wayne's World." But we can always speak of the world of infants—a world that is huge and daunting from the perspective of their inexperienced consciousness, yet tiny and exciting from the perspective of new parents. Then we have the world of music, the civilized world, the planet Earth as our world, and the immense universe in which our own far-flung galaxy is but a speck. At a more inventive level, we have imagined parallel worlds, lost worlds, and worlds in collision. All of these worlds differ not only in extent but in kind. Perhaps one can argue that there is more specificity in the term *globe*, which, along with *global* and *globalization*, has become a term referring to the Earth more specifically than does *world*. But a global command in a computer program refers only to the specifics within that program. For whatever reasons, in English we speak more of the world than the globe, but we label more things as global than as worldly.

The Range of Scales

If the space of the world is big, perhaps the expanse of time covered in world history should also be big. Over what extent of time does one examine world history? Is it all of the twentieth century? All of human history? When I wait on cold Boston evenings for the No. 39 bus on Huntington Avenue, I grumble that "it takes forever" for the bus to arrive. Yet some of my students may write on exams that the Atlantic slave trade was an "event," despite the centuries of its duration. So our notions of what is a long time and a short time seem almost as expandable and contractible as our ideas on the size of the world.

Scale in history is not only about space. It also refers to time and topical breadth. While the debates about geographical space—national, continental, and global history—have stood in the foreground of the discussion of world history, the temporal and topical dimensions of world history deserve equal attention. Nor is it sufficient to define the limits of a study by listing the maximal temporal, geographic, and topical limits of coverage. To describe the scope of a work of history, the historian should indicate the interior as well as exterior dimensions of the study. As I will argue here, the full description of the scale of a historical analysis includes the exterior limits of the space, time, and topics considered, and the interior subdivisions, patterns, and dynamics analyzed by the author for each of these dimensions.

One may get a clearer sense of the possibilities of scale in history by comparison with other disciplines. Astronomers focus on a very wide geographic and temporal scale but a rather narrow set of topics. Chemists focus on the atomic and molecular levels, though chemical engineers extend the work of chemists to an industrial scale. Physicists work with a widely ranging spatial scale, as they make estimates ranging from the mass of subatomic particles to the mass of the universe. Geographers focus on space at several levels; their analyses range widely over time and especially widely over topics. Biologists work at several spatial levels, with specialists working at cellular levels, at the levels of individual organisms, on the behavior of whole populations of a given organism, and on the interaction of various species with each other.

Of all the scales in human society and in the natural world, which are the most basic? Is it the case that the smallest units are the most basic or essential, and that the larger units are built up from them? Are cells the most basic units in biology and atoms the most basic units in chemistry? Is the individual or the family or the nation the basic unit for historical analysis? In a sense this hierarchy of essentiality, going from the smallest unit to the largest, is inescapable. We are certain that multicellular organisms developed out of single cells, and we are certain that civilizations developed out of families, not the reverse.

Yet once we begin to analyze the reproduction and transformation of a world in which there are many levels of existence, no single level can be identified as the essential unit from which all other units and their experience are derived.[1] There

---

1. Peter Novick notes that no less of an historian than J. Franklin Jameson, writing in 1910 while he was editor of the *American Historical Review*, conceived himself to be creating "bricks," monographs that would be the elements for constructing an edifice of historical interpretation (Novick 1988: 56). My view is that this reverential and essentialistic vision of historical monographs still pervades the historical profession, and is reflected in comments that such-and-such is the "definitive" study of a given topic. This essentialism makes it difficult for historians to address either the localized fine structure within a study of monographic scope or the broader context that structures and is structured by the events and processes surrounding monographic analyses. In chapter 17, I develop the terminology of "context effects" and "constituent effects" to address influences beyond and within the principal level of analysis.

are two related reasons why this is so. First, at any given scale (say, the level of molecules in chemistry or the level of an ethnic group in history), the experience at that level is influenced by phenomena at both smaller and broader levels: that is, ethnic groups are changed by families within them and by the empires that command them. Second, at any given scale there exist patterns and phenomena that are unique to that level—patterns that cannot be derived from predictions at another level. The behavior of a horse cannot be predicted from what is known of the cells and organs that make it up; the behavior of a herd of horses cannot be predicted from what is known of individual horses. The global patterns of ecology or society cannot be predicted from what is known of society or ecology within any given nation. Though the global patterns interact with the local patterns, the global patterns must be studied directly if they are to be understood.

For humans who wish to understand the world they inhabit, it therefore makes sense to conduct analysis at many different scales in space, time, and topics. There are reasons to study the Earth as a whole, and to study it for the whole of human history or for the last two thousand years, but such a scope of world history does not have any inherent or essential advantage over other scales of historical analysis.

There may be compelling reasons to focus substantial study at one level or another. The focus of historians on the nation during the nineteenth and twentieth centuries certainly fit with important realities and social needs of the time. Similarly, the need to find a way to make decisions at a planetary level on ecological, economic, demographic, and social issues provides a substantial impetus for historical studies at that same planetary level to provide context for the decisions. Yet in neither case is the cause of historical knowledge served by an automatic decision to focus historical analysis within one or another of these spatial frames. For instance, a study may focus on the interaction among different levels of scale, such as the interaction of volcanic episodes and long-term climate change, or the interaction of local markets and global trade.

Each study in world history must work within some defined limits of space and time. Defining the limits, range, and recommendations for the scope of world history studies is the topic I address within this chapter. There exist solid arguments for many different versions of the scope of world history. The cosmos is a large place, in which interactions retain great importance even though they must span parsecs of empty space. At the same time, no matter how immense and how ancient the reality of the world itself, individual human beings of limited life span apprehend "the world" as a concept. We necessarily cut the world down to size in order to fit it into our minds. Understanding world history must account for these and other poles of reality and perception.

Scales and Patterns of Time

The specialty of historians is the analysis of change over time. Typically, however, historians carry out the temporal dimension of their analysis in an artisanal manner, making their choices without attempting to develop explicit or theoretical statements of how they have handled time. In practice, of course, historians commonly divide their subject matter into periods, and seek to emphasize continuities within each period. Modern history, ancient history, prehistory, Renaissance-Reformation—these are ways of defining historical periods and even the identity of the historians who study them. But to stop here would be to suggest that historians avoid analysis of changes—either from one period to another or within a period.

Instead, historians emphasize change by presenting time in various formats. Periods can be selected to emphasize times of change as well as continuity—hence the "time of troubles." In addition to periods there are trends, as with the expansion of industrial production or the decline in infant mortality. There are cycles, as in weather patterns, and in the politicoeconomic cycles sketched by Andre Gunder Frank.[2] Time can be represented through a continuous narrative, or it can be represented discontinuously, with cross-sectional descriptions at separated points of time, or chaotically, with assertions that there have occurred vast and unpredictable changes within short periods of time.

Fernand Braudel, one of few historians to offer an explicit typology of time, left a terminology distinguishing the *longue durée* at one extreme from *histoire événementielle* at the opposite extreme.[3] Historians use the terms to acknowledge that both extremes exist, but they have developed no consensus on how to balance them. At the level of *histoire événementielle* may be found the chronologies of world history, listing maximal numbers of events with minimal efforts to make sense of them. Yet also at the short-term end of historical time frames is John Wills's *1688*, a book-length study that displays and links the events of a single year around the world.[4] For a time frame of literally cosmic extent, the "Big History" analyses of David Christian and Fred Spier consider processes over billions of years; yet precisely because the authors wish to give some attention to recent human developments, they must mix this cosmic scale with units of time as short as centuries.[5] For the cosmic time-frame, historians have relied on the thinking and the popularized writings of astrophysicists, notably Stephen Hawking.[6] This work, while at a far greater scale than history and involving a much smaller number of variables, shares some of the same conceptual problems as world history: alternative frameworks for space and time, whether to think of time as an "arrow" or a more complex variable, and whether it is possible to develop a unified system for analyzing the world.

A thematically focused example of both extent and flexibility in temporal scope is that of Johan Goudsblom's *Fire and Civilization*. It begins with the "first

2. Those who have seen history as a trend of steady progress have been called "Whig" historians by Herbert Butterfield. Butterfield, *The Whig Interpretation of History* (New York, 1951); Frank 1998.
3. Between the two extremes, Braudel fit the history of the *conjoncture*, or the episodic changes in climate and economic life. Braudel [1949].
4. The work focuses quite systematically on 1688, addressing details of that year in the lives of people in many parts of world. It also provides some background on earlier years and the fates of the protagonists in later years. Wills 2001.
5. Christian 1991; Spier 1996.
6. Stephen W. Hawking, *A Brief History of Time: From the Big Bang to Black Holes* (New York, 1988).

domestication of fire," apparently by communities of *Homo erectus* some 400,000 years ago, and argues that this change was part of a civilizing process that continued thereafter for all humanity. He proceeds thereafter, and using shorter time frames, to explore the use of fire in the development of agriculture, in the ancient Mediterranean, and in industrialization.[7] The long-term interpretation of Jared Diamond, beginning with domestication of cereals some 10,000 years ago, is similar in addressing progressively shortened segments of time as he approaches the present. The two differ, however, in that Diamond's analysis centers on the continuing repercussions of the initial achievements of domesticating plants, while Goudsblom moved his focus to successive developments in a long-term civilizing process.[8]

What is the value of different times? The generally greater space allocated to recent times in history suggests that we value our own era more highly than other times. Despite this appearance, many historians offer strictures against "presentism" (projecting the concerns of the present into the study of the past). In contrast, some historians explicitly support presentism as a value in historical writing.[9]

Thus, while readers of textbooks may conclude that world history seems irrevocably divided into two great periods at the year 1500 or thereabouts, it is easily seen that historians generally and world historians in particular utilize time frames of widely varying shapes and extents to tell their stories. Time can be molded into many different forms. I have found it useful to demonstrate the range of available time frames through exploration of historical narratives. Harvey Green's *The Light of the Home* has served this purpose well: it is an extended commentary on a museum exhibit of material culture in the homes of nineteenth-century upstate New York. This modest but informative volume reveals the range of narratives (each with its associated time frame) that an author can provide in a single work. In the course of this study, Green's discussion of the material culture of nineteenth-century New York homes presents a narrative of stages in the manufacture and use of household implements; a seasonal cycle of household work; a daily cycle of tasks and activities; a life-cycle narrative of individual maturation and aging; a cycle of fashion, as styles once prized become neglected and then rediscovered; an evolutionary narrative of the transformation of industrial society; and a storyteller's narrative of the passage of women into acceptance of limitations on their mobility and then to experimentation with the bicycle.[10]

In yet another formulation, historians often present the past through points in time or episodes. In one sense history is made up of an infinity of such moments, each of significance to those who experience it. In another sense, historians have selected certain moments as defining an era or perhaps transforming history ever after. Jules Michelet wrote in this fashion as the historian recounting the 1789

---

7. Goudsblom 1992a.
8. J. Diamond 1997.
9. Mazlish 1993b.
10. Harvey Green, *The Light of the Home: An Intimate View of the Lives of Women in Victorian America* (New York, 1983).

taking of the Bastille in Paris, and John Reed did much the same as the journalist narrating the 1917 Bolshevik seizure of the Winter Palace in Petrograd.[11]

Scales and Patterns of Space

For world history, the most obvious spatial unit of study might seem to be the planet. But of course data are neither given nor collected at the planetary level. Data come generally from local levels, though for much of the twentieth century data have commonly been aggregated at the national level. To a lesser degree and in more recent times, data have come to be estimated at the continental level. World historians and other writers on global affairs have capitalized on the latter trend, and they explain both past and present in terms of the interactions of Africa, Europe, and Australia. Sometimes the reification of continents goes so far that the author attributes a consciousness and an agency to Asia or to Latin America.

Martin Lewis and Karen Wigen, in an insightful and accessible presentation, have shown the impermanence of the continents in the minds of mankind: they describe how the terms and the regions addressed have shifted and sometimes expanded with time.[12] While identifying the socially constructed nature of geographic space, they seek to develop conventional regions that minimize the inconsistencies within regions. As with the problem of defining periods of time, defining conventional regions raises the question of whether one wants to maximize the consistency within each unit, thus threatening to leave out of account the major changes and inconsistencies, or to define regions that focus on interactions. One move toward the latter approach comes from defining world areas not by continental land masses but by ocean basins.

Another view of the world centers on localities. The great cities of the world have been seen as lenses through which the rest of the world is presented. Babylon, Rome, Baghdad, Constantinople, Paris, London, Tokyo, and many more contenders have gathered the crystallization of human experience within their walls. Not only great cities but cosmopolitan rural areas can serve as lenses through which to view the world, as Donald Wright has recently shown in his analysis of six centuries in the history of the Niumi region of Gambia.[13]

Analysis of continents tends to emphasize common and essential characteristics of large regions, while analysis of localities emphasizes a tiny prism focusing a wide range of human experience. Analyzing of large bodies of water and their surroundings gives emphasis to interactions among distinct localities and regions. The paradigmatic study of this sort is Fernand Braudel's magnum opus on the Mediterranean.[14] This work can be viewed at once as a tour de force in methodology

11. Michelet [1847–1853]; John Reed, *Ten Days that Shook the World* ([1919] New York, 1967). Among the time frames evoked by these studies are periods, flashbacks, cycles (both mechanical and evolutionary), and the contemporaneity of events at different chronological times.
12. M. Lewis and Wigen 1997.
13. D. Wright 1997.
14. Braudel 1995; K. Chaudhuri 1990.

and empirical presentation, and as a meandering and eclectic compilation. Fundamental to its organization, however, is the geographical basin of the Mediterranean and the many distinctions and connections of its waters and its littoral. A similar organization and focus on interactions underlies works that are in other ways quite different: David Chappell's social historical study of maritime workers in the Pacific, and Paul Gilroy's literary and cultural analysis of modernity in the Atlantic.[15]

Just as time cannot be limited to a one-dimensional measure of duration, the space that historians "cover" cannot be limited to a two-dimensional region marked out on a map. For instance, the spatial work of historians reaches three dimensions when they consider the effect of altitude on climate or the elevations of military positions in battle. On the other hand, sometimes space is treated as having less than two dimensions, as when the historian describes cities as if they were points (and thus with no dimensions), or the road between two cities (one dimension).

Scrutiny of the term *coverage* helps reveal the range of shapes of space addressed in historical studies. Historians of France do not generally write about all of France, though there is a French tradition of historical geography that gives the impression of covering all of France by its elaborate mapping of patterns for each of the more than ninety administrative *départements*. Political historians address the full space from which taxes are collected, yet focus their studies on the centers of government or of contestation. Sometimes the "coverage" of historians focuses on routes and pathways more than on places: the range of studies on pilgrimages privileges this approach. Another trick of "coverage" is revealed when one notes how often historians use a point on the map to symbolize a substantial territory: "the Kremlin" for the Soviet Union, "the Sublime Porte" for the Ottoman Empire, and "Wall Street" for the U.S. business community.

At other times the "places" are not earthen but ideological or cultural. What is "the West"? Most users of the term are geographically vague, and when they move toward precision they encounter scores of points of disagreement. Latin American and Caribbean countries are included in most definitions of the twentieth-century "West" but are almost never referred to specifically; Russia and Germany are included or excluded from the "West" according to their changing political regimes.

Another spatial example of the definition of interior scale in world-historical studies is that of the Wallersteinian world-system, in which the world as a whole is broken down into the center, the periphery, the semi-periphery, and the external areas. (The latter allows, in principle, for the existence of more than one world-system, though this is not accepted in Wallerstein's view of the modern world-system.) The time frame for the analysis as a whole is the period since 1500, but the interior time is broken into cycles of expansion and contraction and cycles of hegemony.[16] Equivalently, empires may be analyzed in their entirety, or in terms of the administrative subdivisions within them.

---

15. Chappell 1997; Gilroy 1993.
16. The thematic focus of the analysis is on political economy—the production, commerce, and control of wealth in the economic sphere. Political power, in addition, provides a central aspect of the analysis; in addition, it includes the social structure of classes and leading families. Wallerstein 1974.

A related approach to interior scale is the identification of frontiers within the regions analyzed. On one hand, such frontiers can take the form of sharply defined and administered borders, where crossings are scrutinized by authorities.[17] On the other hand, frontiers can take the form of ambiguously defined zones that are distant from centers of power or that lie between centers of power, where independent influences develop and seek to define their futures.[18] The terms *border* and *frontier*, however, tend to be defined and applied primarily in political terms; the cultural, social, and economic provinces appropriate to other sorts of analyses may not fit neatly with political boundaries.

Territories, however defined, may not be homogeneous. The tendency to draw maps in one color or another does not help in the representation of the many geographic regions containing populations of heterogeneous ethnicity, religion, language, or population. The identification of the spatial dimensions of any historical study, in sum, must account for the internal fine-structure as well as the external limits of the place under study.

Mixtures of Time, Space, and Themes

Spatial, temporal, and thematic frontiers of a study overlap and mix in interesting fashions, so that selecting a "unit of study" for historical analysis involves more than just identifying a place to study. If I choose to study events at two points in space occurring simultaneously, I must still account for the fact that people at each point experience the event at the other point as having occurred later, because they find out about it later. (The term "light years" describes interstellar distances, relying precisely on this effect.) In more practical terms, historians usually choose between emphasizing a chronological or regional organization of a study. They choose to focus on a single period for a wide area, or they focus on a single region for a long time. Do the interactions of history move more easily across space within limited time, or do they move more easily across time in a limited space? Those who prefer chronological analyses generally believe they can find more interaction in restricted time periods; those who prefer regional organization generally believe that the interactions are stronger within regions than within time periods.[19] Perhaps there is no general pattern: the realities of past processes may have developed more readily within time frames in some cases and within regional frames in others.

17. Asiwaju 1976.
18. The frontier literature is immense, but see, for instance, William H. McNeill, *The Great Frontier: Freedom and Hierarchy in Modern Times* (Princeton, 1983).
19. Spengler gave primacy to the region or civilization, though he argued that each civilization had a fixed lifespan associated with it. Toynbee too gave primacy to space, though he ranked civilizations according to whether they were original or derivative. Braudel's study of the Mediterranean also privileged space. McNeill, while conducting his analysis in terms of regionally based civilizations, nonetheless gave primacy in his analysis to the temporal evolution of the ecumene. Spengler [1918–1922]; Toynbee 1933–1961; Braudel [1949]; W. McNeill 1993.

This review of the scale of world historical study may seem indecisive to the reader, in that it accommodates to applying the label "world history" to studies with a temporal from as short as a single year or as long as all of time, and to studies with a geographical scope as constricted as a single city or as broad as the entire planet and beyond. Perhaps the way out of this uncertainty is to distinguish between the *proximate* frame of analysis and the *ultimate* frame of analysis. In world history, the ultimate frame of analysis is human society in general, and the proximate frame of analysis is the region, time frame, or theme one studies to gain insight into the world. An analogy to national history may help: national historians of France or Japan may investigate local communities or even international institutions with the objective of throwing light on the national experience. This formulation of proximate and ultimate frames of analysis, while helpful to a degree, also has its limits: the reader will already have noted that those who call themselves world historians are not necessarily trying to explain the world as a whole.

As elsewhere in this book, we are caught between two sorts of reasoning. On the one hand, I seek to develop a systematic picture of the distinctive and appropriate characteristics of global historical studies, especially the emphasis on connections, multiple perspectives, and large-scale patterns. On the other hand, I seek to emphasize the substantial overlap of world history with all the rest of historical studies, especially in the artisanal techniques and philosophical distinctions to be utilized in research and interpretation. Within the first perspective, world history is a distinctive field that needs to be supported and developed for the new insights it can bring to our understanding of the human past. From the second perspective, world history is not so much a distinct field as a critique of national history. The problem is not with studies of nations but that the national framework constrained twentieth-century historians to limit their research and writing, granting recognition to the families within nations and the diasporas overlapping nations only insofar as they contributed to understanding the national experience. In the latter sense, "world history" means simply the removal of the national constraint.

Conclusion

Both for analysts of history constructing an interpretation and for readers of history absorbing and assessing an interpretation, the scale of the study defines its most basic character. The scale of a historical study is set by the range of space and time to be enclosed and by the dynamics assumed for interaction within those categories. Through an extension of the same logic, the topical breadth of a historical study adds a third dimension to its scale. As with space and time, the topical scale of a historical interpretation includes not just the range of topics, but the relative emphasis on topics and the assumed dynamics of their interaction. All studies in world history are big, but each study is big in its own way.

Chapter 16

Modeling Frameworks and Strategies

H istorians, because of their desire to retain contact with the complexity
and nuance of the past they re-create, commonly decline to write in neatly
formalized fashion. The historical narrative is not generally a presentation in standard, orderly steps. The artfulness of historical writing may disguise the logic with
which the author constructs the text, yet the results of the author's logic must
still affect the priorities and the conclusions of the interpretation. The past generation's debates on modeling in history have shown clearly that historians use
a range of logical structures in developing and presenting their interpretations.[1]

In principle, the historian begins with a research agenda. Most generally, these
are the questions one asks out of a desire to understand the world of today and
its antecedents—and to understand the world in broad and interconnected
terms. The questions stem from intellectual curiosity and from social or natural
crisis. Will global warming continue? Will inequality continue to deepen? Is
war inevitable? Is progress dependable? These questions launch and redirect
campaigns of reading, classroom inquiry, and research.

I encourage world historians to develop a formal statement of their research
design in any sizable study, and I propose the contents of such a statement in the
course of this chapter. In building up to this point, the chapter distinguishes
among the various elements of framework and strategy in analyzing history. The
historian's choice of how to combine these elements is contingent on the topic, the
discipline, and the question under study. No single element of these frameworks
or strategies is brand new, nor is it even new to combine them. But the systematic
effort to compare and contrast, to link, and to explore historical systems, to
develop coherent strategies for analysis and for presenting the results—these,
I argue, yield a substantial advance in the methods of world history.

1. P. Burke 1992. The classical manuals of historical method are: Marc Bloch, *The
Historian's Craft*, trans. Peter Putnam (New York, 1953); Collingwood [1946]; Carr
1961; Hexter 1971.

The term *analysis* can be used in so many different ways to discuss history that I will use it only at a general level. *Analysis* can refer to everything from scrutinizing an individual document to organizing an overall interpretation, so I will use the noun form to apply to the entire project of historical analysis, and will use the adjective *analytical* at any level to distinguish analytical from descriptive thinking.

In the terms I will use in this chapter, *frameworks* are ways to set up historical problems, *strategies* are ways to solve the problems, and *interpretations* are the solutions. A historical framework is a lens through which to scrutinize the past and to collect basic data on it. An analytical strategy is a plan for interpreting the past, turning the data into statements about patterns, dynamics, and meaning in the past. An interpretation is the result of successful implementation of the strategy.

Frameworks for Global Historical Analysis

The framework of historical analysis, as I define it, includes the *objects* of study and the *procedures* for study. The objects of study are usually known as cases.[2] A case is generally understood to be a coherent and localized unit such as an individual or a polity, though I shall also argue that a network provides an interesting and relevant sort of case. In selecting an object of study, one defines its boundaries.[3] Thus, national history focuses on nations as cases; comparative history may also focus on nations as cases; and world historians have options on whether to view the world as an aggregation of national cases, as a single planetary system, or otherwise.

The procedures for study in history include, most basically, descriptions of the elements and the relationships within the object of study. Going beyond description, the basic analytical procedures are comparison (looking for similarity and difference), linkage (looking for connections), and chronology (ordering elements in time).

Cases, networks, comparisons, and linkages thus provide the basic grammar of our review of historical frameworks and the object and procedures for study within them.[4] At a more general and complex level, the objects of study are *systems*, and the procedures of comparison, linkage, and chronology can be used to catalogue any system and its constituent subsystems. The systems framework, along with its application to historical studies, is explored in detail in this chapter.

In the concluding section of this chapter, I discuss strategy as an envelope for a range of analytical terms—paradigm, model, metaphor, theory—that point toward proposed answers to questions about the past. In general, the historian's strategy consists of an analytical objective and a plan for how to get to it. To use the

---

2. Also known as units of analysis.
3. The previous chapter has addressed ranges in temporal and geographic scale; here we consider the framework set up for description and analysis within a given scope.
4. These are the inner boundaries, segments, and structures of historical analysis, set by the historian. They are related to the boundaries and structures assumed to exist in the historical situations. The outer boundaries have been addressed in chapter 15.

metaphor of a road map leading to the answer to the historical question, a paradigm can be seen as a key to the map but not the map itself. A model or relationship among variables can be seen as the relationship among roads on the map. A metaphor would represent the road map in symbolic terms. A theory includes both the paradigm and the model.

Interpretations and narratives are the actual solutions developed by historians to the problems they have set. Debates, as we will see, may focus on the framework, the strategy, the interpretation, or a mixture of these three aspects of analysis.

In seeking to focus on the framework and strategy of historical analysis, I emphasize the lens through which we view the past. But as will be apparent as soon as we get into examples, one can never maintain the distinction between the lens and the situation that is observed through it. In treating a set of port towns as historical cases, the historian will almost always find it difficult to persist with thinking about the cases as analytical abstractions for exploring the past (for that is how they were chosen) and will begin to think of them as real and complex social situations. Of course, the latter represents the history we seek to reconstruct, but it is best not to forget that the formerly living past is quite distinct from the documents and categories through which we analyze it.

Units of study include cases and networks. The case study is a time-honored form of historical analysis and presentation. From the biography and the family history to the study of a locality, monarchy, or nation, historians have developed the skills of researching and telling an individual story in relative isolation from other stories. These stories are expected to be relevant to the broader study of history, and to draw upon the lessons of other studies, but the connections are neither explicit nor formal. Edward Gibbon's *Decline and Fall of the Roman Empire* was clearly understood by its readers to be offering a parallel between the situations of ancient Rome and eighteenth-century Britain, but that parallel operated between the lines of text.[5]

The inherited idea of the historical case study is that each history is worthy in its own right and that the research and telling of each history is mostly a matter of locating the documentary evidence and assembling it into an orderly and logical narrative. This vision of historical writing, which can be traced far back in time in the European tradition as doubtless in other traditions, received a particular sort of reinforcement in the era of national history. Professional history, under the inspiration of Leopold von Ranke, focused primarily on location and analysis of textual documents (especially diplomatic, in Ranke's case), and on reconstruction from them of the past, to present the reader with history "wie es eigentlich gewesen." That Ranke himself went on to launch a study of world history, of which he completed eight volumes before his death, suggests that he had a vision of history that went beyond the case study.[6]

But the term *case study* can have more than a single meaning. In the social sciences, particularly sociology, the case study has a role somewhat analogous to that of the laboratory test in the experimental sciences. That is, a positivistic

5. Gibbon 1776–1778.
6. Ranke 1883–1887.

theory may be tested with a group of case studies that may show, in the aggregate, whether the theorized relationship can be observed in history. Thus a case study might be an isolated inquiry, or it might be seen as one of many possible observations that might add up to a larger pattern. As a twist on the latter sort of reasoning, Immanuel Wallerstein's analysis of the modern world-system is presented as the study of a single and unique case, as it is the first world-system and encompasses the whole world.[7]

In the literature on Atlantic slavery, Peter Wood's *Black Majority* provided a study of a single region, South Carolina, establishing its uniqueness as a rice-growing colony, yet also emphasizing its connections to Barbados and to Sierra Leone. Eugene Genovese's *Roll, Jordan, Roll* made a single case out of the whole antebellum U.S. South, combining wide-ranging data into a deeply nuanced but nonetheless singular and cross-sectional view of the character of U.S. slavery.[8]

Still, for all its importance in historical studies generally, the approach of the case study cannot be of much importance for world history. Or, to be more cautious, case studies cannot be of importance in world history until we have a more fully tested conception of the overall structure of world history and how cases might appropriately be defined within it. Otherwise, case studies tend to reify the isolated vision of self-sufficient national and regional histories.

Cases in history need not be limited to individuals, regions, and nations. The network, defined as a transregional grouping of individuals, families, or other organizations, is of particular interest for the study of world history. The network is a particular sort of case. Hugh Thomas's *Atlantic Slave Trade*, in many ways a very old-fashioned, biographical study of the topic, can be said to focus on the commonalities and connections among slave merchants around the Atlantic. In work on other topics, Adam McKeown's work on the Chinese diaspora relies heavily on the notion of the network.[9]

The notion of the network has considerable metaphoric power, and this will be explored in a later section on strategies of analysis. To anticipate that discussion, networks can be thought of as connective social tissue holding the world together. Networks can be seen as a space, though not a locality. Migratory networks make up a space of global circulations. Viewed in this way, migrants are not just moving among settled people; they form a unit unto themselves. Networks can be seen as having a structure that, in mechanical terms, consists of strings and knots. Languages can define networks of communication, especially as they link to

7. Wallerstein 1974.
8. Peter H. Wood, *Black Majority: Negroes in Colonial South Carolina from 1670 through the Stono Rebellion* (New York, 1974); Eugene D. Genovese, *Roll, Jordan, Roll: The World the Slaves Made* (New York, 1976).
9. Hugh Thomas, *The Slave Trade: The Story of the Atlantic Slave Trade: 1440–1870* (New York, 1997); McKeown 2001; Roger Anstey *The Atlantic Slave Trade and British Abolition, 1760–1810* (Atlantic Highlands, N.J., 1975). William McNeill, in the preface to the Frank and Gills volume on the world-system, argues for an emphasis on networks as a basis for analysis. W. McNeill, in Frank and Gills 1993, xii.

migrants. The early modern Portuguese and Dutch empires, which fit poorly into the analysis of empires as land-based polities, can perhaps be seen more successfully as military and commercial networks. The same may be said for the commercial networks of Armenians and Bugis.[10]

We turn now from units of study to procedures for study and, therefore, comparison. In a world of case studies, any sort of comparison is cosmopolitan and, for that reason, daunting. By the logic of case studies, and assuming one were comparing two national or local units, one now had to master two sets of archives, literatures, and perhaps languages. The Wisconsin program in comparative tropical history (later comparative world history), in which I studied during the 1960s, was explicitly comparative in this fashion. One was expected to master a home territory, then develop historical questions that could be explored by comparison with other regions. One learned to absorb the secondary literature on the more distant region and to dip into its primary sources as time allowed.[11]

Studies of slavery have been a particular center for comparative studies. Frank Tannenbaum's classic *Slave and Citizen* (1946) compared slavery in Protestant and Catholic countries. Carl Degler followed this up with a comparison of slavery in Brazil and the United States, and Herbert Klein began his work with a comparison of slavery in Cuba and Virginia. Even Curtin's *The Atlantic Slave Trade*, while it yielded global conclusions, was organized as a comparative analysis of the volume of slave exports for each of the European powers. Extending this approach from slavery to racism, George Frederickson and John Cell published comparative studies on the United States and South Africa. This sort of comparative approach did lead the authors to wide reading and to development of interpretations on a transnational scale, but it did not explore in detail the variations in approach included within the term *comparison*.[12] The novelty of transnational studies was such that even into the nineties, it was common to hear historians use the term *comparative* to refer to all studies beyond the national level.

But the term *comparison*, as is usual for important words, has multiple meanings. To compare is to bring two or more things together (physically or in contemplation) and to examine them systematically, identifying similarities and differences among them. *Comparison* has a different meaning within each framework of study. Any exploration of the similarities or differences of two or more units is a comparison. In the most limited sense, it consists of comparing two

---

10. Curtin theorized Armenians and others as a "trade diaspora." It would be of interest to compare in detail the models of trade diaspora and network. Curtin 1984b.

11. Philip D. Curtin, "World Historical Studies in a Crowded World," *Perspectives* 24 (January 1986), 19–21; Adas 1979.

12. Frank Tannenbaum, *Slave and Citizen: The Negro in the Americas* (New York, 1946); Carl Degler, *Neither Black nor White: Slavery and Race Relations in Brazil and the United States*(New York, 1971); Herbert S. Klein, *Slavery in the Americas: A Comparative Study of Virginia and Cuba*(Chicago, 1967); Curtin 1969; George Frederickson, *White Supremacy: A Comparative Study in American and South African History* (New York, 1981); John Cell, *The Highest Stage of White Supremacy: The Origins of Segregation in South Africa and the American South* (New York, 1982).

units isolated from each other. These two separated units might be undergoing some sort of influence causing them to change: studies of the responsiveness of colonized societies to the impact of metropolitan rule fall within this framework, as do the studies on comparative frontiers in history. In such comparative studies, one of the main choices is whether to give greater emphasis to the similarities or the differences among the cases.

These units under comparison might be seen as parts of a larger system. The studies of individual colonized societies might be seen as contributing to the understanding of a larger imperial system, including colonies and metropole. Or the individual civilizations or societies examined by Spengler and Toynbee in their grand reviews of history may be seen as elements of the larger evolution of human history. For all of these uses of the term *comparison*, however, the general assumption is that the objects of comparison do not influence each other. That is, the colonies, frontiers, and civilizations in the examples each other might influence and be influenced by historical processes overall, but one colony or frontier is not assumed to have influenced others.[13]

It is a third and more specialized meaning of *comparison* that refers to units that are in contact with one another and that influence one another. The political systems of France, England, and Germany, while distinctive as any static comparison of them would show, have nonetheless influenced each other greatly through direct intervention, emulation, or alienation. The comparison of Japan, Korea, and China is not complete until it includes, in addition to their similarities and differences, the list of ways in which each has influenced the other.

If we think of the work of economic and demographic historians, in addition to the work of social and civilizational historians, another distinction of approaches within comparative history becomes possible. Historical studies that collect multiple observations on similar cases, as would be the case for trade statistics in economic history, provide a basis for hypothesis-testing—which may in this sense be seen as a technique for developing conclusions in comparative history. These might be labeled as "micro-comparisons," in contrast to the "macro-comparisons" of large and complex units, which address many more variables and whose conclusions are developed by informal inspection rather than through a formal procedure.

If comparison contributes to the framework for studying world history by addressing units assumed to be relatively autonomous, then *linkage* contributes to that framework by assuming that units are connected and interdependent. In attempt to maintain the distinction between the historian's present-day analysis and the past patterns of the historical situation, I shall use *linkage* to refer to the historian's effort to locate past interdependence, and *connection* and *interaction* to refer to past ties. *Interaction* is the term usually used in scientific discourse, but Alfred Crosby steered away from it to the vaguer and more general term *connection*,

---

13. Spengler [1918–1922]; Toynbee 1933–1961; Walter Prescott Webb, *The Great Frontier* (Boston, 1952); Howard Lamar and Leonard Thompson, eds., *The Frontier in History: North America and Southern Africa Compared* (New Haven, 1981).

and others have followed him. The wisdom in the choice may be that *interaction* refers to connections linked by positivistic and deterministic theory, while *connections* refers to links that may not be reducible to cause and effect. Jerry Bentley, on the other hand, has used the term *cross-cultural interaction* to refer to major connections in world history without causing confusion.[14]

The terms *connection* and *interaction* both have the advantage of being more general than that other term commonly used by historians, *diffusion*. There are many different models or types of connection in history, of which *diffusion* is only one, and a very basic one at that. Diffusion is a one-way interaction in which an item leaves place A and spreads to place B and many other places, implanting itself in the new places without changing. (The only simpler model one could imagine is that the item leaves place A and moves only to place B—this model might be called *transference*.) For instance, a food crop might move from one region to others without any hybridization.

It should be noted that the connections in history are not only those across space, but also those over time and across themes. To give an example of the latter, it is now a commonplace that changes in technology bring about changes in the look of clothes and the sound of music, and therefore in cultural practice; some changes in the reverse direction may also be imagined.

For all historical connections, whether across space, time, or theme, it is not enough to assert a connection: one must specify or model how that connection works. A few other models for types of connection may be mentioned, to give an idea of their range.[15] As an item or an idea moves from place A to place B, it might change while en route. If the potato itself remained unchanged in moving from Peru to Europe and East Asia, the ways of cooking it might have changed with its travel. I prefer to call this *revision*.

The term *syncretism* is widely used, especially with regard to religious belief. In this case, a group of believers retains ideas from its own past but also adopts religious beliefs from one or more additional traditions. The usual implication of *syncretism* is that the two or more sets of preexisting beliefs, while combined, are not really integrated, and the new beliefs may be linked with their domestic or imported antecedents. A different outcome of such connection can be labeled with a musical term, *fusion*. If the preexisting and new ideas (or two sets of new ideas) bring sufficient inspiration, they might fuse into a new cultural creation

---

14. In his 1993 book, Bentley referred to "cross-cultural contacts"; in his 1996 article, he spoke of "cross-cultural interactions." Bentley 1993; Bentley 1996.
15. In formal terms, interactions and their results are of several sorts. The simplest sort of interaction is a one-way displacement of power or an innovation from one society to another (followed by a one-way diffusion of influence from one place to several other places). Influences between societies may also be a two-way affair. In more complex two-way interactions, an influence sent from one society to another may be transformed as it reaches another society. In still more complex interactions, the influences from two societies may produce new forces or new phenomena; these phenomena may be observed at the level of each society but actually have their existence at a level including both societies.

that, though its links to its antecedents might remain clear, nonetheless has a logic and consistency of its own.[16]

Such modeling of types of connection could go on at length, and productively so. For instance, the revision of a tradition might take place over time in a given place even without outside influence, or the old tradition might persist; and persistence of a tradition, once it has moved a distance, is often called survival. The point is not to make the longest possible list of models, but rather to affirm that the study of connections in history opens up a discussion of the various dynamics of connection and requires attention.

My overall point is that the analyst should be clear on the range of tactics and models available for making comparisons and linkages in historical evidence, should choose appropriately in order to prepare a coherent and consistent interpretation, and should provide readers with a clear statement of the research design for each study. World historians, if they are to look at more than just a tiny bit of world history, must be prepared to address a wide range of analytical standpoints and interpretive techniques. This breadth should make world historians cosmopolitan, but it risks leading them to indecisiveness or to arbitrariness in choosing their own interpretive approach. The indecisiveness is certainly apparent. World historians can see all the different topics and methods. When they start to pursue a given interest and a given methodology, they can always see other approaches and dimensions. So they tend not to polish their approach within one framework but to alternate among frameworks. At the strategic level, authors should identify the framework of each work as a case study, comparison, network, or system. In practice, authors do pretty well at this. At the tactical level, authors should distinguish among the varying structures of comparison and models of connection they utilize. And they should be explicit in stating the procedure used for confirming the main line of argument of the work.

The complete structure of a research design in history is rather complex, and for world historians it is far-flung in addition. But even for world historians, to produce studies of lasting value, it is important to maintain contact with bedrock—with a clear handling of the comparisons and linkages in our historical analyses, and with a clear indication of how we are verifying the interpretive statements we offer. Attention to detail at this level shows the various types of comparison, linkage, and even case studies. It shows as well that comparison is a part of any study of connections and that case studies are part of any comparison.

16. See the reference section on "Concepts" in Manning et al. 2000.
    The term *hybridity* has come to be widely used to refer to cultural mixtures, for instance in Gilroy 1993. My impression, however, is that *hybridity* is generally employed as a one-size-fits-all term, lumping together the numerous types of connection and thereby vitiating the very precision in analyzing cultural mixtures that authors claim to seek.

Strategies: Research Design

Having reviewed in earlier sections the units of study and procedures for study, I turn now to the strategy of analysis: the historian's analytical objective and plan for getting to it. Analytical strategies differ widely among individual historians and among fields of history, so the terms used to describe the strategic toolkits of historians include *typology, paradigm, mechanism, theory, metaphor*, and more. *Paradigm* and *theory* are usually treated as broad and general terms, while *mechanism* is usually applied in rather specific terms. *Typology* is utilized at levels from the broadest to the most specific. (Note that sometimes analysis goes no further than description and typology. In that case the historical dynamic is only hinted at, rather than explored explicitly. In fields such as literature, in which the number of interacting factors is so vast, analysts sometimes feel they do better by identifying major distinctions and leaving it at that.)

Comparison is often treated (though rarely explicitly labeled) as a strategy for historical analysis. Quite aside from the many acts of comparison conducted in the procedures for any historical study, the comparison of historical cases is adopted in some cases as the main purpose of the study. This is "comparative history" as emphasized by Philip Curtin and Michael Adas, and it is the comparative approach in the sociology of Barrington Moore and Theda Skocpol.[17] In the discussion that follows, I mean to include this sort of research design along with those that focus more on interactions or on global patterns.

Typology. A typology is a classification of phenomena. It is any organized terminology. Typologies include the division of the world into civilized societies and barbarians, or into social classes of peasants, pastoralists, wage workers, shopkeepers, and elite classes of landowners, entrepreneurs, priests, and officials. The division of historical time into ancient, medieval, and modern is a typology, as is the division of world-systems into center, semi-periphery, periphery, and external areas. We use typologies to describe categories of age, gender, family, and areas of the arts.

Making the categorical distinctions in a typology can be a very important analytical step and can at times unravel incorrect interpretations. The typology may be part of a larger and more complete system of analysis, or it may imply further relationships among the categories that are not made explicit. But the typology itself is no more than a set of categories, and it does not generally include logical interrelations or patterns of change affecting the categories.

Paradigm. A paradigm makes an explicit statement of the framework for a study but leaves implicit the dynamic of the events and processes studied within that

---

17. For a critique of the comparative method as a strategy of social-science analysis, see Khaldoun Samman, "The Limits of the Classical Comparative Method," *Review* 24 (2001), 533–73.

framework. The world-system paradigm, for instance, assumes that a world-system can exist once there has been created a world-economy exceeding the limits of any single empire, and the paradigm urges one to explore the interrelations among center, semi-periphery, and periphery in the transformation of this system.

The *network* can be treated not only as a type of case, but as an analytical paradigm. One can define a network as a space (though not a locality) that has nodes but probably not a center. One's participation in a network is determined by exchange—of funds, goods, or information—so that ceasing to exchange means leaving the network. The theory of networks should focus on their principal dynamics: their creation and expiration, flows within networks, control and facilitation of flows, and the function of networks in wider society. Networks can be seen as systems that are open or closed.

In economic history, microeconomic theory and the neoclassical paradigm provided Robert William Fogel and Albert Fishlow with competing versions of a model with which to study the place of railroads in U.S. economic growth.[18] The model was that railroads increased U.S. capital investment and reduced transportation costs, but only to the degree that transport costs by rail were lower than those of an alternative system based on canals. The result of this analysis led to a revised outlook by economic historians and a decline in the tendency to celebrate railroads as the principal source of economic growth. Our concern here is to focus on the model, theory, and paradigm rather than leap to discussion of the results of the analysis.

Mechanism. A mechanism is an explicit statement of the dynamic of a process, without an explicit statement of the framework within which the process is studied.[19] In commerce, the idea of buyers and sellers haggling over prices is a mechanism: if the price is too low, a surplus of goods will result, but if the price rises, supply will increase and demand decrease until a point where the market is cleared. Demographic analysis assumes a number of mechanisms: one is that when famine or disease leads to a peak in mortality, the aftermath brings an increase in births. The reactions of one generation against the tastes of the preceding provides one mechanism for describing the changing styles of dress. The notion of cultural diffusion is a simple mechanism: in it, an innovation is created in one place and simply spreads to other places. More complex mechanisms are often used in study of music, where combinations of multiple musical traditions lead to a mix of new traditions and occasionally to fusion, the creation of a new

---

18. Fogel 1964; Albert Fishlow, *American Railroads and the Transformation of the Antebellum Economy* (Cambridge, Mass., 1965). Thomas Kuhn popularized the term *paradigm* in Kuhn 1962; I am using the term in a slightly narrower fashion.

19. The *mechanism* need not be mechanical. Oswald Spengler's organic metaphor for the rise and fall of civilizations is a sort of mechanism, defining the dynamics of historical change. The term *model* is perhaps more often used, but I have sought to avoid the confusion that would be brought by using the same term in two ways, since I am using the term *modeling* to refer to the overall process of formalizing a historical conceptualization. *Modeling*, as I use the term, includes typologies, paradigms, mechanisms, theories, and metaphors.

tradition. A mechanism gives analytical attention to the dynamics of change and interaction, but abstracts from the boundary conditions: the statement of the mechanism itself does not indicate the framework that might restrain, reinforce, or transform the dynamic it emphasizes.

Theory. A theory is the most fully developed of analytical statements: in it both the framework and dynamic of the processes within the framework are modeled explicitly. The theory, in the case of microeconomic theory, is a well-developed set of assumptions, variables, and relationships among them, which makes it possible to predict the value of major variables once sufficient data are known. Contending theories can differ in any of these areas: neoclassical economic theory and Marxian economic theory, for instance, differ not only in the variables they identify but in which factors are variables and which are parameters set for the duration of the analysis. That is, social composition and activism are variables in the Marxian analysis, but they are taken as given in neoclassical analysis.

The meaning of the term *theory* in literary theory is quite different. In the earliest days of deconstructionism, in literary theory, the emphasis on identifying the perspective of the author was central. This challenge to the universality of the author's voice was generalized to give attention to the perspective (and by extension the agency) of individual participants in any historical situation. This was not theory with discrete variables or quantitative relationships. In fact, the theory consisted principally of typology: identifying various standpoints from which an author might be writing, then developing techniques for revealing the author's standpoint from his or her writings. Thus "standpoint theory" is a set of categories, and "microeconomic theory" is a set of relationships—so that (despite my effort to develop a consistent typology for the toolkits of historians' strategies) the meaning of the term *theory* remains situational.

Metaphor. Metaphor provides another basis around which to build an analytical strategy—one allowing for more imagination and avoiding the explicit, dynamic assumptions of a model or theory. Eric Wolf, in a well-known interpretation of world history, incisively picked out the limitation of history based on national units by recasting his argument in terms of a skillfully chosen metaphor. He noted that typically, in the tradition of national history, one interprets the history of Sweden or Japan in terms of internal influences, and one interprets international affairs in terms of national character or of the decisions of national political leaders. As Wolf argued, "By endowing nations, societies, or cultures with the qualities of internally homogeneous and externally distinctive and bounded objects, we create a model of the world as a global pool hall in which the entities spin off each other like so many hard and round billiard balls. Thus it becomes easy to sort the world into differently colored balls, to declare that 'East is East, and West is West, and never the twain shall meet.'"[20]

Wolf's metaphor, the global pool hall, is a potent critique of national history: he attributes this model to those who write the history of nations, then reveals the limits on their thinking. But he should have explored the metaphor more thoroughly to reveal more of its properties. If he had, he would have discovered implications

20. Wolf 1982: 6–7.

that magnified the potency of his critique. First, the billiard table is flat; second, it is bounded. That is, the billiard table metaphor reminds us that the approach of national history utilizes flat maps and tends to assume a flat earth, and that it emphasizes the borders around each country or, for instance, the limits of the European continent. Within this bounded region, certain areas appear to be central. Thus it is that world history, as written by scholars who have not escaped the national paradigm, ends up focusing simply on the powerful nations of Europe, though with the United States, Japan, and the Soviet Union eventually gaining some attention.

Let us think, in contrast, of the surface of a spinning globe—the metaphor with which H. G. Wells began his *Outline of History*.[21] Wells, too, left his metaphor undeveloped, but with a little effort we can add to its details and implications: it has no edge, no firm boundary; no way to leave it, and no way to exclude people from it. Events and processes—even if they were billiard balls—spread across it and encounter each other in a different pattern. Of course the surface of our globe is not uniform. It has varying environments, and most of its surface is water; it has concentrations of population and resources, but these areas of concentration can never be isolated from each other. This world has no center, nor does it have the flat and linear surface of a national map. Thus, to keep in mind the image of the globe is to take a step toward thinking of the world community as an organization of society that is global, interactive, and multipolar.

The benefit of using metaphors to aid in our thinking is that they convey a simplified logic. The problem with metaphors is that they simplify too much: they tend to leave out, as historical factors, physical dimensions, time, interactions, and the workings of the human mind. Our problem, as historical analysts, is to find metaphors simple enough to keep in our mind and that accurately reflect global historical interaction.

The analyst's game in conceptualizing global society is therefore to pick the best metaphor while keeping others handy, and to use each where it is most helpful in explaining the forces at work. Let me describe this game through a metaphor of metaphors. Consider the observation and interpretation of a complex organism—say, a human being. This human being is not only the subject, the particular case in question, but also the unit of analysis. That is, the physician (if well advised) seeks to diagnose and cure the person as a whole, though most attention may be focused on an ailing stomach or heart. A physician can view the human subject through a variety of lenses or other instruments: these lenses range in magnification from those permitting a view of the whole subject at once, to those focusing on a given organ, to those focusing on the cellular or even microcellular level. Different lenses are appropriate for viewing different levels of the subject's existence. Still, the judgment of the physician is itself a factor, for it is the physician-as-analyst who decides which is the best level of analysis for understanding any given issue, as well as what is the best lens for use in study. The physician's written report, in this case, corresponds to the history book.

21. "The earth on which we live is a spinning globe." H. Wells 1920: 3.

This metaphor, as all others, is imperfect. But in its imperfect way it distinguishes among six essential elements: the unit of analysis (in our case, it is the whole world community we seek to explain); the subject itself (that is, the reality of the world community and its past experience); the phenomena under study (the problems or historical issues within the subject that have drawn our attention); the instruments available for its study (the methods of historical study, the paradigms or frameworks of analysis, the theories and metaphors for human social processes); the analyst, who focuses on a problem and selects the appropriate analysis (identifying the themes and dynamics of world history); and the interpretation of the subject (writing descriptions and interpretations of the past).

The metaphor of the spinning globe and the metaphor of the global pool hall are mechanical images for the human community. Let me distinguish three orders of images out of which to draw metaphors to explain our human community: I will call them mechanical, organic and societal.[22]

In mechanical models the oppositions and interactions are presented as action and reaction, force and inertia. Mechanical images of the march of progress assume the straightforward working out of a process according to a set of rules. These images are used, not uncommonly, by historians of technology to describe a logical and relentless development of human civilization in recent centuries. Economic historians, in focusing on the expansion of markets, often reason with economic variables in what can be treated as a mechanical model of historical change. Yet another mechanical image is chemical: one can imagine the progressive crystallization of the world community, first precipitated in key regions by certain local circumstances, then propagated along lines of trade and contact to other areas, and finally filling in all the spaces. Even within mechanical models, therefore, we can find a range of images with which to interpret global history. For all mechanical models, however, interactions among units tend to be very simple: collisions and displacements, in the case of billiards.

Second, there are organic images of the birth, maturation, and death of human societies. In organic models, the oppositions are life and death, growth and contraction, parent and offspring. Historians of empires and civilizations often resort to organic metaphors in referring to the birth, maturity, and ultimate extinction of their historical subjects. As with mechanical images, there are several distinct types of organic metaphors. One of these is biographic (as Oswald Spengler calls it), focusing on the life cycle of an individual: the schemata of Spengler and Arnold Toynbee represent world history as the birth, maturation, and fall of distinct civilizations. Another organic metaphor for the global community is that of the family: in it, various societies are treated as more or less distantly related clans and lineages. Still another organic metaphor, utilized by Sigmund Freud, is that of species evolution, in which human society is seen to be gradually and unconsciously evolving and adapting to its conditions. Since organic models focus on the

---

22. In fact, there is a fourth order of images that I cannot fairly exclude from the discussion: the mystical. In this case the governing factor is assumed to be the working of the mind of God rather than the working of a mechanism, an organism, or a society. Kearney 1971.

integrity of each organism, they tend to include little discussion of interaction among organisms: one exception is the notion of infection by foreign organisms.[23]

For more recent times, a further sort of a biological metaphor can be proposed. That is, the West can be seen as a parent civilization for several new civilizations: the emergent, modern cultures of Africa, Latin America, South Asia, and East Asia can be seen as having a European father and a local mother and to have been conceived during the time of European colonization. To pursue the image, the polygamous European father, often absent, remains the authority figure, while the local mother provides nurturing for the offspring. By the same logic, the new cultures have lives of their own; they neither resemble their parents nor obey them, much less express gratitude to them. This image, however intriguing for some cases, is not helpful in explaining the evolution of European society. Nor does it fit the story of the creation of the world community. For the world community is not created by a mating of Europe with the generic other; further, the offspring in this case, the global society, envelops and absorbs the European parent.[24]

The third or societal type of image involves the human mind. That is, change takes place because of conscious and unconscious human reflection, and not simply because of automatic processes, be they mechanical or organic. In the case of societal models, it is easier to think of interaction in terms of the exchange of ideas, and in terms of the formulation of new ideas based on elements drawn from another group. In the following discussion, I list four possible metaphors for the world community, each of which is based on the action of human mind: the family, the tribe, the church, and the neighborhood.[25]

The term *the human family* is applied not uncommonly, and we do speak of one another as brothers and sisters. But we apply these terms in a fictional sense. Our scholars and our religious leaders each affirm that we are a single family, but we do not act like one. (Note that I have introduced the family in two ways: as an organic metaphor, in which we emphasize common biological descent of humanity, and as a societal metaphor, in which we emphasize the conscious interaction among family members.)

The image of the tribe can be seen as a special case of the family. The tribe is a group of people who claim to be descended from a common ancestor and who share common traditions as a result of that ancestry. In fact, the study of historical "tribal" communities usually shows that many members of the community immigrated as individuals (wives taken from other tribes), or that whole groups joined the community and developed a fictional descent from the founder. This image of the tribe fits with the vision of national history and with the idea of the rise and fall of distinct civilizations and, further, with racial history. That is, the metaphor of the tribe assumes that each definable "tribe" (or nation, or civilization) has a distinct ancestry, a distinct outlook, and that the traditions of each are inconsistent with the others. Thus, the peoples making up "Western Civilization" are often

23. Freud 1930.
24. I have toyed with this metaphor in Manning 1988.
25. Within the societal metaphor there arises the important question of the roles and significance of the individual and the group in the society.

thought of in tribal terms, on the assumption that belonging to this group separates its members in an essential fashion from all other peoples or tribes. One basic difficulty with the tribal metaphor is that it provides no basis for interpreting common interests and recurring patterns, or for studying interactions across "tribal" lines—except through of conquest and assimilation. The logic of the tribal metaphor further defines members of the tribe as "us" and all others as "them," based on assumed ancestry rather than any other distinctions. It therefore emphasizes the divisions in the human community rather than its commonalities.

The metaphor of the church invokes the idea of a community united by its shared beliefs rather than by ancestry or proximity. The great world religious communities—Christian, Muslim, Judaic, Hindu, Buddhist—are the most obvious applications of this metaphor. But Marxian philosophy or the doctrine of free enterprise can also be thought of as churches in these terms. At a more general level there may exist, throughout the human community, shared beliefs in the destiny of man and, more recently, in the desire for equality, which form the basis for a secular church of human rights. Meanwhile, although a number of prophets have proposed a universal religion, none of them has yet been universal enough to encompass the whole human community.

The metaphor of the global community as a neighborhood provides many advantages for the historian. Inhabitants of a neighborhood share a common space; they interact. They do not claim a common ancestry, and they define themselves by the space they share rather than in opposition to other collectivities. They share a common fate rather than a common faith. They may be divided and distinct from each other and in conflict with each other, but they do recognize their commonality. Neighbors can and do withdraw into their own households and families, but they cannot escape the fact that they live together. The history of a neighborhood responds to the mechanical effects of economic necessity and to the organic effects of its birth and maturation, but ultimately the path of change depends on the debates, battles, and conscious decisions of the neighborhood's residents.

Systems

The term *comparison* has already led to considerable methodological discussion among historians. The term *system*, in contrast, has become very important in many other fields of study, but its discussion by historians has mainly been limited to one specific issue: the "modern world-system," as defined by Immanuel Wallerstein, and the question of whether "world-systems" (hyphenated or un-hyphenated) existed at earlier times.[26]

What requires discussion by historians, I argue, is the notion of "system" in general. As a general framework for analytic thinking and as a framework for historical analysis, it presents a logic distinct from the case studies, comparisons, and connections described earlier. Quite aside from the specific hypotheses of

26. Wallerstein 1974; Frank 1991.

Wallerstein, Abu-Lughod, Gills and Frank, and others, the development of systems thinking and its relevance to history should be part of the methodological preparation of any world historian.[27] A successful practical example of systems logic in the history of slavery is Joseph Miller's *Way of Death*, in which he traces the eighteenth-century commerce in slaves in Angola, Brazil, and Portugal, showing how the connections of regional specializations led to a system of slave trading that was very different from the system drawing slaves from West Africa.[28]

Systems theory came into being in the same post–World War II era that was crowded with the growth of area studies and the development of so many other sorts of theory in natural and social sciences. The principal focus of systems theory is in understanding the interactions of the various elements of a complex system: it is an approach that focuses on modeling complex wholes rather than on breaking large problems into smaller problems for separate solution. Two key figures in the formalization of systems theory were Ludwig von Bertalanffy, a biologist, and John von Neumann, a mathematician. Von Neumann gained the greater fame for his systems-based work in cybernetics, while von Bertalanffy, with no great individual achievements to his credit, carried on as the leading propagandist for systems theory in general.[29] Systems theory, as elaborated through the work of these and other figures, has had significant impact in engineering, computer science, psychology, and business management. The description to follow, introducing some of the issues and principles in the systems approach, is presented at some length, as it is unfamiliar to most historians yet important for the conceptualization of world history.

If physics was the most prestigious branch of scientific study during the twentieth century, especially through the impact of Einstein's formulations on relativity, it remained the case that theory in physics (as in most other fields of study) proceeded mainly by breaking problems into segments and seeking to account for the working of each segment. Einstein's search for a unified theory, combining various branches of physics, eluded him, and this check to his efforts may serve as a symbol of the limits reached by positivistic, segmented research design and theory in his and other fields.[30] Von Bertalanffy emphasized the limits of "classical" science in modeling interactions: "The method of classical science was most appropriate for phenomena that either can be resolved into isolated causal chains, or are the statistical outcome of an 'infinite' number of change processes.... The classical modes of thinking, however, fail in the case of interaction of a large but

---

27. Wallerstein 1974, Abu-Lughod 1989, Frank and Gills 1993.

28. Miller 1988.

29. John von Neumann, "The General and Logical Theory of Automata," in L. A. Jeffries, ed., *Cerebral Mechanisms in Behavior* (New York, 1951), 1–31; Ludwig von Bertalanffy, *General System Theory* (New York, 1968). Von Neumann had already become well known for his development of game theory; both game theory and cybernetics each found a place within the larger rubric of systems theory.

30. Peter Novick has identified parallels of thinking in history, philosophy, and science in the United States in the nineteenth and twentieth centuries. For the influence of Einstein's notion of relativity during the interwar years, see Novick 1988: 133–67.

limited number of elements or processes."[31] In contrast, he wrote optimistically about the progress of "contemporary" science in analyzing the interactions within systems: "While in the past, science tried to explain observable phenomena by reducing them to an interplay of elementary units investigatable independently of each other, conceptions appear in contemporary science that are concerned with what is somewhat vaguely termed 'wholeness,' i.e., problems of organization, phenomena not resolvable into local events, dynamic interactions manifest in the different of behavior of parts when isolated or in a higher configuration, etc.; in short, 'systems' of various orders not understandable by investigation of their respective parts in isolation."[32]

Based on a generalization of his approach to biological systems, von Bertalanffy proposed the formulation of a general systems theory, whose principles could be applied to a wide range of fields: "Its subject matter is formulation of principles that are valid for 'systems' in general, whatever the nature of their component elements and the relations or 'forces' between them."[33]

General systems theory posits "the appearance of structural similarities or isomorphisms in different fields." These structural similarities, observed empirically and through the parallels in theories applied to them, are taken to be characteristics general to all systems, somewhat independently of their specific properties. One early and important area of development of systems theory was in study of organizations, especially business firms. Kenneth Boulding's *Organizational Revolution* was a key early text in the field, on which von Bertalanffy draws to emphasize the notion of the hierarchy of systems, within which are posited repeated appearance of structural similarities. These range from static structures such as molecules, to machines, to lower organisms, to humans, to sociocultural systems, and finally to symbolic systems.[34]

As von Bertalanffy notes, the concept of organization "was alien to the mechanistic world.... Characteristic of organization, whether of a living organism or a society, are notions like those of wholeness, growth, differentiation, hierarchical order, dominance, control, competition, etc. Such notions do not appear in conventional physics."[35]

More specifically, von Bertalanffy's interest focuses on "open systems" rather than "closed systems." Conventional physics and thermodynamics relies on the model of the closed system, in which the unit under study is entirely closed off from the rest of the world. The assumption of the closed system underlies every

31. Von Bertalanffy 1968: 35.
32. Ibid., 36–37.
33. Ibid., 37. Sub-fields and approaches within the systems approach, as listed by von Bertalanffy, included mathematical systems theory, computerization or simulation, compartment theory, set theory, graph theory, net theory, cybernetics, information theory, theory of automata, game theory, decision theory, and queuing theory. Ibid., 19–23
34. Ibid., 33; Kenneth Boulding, *The Organizational Revolution: A Study in the Ethics of Economic Organization* (New York, 1953).
35. Von Bertalanffy, *General Systems Theory*, 47.

"classical" theory, from physics to economics. Every living organism, meanwhile, is an open system: "It maintains itself in a continuous inflow and outflow, a building up and breaking down of components, never being, so long as it is alive, in a state of chemical and thermodynamic equilibrium but maintained in a so-called steady state which is distinct from the latter.... Living systems, maintaining themselves in a steady state, can avoid the increase of entropy, and may even develop towards states of increased order and organization."[36]

The greatest part of von Bertalanffy's work on systems, therefore, is the exploration of open systems and their characteristics. Open systems, surely, are the type of systems relevant to the study of history. Groups of humans, whether defined by the limits of families, nations, or continents, live not in isolation from each other but by drawing and sending human and physical resources (and wastes) across the borders separating them from their neighbors. Systems theory thus provides hope of clarifying connections among human groupings.

Open systems, which can survive only by establishing some balance of the inflows and outflows to and from their surroundings, tend to develop adaptive patterns or behavior. Von Bertalanffy focuses on two models of adaptive behavior that have gained currency in analyses of open systems. First is "equifinality," the tendency toward a characteristic final state from different initial states in an open system attaining a steady state. One might argue that a "characteristic final state" for historical systems is one with a social hierarchy that results from intersocietal contacts whatever the initial state of a social grouping. Second is "feedback," in which circular causal chains help to maintain a steady state within the system. Perhaps "brain drain," as a type of migration causing universities to expand or contract their academic offerings, can be seen as an example of feedback in operation.

Von Bertalanffy shows particular interest in civilizational and world history. He restates Peter Geyl's portrayal of the numerous and contradictory interpretations of Napoleon, and in so doing provides a concise refutation of the common dichotomy as to whether historical studies are most appropriately "nomothetic" (history results from the workings of historical laws) or "idiographic" (history results from individual choices). He uses a chemical image to provide a parallel, distinguishing "molar" (large-scale) from "molecular" (individual) approaches to history.[37] "In the light of modern systems theory, the alternative between molar and molecular, nomothetic and idiographic approach can be given a precise meaning. For mass behavior, system laws would apply which, if they can be mathematized, would take the form of differential equations of the sort of those used by Richardson.... In contrast, free choice of the individual would be described by formulations of the nature of game and decision theory."[38] The point, in between these dense references to numerous fields of analysis, is that systems analysis is

36. Ibid., 39, 41.
37. Ibid., 110–11.
38. Ibid., 114. Immanuel Wallerstein has devoted a major research project, with support of the Gulbenkian Foundation, to the evolution of methodology in the social science disciplines, with particular attention to the distinction between nomothetic and idiographic approaches. Wallerstein et al. 1996.

proposed as a framework allowing for encompassing and linking both mass and individual behavior. For it to provide specific and useful results would be an additional benefit, but at least it treats the problem of linking multiple levels as being within the purview of the analysis.

The analysis of open systems addresses another issue of great importance to historians, that of teleology. The label of "presentism" is often thought to disqualify a historian as a qualified commentator when it can be shown that he or she is interpreting the past with particular reference to the concerns of the present. Then again, the label of "teleological" is often used to dismiss the validity of a historical interpretation, when that interpretation focuses on achieving a certain destiny. My favorite example of teleological history is studies of the antebellum United States, which tend to evaluate every event according to whether they contributed to the Civil War, as if people had nothing to do for thirty years but get ready for that conflict. Nevertheless, von Bertalanffy's analysis of the open system makes clear that the critiques of presentism and teleology are commonly made from the philosophical standpoint of the classical, atomistic viewpoint, which assumes that the world and history are divided into independent units.

> In the world view called mechanistic, which was born of classical physics of the nineteenth century, the aimless play of the atoms, governed by the inexorable laws of causality, produced all phenomena in the world, inanimate, living, and mental. No room was left for any directiveness, order, or telos ... notions of teleology and directiveness appeared to be outside the scope of science and to be the playground of mysterious, supernatural or anthropomorphic agencies; or else, a pseudoproblem, intrinsically alien to science, and merely a misplaced projection of the observer's mind into a nature governed by purposeless laws. Nevertheless, these aspects exist, and you cannot conceive of a living organism, not to speak of behavior and human society, without taking into account what variously and rather loosely is called adaptiveness, purposiveness, goal-seeking and the like.[39]

As a result, it appears that the logic of the open system may help reopen, in a helpful fashion, questions of purpose and teleology in the interpretation of human history.

The actual applications of systems theory by historians, however, have not been greatly successful, and it is surely for this reason that historians have not paid much attention to systems theory. The most successful such attempt has been the "modern world-system" of Immanuel Wallerstein. However, while Wallerstein used the term *system* and some of its most elementary principles, his approach relied far more heavily on the analyses of Marx and Weber and on the debates in political economy of the 1960s and 1970s. Those who have followed him in this analysis, including Christopher Chase-Dunn and Thomas Hall, have similarly focused more on political economy than on systems theory.[40]

In the era of greatest excitement and publication on systems, during the 1960s and 1970s, authors seeking to show the general relevance of the approach tended

---

39. Ibid., 45.
40. Wallerstein 1974; Chase-Dunn and Hall 1991.

to include chapters showing the systems approach to world history. These ended up being parodies of national and civilizational history, providing crude evolutionist interpretations of change over time. The authors violated the assumptions of the open system, treating each civilization as a neatly sealed package and neglecting what might be thought of as the potential of the systems approach to focus on interactions within and across social systems. They failed to pay sufficient attention to the subsystems within any historical system.[41]

Perhaps these authors were victims of the too-simplistic social assumptions of their time, or they had not completed sufficient historical reading. But we must also consider whether the limitation was in the systems approach they were applying. For the systems theory of von Bertalanffy and his successors, while it focuses on identifying complex interactions and hierarchies of systems, also focuses on finding very simple rules of their behavior. The notion of equifinality (the open system equivalent to the equilibrium of a closed system) provides a useful approach to the study of "steady states" in society, but it does not encourage the analyst to seek out complex interactions within or among societies. Systems theory may be of significant use in the study of world history, but only if versions of the theory can be developed which give attention to the fine-structure of historical processes as well as the overall results.

Results: Interpretations, Narratives, Debates

The author's decisions in selecting units of study, procedures for study, and analytical strategies all serve to point historical analyses in one direction or another. Studies structured in terms of interacting networks, analyzed through social anthropological theory, will almost surely come out differently than studies of single case studies analyzed through a Spenglerian organic metaphor.

Yet the results of a historical analysis depend not only on the analytical framework but on the historical data and on patterns that emerge when data are analyzed in the framework. These resulting patterns constitute the author's interpretation. The logic of the author's interpretation is to be presented in the way most convincing to the reader, and this is the reason for all the work put into presenting a clear narrative to convey the interpretation.

Debates in history address the full range of these factors. Readers and reviewers may debate the data used by the author—"the facts." Or they may debate the author's analytical framework—the units of analysis, the analytical comparisons and linkages, and especially the model, theory, or metaphor underlying the analysis. Finally, reviewers may dispute the author's interpretation, even if they find the data and the framework of analysis to be appropriate.

41. World historians, if using the systems approach, must pay explicit attention to the varying scales of historical processes, rather than focus on a single scale. These scales are known by such terms as *local, regional, civilizational,* and *planetary,* but also as *social, cultural,* and *political.* In general, a systems model for any historical system must identify its subsystems and also the larger systems of which it forms a part.

Debates in history thus require decoding, and if they are not decoded thoroughly they may add more confusion than clarity. Debates on the facts can be taken at face value, except that people identify factual errors more readily when the erroneous data tend to undermine their chosen interpretation. Debates over analytical framework often take up a great deal of space, so that critiques of world-system analysis in general tend to be more voluminous than critiques of the interpretations developed through the world-systems framework. Yet it is equally limiting to debate contending interpretations—as with the dominance of Western Europe or East Asia in the world economy of the eighteenth century—without including the frameworks of analysis in the debate.

Debate in history goes further, of course, than assessing a single book or article. It also goes beyond attempting to set up any single framework or interpretation as correct, and others as inappropriate. No single vantage-point can be taken as definitively privileged. Attempts in the past to determine that a single vantage-point is correct and others are invalid become steadily more difficult to sustain. This leaves the problem of how to reconcile or aggregate the various distinct views and interpretations.[42] This problem—the work of compiling or selecting a broad statement of historical interpretation out of a mass of smaller and incommensurate studies—is a problem we cannot yet solve, and only by posing the problem directly and repeatedly will we develop a range of solutions.

Conclusion: Rigor in History

I believe that world historians should emphasize logical rigor in their analyses. By that I do not mean that historians should don some analytical straitjacket or give precedence to theory over empirical study of the past. But I do mean that historians' research design and interpretations should be well thought out and internally consistent, and it should address all parts of the issue under study. In this chapter, I have described in general the elements of historians' frameworks and strategies. This review of method provides a checklist that can be set next to the details of a plan for research in world history and can minimize the logical gaps in articulating the research design and putting it to work. The "framework" point on the checklist reminds the analyst to distinguish the objects of study (usually cases or networks) from the procedures for study (especially comparisons and linkages). The "strategy" point on the checklist reminds the analyst to specify in detail the plan for developing and documenting an interpretation, and to identify the types of tools employed in the analysis, be they typology, paradigm, mechanism, theory, metaphor, or system. With careful attention to framework and strategy, it is more likely that the results of the analysis—the published interpretation, the assessment of the work by readers, and the debate over the results—will actively advance the understanding of world history rather than retire to a quiet spot on a shelf.

---

42. For examples of this issue in the study of cultural history, see chapter 13.

Chapter 17

Verifying and Presenting Interpretations

How do readers know whether to accept the descriptions and interpretations of world historians? Skeptical and curious readers will wonder whether a work in world history conveys knowledge about the past or just a tissue of the author's imagination. Even when readers find the interpretations largely convincing, they will wonder whether alternative interpretations might work just as well.

Historians have long and justifiably prided themselves on their advanced skills in confirming the factual evidence they present. The facts, however, can be assembled in multitudinous ways. Historians in recent decades, cognizant of this issue, have addressed with great success the construction of stronger assemblages of the facts and broader interpretations of the past: in earlier chapters I have identified revolution after revolution in the analytical accomplishment of historical studies. Yet when it comes to verifying historical interpretations, I can say only that a revolution is due in this area. All too commonly, historians develop an interpretation of their materials, demonstrate its plausibility, and then stop short of any attempt to verify its validity.[1] Verification of interpretations has been a high priority only for issues on which intense debate developed among contending perspectives, or (within certain sub-fields of history) in a campaign of quantitative hypothesis-testing.

---

1. Here are three levels at which authors of historical works commonly stop short of attempting to confirm an interpretation. First, there are many studies that provide descriptions of the past, with no explicit attempt to develop an overall interpretation—a history without a thesis. Second, there are studies in which the author presents a clear interpretation, then deploys enough reasoning and evidence to make their interpretation plausible but takes no steps to confirm the interpretation, not even comparing it to other possible interpretations—a thesis defended more by logic than by evidence. Third, there are works of history that give primary emphasis to developing and explicating an analytical paradigm but do not take the next step of actually applying it in detail—a research design but not a history. (As an addendum, one may also note studies in which an interpretation is illustrated, but the illustration is not shown to be representative.)

Readers of world history, however, are too skeptical to allow the need for verification to be swept under the rug. For world history, it is necessary to make verification a regular part of the analysis rather than an optional exercise. World history will develop fully as a field of scholarship only if its analysts become effective in addressing verification of their interpretations. Readers should be encouraged to be skeptical of studies in world history, and authors should address the skepticism directly.

World historians, it appears, may have to play an active part in clarifying and strengthening the practice of verifying historical interpretations. World historians have a great need for improved techniques of verification, in order to confirm the credibility of their own studies, but they may also bring particular strengths to the problem. The breadth of world-historical study works against an oversimplification of the issues, and the globalist focus on connections may locate new strategies for verification. This chapter identifies the issues in verification, and proposes techniques for assuring readers that world-historical statements about connections and changes in the past represent knowledge and not just imagination. The chapter also addresses the related issue of writing up an analysis in convincing style. These stages of verification and presentation, though summarized concisely in this chapter, are central to the success of world history as a field of scholarly analysis.

The Scope of Verification in World History

The scope of verification is set by the assumed structure and dynamics of the historical situation. The elements of verification and the procedures of verification lead the analyst through a review of research design and into the alternative scenarios against which the author compares the record of the past. Verification in world history is more than a simple matter of truth or falsity in historical interpretations. It involves checking facts, checking the logic of interpretation, and, most importantly, checking the interpretation against the evidence according to a clear procedure. Since world-historical studies often span long times and major transformations, the assumed structures and dynamics may change in the course of the study, thus adding an additional complication.

The elements of verification begin with categories of evidence. As a terminology for addressing this range of factors, I will utilize the following: *objects* are the persons, things, and events being described; *relations* are the processes and other connections that link objects, and that the historian also describes. The *level* of a study refers to the level of aggregation at which the principal analysis is conducted, though it is possible for an analysis to be focused principally on the interaction between two or more levels of aggregation. *Constituents* refers to objects and relations at a level more specific than the principal level of study; and *context* refers to objects and relations at levels more general than the principal level of study.

Verification in world history takes place on multiple levels. As I argued in chapter 15, the varying levels of breadth in world-historical study make world historians reluctant to identify certain phenomena as fundamental or essential. There are no "atoms" of world-historical analysis: we define elements, but define them situationally, and treat each element as problematic. "Nuclear" families, no matter now nuclear, are influenced at once by the individuals within them and by the broader structures of society around them; patterns of the individual human life course are affected by the short-term impact of bad weather and by such long-term influences as global warming. So world-historical analysis involves keeping track of relevant evidence at several levels of generality. World history, seen from this perspective, does not make the choice between microstudy and macrostudy but conducts both at once.[2]

The elements of verification continue, second, with the models of structure and dynamics of the historical situation. For this discussion of verification, I emphasize models of structure and models of dynamics in history. For models of historical structure, here are three main categories:

- Models containing a single population of roughly comparable objects (or cases). Electoral studies use this sort of model. The analysis is expected to reveal a population-wide pattern of behavior.[3]
- Models containing systems that interact with each other, though each operates by different rules. Analyses with these models tend to be eclectic and open-ended.
- Models containing a single system, with its constituent subsystems. This is the structure of world-system analysis. The analysis is expected to reveal the patterns of operation and evolution of the system.

For each of these types of model, verification of the interpretive results must be organized differently, to be consistent with the structure of the evidence and the analysis. Second, I offer two categories of models of historical dynamics, which I label "positivistic models" and "postmodernist models." Positivistic modeling emphasizes solving analytical problems by dividing them into pieces, classifying and labeling the basic elements of nature as independent elements. Positivistic reasoning is that of cause and effect, as independent variables have their effect on dependent variables. Truth, in this case, is a statement about the effect of independent variables on the variables under study. Short of certain knowledge, one offers a hypothesis as a tentative statement of that truth and tests its fit with known data against other hypotheses. Postmodernist modeling, in contrast, emphasizes linkages among factors in nature and society. While individual elements may be defined

2. Microanalysis tends to assume that one is analyzing at the most basic level. In economics, firms and households are not broken down for more detailed study, but analysis may include accounting for macrolevel influences on the microsituation. With macroanalysis, it is assumed that one is working with the largest relevant unit; analysis may include accounting for microlevel influences on the macrosituation. World-historical analysis, in general, includes both of these, accounting at each level both for the more specific and more general influences on the situation under analysis.

3. One could also add models containing a single population of roughly comparable cases, each of which is itself a system (containing objects and relations) for which the analysis is expected to reveal patterns of behavior of the individual cases. This is the case, for instance, of medical diagnosis of individual health based on study of a whole population.

provisionally, they overlap and interpenetrate each other. The reasoning is not that of cause and effect, but of linkage and mutual definition. Truth, in this case, is a statement about the linkage of various influences under analysis. Because statements within the postmodernist outlook are complex and nuanced, an equivalent to the statement and testing of hypotheses has not developed—at least not yet. These contrasting approaches to analysis lead naturally to differing strategies of verification.

Further, the elements of verification include the set of alternative pasts that the historian imagines as having been conceivable. These have come to be known by the term *counterfactual*. This term, popularized if not coined by Robert William Fogel, refers to the historical "might have been." In contrast to the outcome or the narrative of history as we understand it to have occurred, the counterfactual is the set of events that might plausibly have taken place but did not.[4] It is the early modern economy if no silver had been found in the Americas. It is the Atlantic world if Africans had crossed the ocean as free rather than slave laborers. It is the early twenty-first century if the Soviet Union had not dissolved, or if the HIV virus had not emerged. In assessing the conquests of the Mongols, one can focus either on the question of how they got so far or on why they did not also conquer Java, Japan, Egypt, India, and Germany. The same historical data may be assessed quite differently when compared to different counterfactuals.

My point is not to give an extravagant list of possibilities that did not occur but to argue that, of logical necessity, the historian has a counterfactual in mind when making any interpretive statement.[5] Selection of a relevant counterfactual and comparing it to the historical record provide a systematic basis for evaluating interpretations of the past. To put it another way, many of the disagreements among historians are not disagreements about the facts, but disagreements about the counterfactual. When making a judgment on the past, it is helpful to give an explicit response to the question, "Compared to what?"

Checking Facts

Historians have a well-deserved reputation as scrupulous scholars, which results especially from their care in checking facts. For world history, however, the nature

---

4. Fogel developed the notion of the explicit counterfactual in a study of U.S. railroads and economic growth. If one compares the U.S. railroad network of 1890 with the transportation network of 1840, it appears that the contribution of railroads to economic growth was immense. But, as Fogel showed, if one compares the U.S. railroad network of 1890 with what could have been the canal network of 1890, in the absence of railroads, then the contribution of railroads to economic growth seems to have been much smaller. Fogel 1964. For some of my attempts to articulate counterfactuals in African and Atlantic history, see Manning 1982: 49–50, 225–29.

5. In simplifying here, I have combined two sorts of counterfactual: an alternative explanation of the path that history took ("alternative hypothesis"), and an alternative path that history might have taken ("hypothetical alternative"). Here I discuss both but emphasize the latter. The distinction will grow more important if historians begin to address verification in more detail.

of "facts" is broadened somewhat, as compared with local or national history. First, as usual, it means confirming the validity of information on objects and events. Second, and because world history focuses on connections, it means confirming that descriptions of relations between objects or people are valid. Thus R. J. Barendse, in *The Arabian Seas*, describes not only the character of the Portuguese Estado da India and the English East India Company, but the relations of each firm to its employees, its clients, and the commodities in which it traded.[6] To be valid, both these types of factual descriptions must be precise, and they must be shown to be either representative or unique. Two more sets of fact-checking complete this picture. First, in addition to the main objects and relations being described, the historian needs to confirm the facts of objects and relations at a more specific level (the constituents) and at a more general level (the context). For instance, in a national analysis, one must check on relevant local facts and on relations among them; in analysis of an empire, one must check on facts and relations of the global and interactive context surrounding the empire. Second, one needs to check facts on the paths of change for the persons and objects under analysis, and also on paths of change for the relations among objects. Overall, before verifying an interpretation, the world historian must check all sorts of facts: facts on big objects as well as small, facts on big relationships as well as small, and facts on paths of change.

Verifying Interpretations

I now turn to the main business at hand: the process of verifying world-historical interpretations. A world-historical interpretation, as I have used the term, is an explanation of change over time, based on the interaction of historical factors according to the model adopted. There are three basic steps to verification (though there are various procedures relevant to each step): first, the interpretation must be shown to be logically consistent; second, it must be a more effective explanation than relevant alternatives; and third, it must be consistent with the descriptive evidence as it is organized into the model. Each of these aspects of verification is explored in the paragraphs that follow.

To begin with logical consistency, each model in world history has its own logic, as expressed by the author. For verification, the model needs to be articulated and reviewed to see whether it contains logical inconsistencies. Rather than go through the full process of logical review here, I will outline the first step, identifying some of the categories of logical review. Most models of the structure of historical situations can be classified into one of the three categories I have identified: models containing a single population of roughly comparable cases (or objects); models containing systems that interact with each other; and models containing a single system, with its constituent subsystems. In each of these models, one must specify the levels of analysis: for instance, local, national, global, or some specific combination of levels. One inconsistency to watch for in verification

6. Barendse 2002, 299–363, 424–52.

is cases in which the analysis applies techniques that are appropriate for one type of model to another where they are inappropriate. For instance, formal statistical techniques can be applied to the first type of model, but the only way statistical techniques can be applied appropriately to the third type is through multiple observations of the system over time. A second issue in logical consistency is the dynamic assumed within the historical situation, including the dynamic of connections among levels of analysis. For this discussion of verification, I will identify four types of dynamics: cause-and-effect dynamics in which, at a given level, some factors are seen as causal and others are seen as effects; feedback dynamics in which, at a given level, factors influence each other through interactions; constituent effects in which factors at a more specific level affect patterns at the level under analysis; and context effects in which factors at a more general level affect patterns at the level under analysis.

The second step in verification is to consider the alternatives: that is, the counterfactuals. If the counterfactuals are defined clearly, as indicated earlier, then the task of verification comes down to showing clearly that explanation proposed is superior to the alternative explanation. Since a world historical analysis usually ranges across levels of generality, yet focuses its interpretation on a certain level, verification of the analysis requires demonstrating that the evidence on the context and constituent effects remains consistent with the interpretation. The specific techniques for verification of these effects have yet to be developed, but I think it is clear that these are an important issue in verification of world-historical interpretations.

The third step in verification is consistency of the interpretation with available evidence. I will discuss general procedures of verification: articulating the plausibility of the interpretation, inspection of the evidence, hypothesis-testing, feedback-testing, and testing through sustained debate. This is not an exhaustive list of approaches, but it indicates that there is more than one approach, yet that each approach has its logic.

Too often, historians defend interpretations of the past simply by showing the logic of their argument and illustrating it with cases and examples. Such a procedure may be labeled a conclusion but not verification because it fails to go beyond the assertion of plausibility to show that a preponderance of evidence supports this interpretation more than an alternative. Establishing the consistent logic of an interpretation is an important preliminary step. But before an interpretation in world history can be adopted by critical readers, it needs to be tested against an organized body of evidence.

To verify an interpretation by inspection of the evidence is to conduct an informal exploration of possibilities within the data under analysis and comparison of these possibilities to other data. Exploratory verification can almost never confirm the validity of an interpretation, but it can be effective in rejecting weak interpretations as well as in suggesting new relationships to consider. Thus, in an attempt to confirm patterns in one migration, the analyst may be assisted by seeing whether similar patterns appear to hold for other migrations: finding parallels in exploratory analysis tends to validate the pattern. In exploratory analysis, one still has to validate descriptions, but there need be no effort to verify the interpretation.

Most studies in world history will include both exploratory and confirmatory analysis. John Tukey, who has developed the informal and exploratory analysis of quantitative data to a fine art, has made an effective case for the importance of exploratory study, emphasizing the benefits of its breadth.[7] Nonetheless, the ultimate purpose even of exploratory verification is to become clear about which historical propositions are tentative assertions by individual observers and which can be advanced with real confidence.

With the procedure of hypothesis-testing, in contrast, the techniques for verification have been developed in great detail. This procedure applies to models of a single system with a population of roughly comparable cases, and in which the cases are assumed to act independently of each other. The formal reasoning of hypothesis-testing provides an elegant and logically consistent, if sometimes limiting, framework for analysis. It is based on positivistic, cause-and-effect reasoning and requires explicit statement of the hypothesis, the alternative hypothesis, the nature of the data, and the logic of relations among the variables. The analysis requires comparison among a number of cases, where the cases could be independent experimental observations, individual cases drawn from a population, or repeated annual observations on the same place.[8]

One begins by identifying an analytical question, a body of data on which to test it, and a theory that defines variables and proposes relations among them so that it can yield predictions. Relying on the theory, the analyst proposes a hypothesis— a predicted relationship between independent variables (the cause) and a dependent variable (the effect). To conduct the test, the analyst must also propose an alternative hypothesis (the counterfactual) and see whether the data fit more fully with the hypothesis or the alternative. Data are then analyzed for each of the cases to see if they fit significantly better with the hypothesis than with the alternative hypothesis.

Typically, the alternative hypothesis is the "null hypothesis," asserting that there is a random relationship between the independent and dependent variables. The

---

7. "Once upon a time, statisticians only explored. Then they learned to confirm exactly— to confirm a few things exactly, each under very specific circumstances. As they emphasized exact confirmation, their techniques inevitably became less flexible. The connection of the most used techniques with past insights was weakened. Anything to which a confirmatory procedure was not explicitly attached was decried as "mere descriptive statistics", no matter how much we had learned from it.... Today, exploratory and confirmatory can—and should—proceed side by side." John Wilder Tukey, *Exploratory Data Analysis* (Reading, Mass., 1977).

8. Numerous manuals describe this procedure and related quantitative work in detail. Early historical manuals for computers and quantitative studies included Edward Shorter, *The Historian and the Computer* (Englewood Cliffs, N.J., 1971); Roderick Floud, *An Introduction to Quantitative Methods for Historians* (Princeton, 1973); Charles M. Dollar and Richard J. Jensen, *Historian's Guide to Statistics: Quantitative Analysis and Historical Research* (New York, 1971). For more recent guides, see Janice L. Reiff, *Structuring the Past: The Use of Computers in History* (Washington, 1991); Daniel I. Greenstein, *A Historian's Guide to Computing* (Oxford, 1994); and Charles Harvey and Jon Press, *Databases in Historical Research: Theory, Methods and Applications* (New York, 1996).

shape of a random distribution is known mathematically for the various types of data, as is the probability that any case may vary from this distribution. If enough cases are more like the hypothesis than the alternative, the test has shown that a relationship exists. At the conclusion of the test, one may say whether the hypothesis has been confirmed or rejected in comparison with the alternative.

For hypotheses on quantitative data, a single general framework of hypothesis-testing is applied with slightly different procedures to each of three categories of data: interval, ordinal, and categorical. The statistics that can be calculated differ for the various types of data.[9] The analytical tests that one performs to establish relationships among variables differ with the type of data: linear regression is the principal technique for analyzing interval data, and cross-tabulation is the principal technique for analyzing ordinal and categorical data. Additional tests of relationships among variables include factor analysis for interval data and analysis of variance for linkages between ordinal and categorical variables. These results will indicate the strength of the relationship between the independent and dependent variables—how much the dependent variable will change for any given change in an independent variable. Once the analytical tests are completed on the available data, one may conduct statistical tests on the results to see whether they indicate that the hypothesis is confirmed—typically, one requires that the statistics show a 95 percent confidence level that the hypothesis, rather than the alternative hypothesis, is valid.[10]

As one works with individual-level data—for instance, observations on individual households rather than citywide summaries—one encounters the important distinction between individual-level data and aggregate data. The variation in individual-level data may be much greater than that in aggregate data, because the individual-level variations are averaged out in aggregate statistics. So the relationship between religious affiliation and occupation, as measured through a comparison of citywide aggregate statistics across a nation, is likely to appear far more regular and determined than if it were measured through comparison of individuals or families in the same cities.

Despite the differences among interval, ordinal, and categorical data and the differences among individual and aggregate data, the procedure of hypothesis-testing yields an orderly analytical process and a clear result: the hypothesis is

9. For *interval* data, the value of a given variable can be attributed a number, as with height or weight of individuals. For *ordinal* data, the values of a variable can be ranked but not necessarily given numerical weights, as with first and second class seating on a railroad. For *categorical* data, the values of a variable can be distinguished but not ranked, as with the different religious affiliations. One can calculate the average height of a group, but one cannot calculate its average religion—though one can deduce the modal religion (the one with the largest number of adherents).

10. Here a substantial difference shows up between interval data and categorical data. For interval data, as few as twenty to thirty observations can show a relationship with 95 percent confidence. For categorical data, roughly a thousand independent observations are required to demonstrate that one may have 95 percent confidence in distinguishing between a hypothesis and an alternative hypothesis. Social historians have to work harder than economic historians to collect the data that will confirm an analysis.

confirmed or rejected for the data on which it is tested. Debate need not halt, however, at the conclusion of a test. One may reformulate the hypothesis, proposing different relationships. Or one may compare the same hypothesis to a different alternative. These variations and their combinations enable the testing to continue until one obtains a progressively improved fit of theory and data.

Hypothesis-testing may be difficult to carry out in practice or may even be irrelevant for many historical analyses. Quantitative data may be lacking; the question may not be formulated in terms of precise variables, and so forth. But the logical coherence of the hypothesis test and the identification of the elements of the test provide a useful if abstract standard to keep in mind when seeking to confirm the validity of any historical proposition.

A third category of evidence verification is feedback-testing. While it has not to my knowledge been formalized, I believe that a process of feedback-testing is in process of development in systems analysis and in other frameworks of analysis emphasizing connections, which I have labeled as postmodern. At a hypothetical level, the analyst follows the flows (of biological elements for Von Bertalanffy, of power and knowledge for Michel Foucault) from segment to segment of the system for several repetitions, until the hypothesized interrelations correspond to a steady state.

In those areas of engineering and psychology in which systems approaches have gained currency, it is my impression that interpretive statements and professional practice are chosen based not on a formal test but on informal "fit." Reliance on feedback and repetition of work means that analyses are continually being reformulated and replaced.[11]

For world historians working with a postmodern outlook and emphasizing mutual determination of a range of interacting factors, the logic of verifying the analysis by using evidence to confirm the feedback among the interacting factors would seem to be appropriate.[12] The tendency so far has been to verify the interactions by inspection of a few examples, stopping short of a more systematic test. Perhaps it will take debates among contending postmodern interpretations of global issues to develop the habits of verifying large-scale interpretations.

Finally, let me mention briefly the testing of interpretations by sustained debate. This represents a forum though not a procedure for verification: it is the pragmatic approach that has served historians best. Analysts present contending perspectives on a single issue, often using quite different sets of evidence and

11. In the interdisciplinary work of William Durham on co-evolution, a systems approach leads to a number of new hypotheses, but again there is no great emphasis on the standards for evaluating them. Durham 1991.
12. Here is a positivist equivalent to this sort of argument. In an economic analysis of colonial Dahomey, I developed a formal input-output model tracing the links among all the sectors of the regional economy. While the numbers in the input-output matrix were all estimates, successive estimates to make them consistent yielded a tool of considerable use. The results showed, for instance, that manufacturing (handicrafts) and transport were the sectors most linked to other sectors, and tree crops (mostly palm oil) the least so. Manning 1982: 83–84, 301.

frames of analysis. They argue eclectically about strong and weak points of each interpretation, and the result over time is steady reformulation of both the data and the interpretations. Among the issues that have been debated in this way are the causes of World War I, the transition from feudalism to capitalism, the wages of industrial workers, and the causes of the French Revolution.[13] At the least, this approach requires multiple scholars to work on a single issue.

Beyond selection among these basic procedures for confirming interpretations, there remain other issues in verification. A rather subtle set of problems lies in the distinction among an author's hypothesis, the work's conclusion, and a verified interpretation. In strict positivistic terms, a hypothesis should be posed before data are collected. Then once the test is performed, the result should be a verified interpretation. More likely, the author reads through the data, then develops an interpretive statement. The book is written up as a test of the interpretive statement—and not surprisingly, the proposition usually passes the test. In strict terms, one should call the interpretive statement a hypothesis, and one should verify it through other data. But of course such data do not exist if the historian is an energetic researcher. To take a step away from this positivistic reasoning, one may say that there is a good deal of feedback in the development and the assessment of interpretive statements in history.

Examples of Verification

To show the practical application of the techniques of verification I have described, here are four brief discussions of the verification of interpretations of major works in world history.

Jerry Bentley's *Old World Encounters*, a study synthesizing social, political, economic, and cultural patterns, poses the question: what were the processes and changes associated with "premodern encounters between peoples of different civilizations and cultural regions"?[14] The answer, he argues, is that patterns of cross-cultural interaction, especially in trade and migration, evolved past various turning points to create six distinct periods in the history of the Old World. His objects are civilizations and peoples, and his relations are three patterns of social conversion: voluntary association, social pressure, and assimilation of minorities. The model in which the argument is set is that of multiple systems, with the interactions among them leading, in the long run, to expansion of the main civilizations. Bentley sets his reconstruction of the past against two alternatives: the isolation of cultures and societies, and the alternative of diffusion, in which cultural practices spread across borders as readily as do technical innovations or disease vectors. He emphasizes practices of cultural resistance and syncretism to argue that cultural influences spread from centers of civilization but at a rather

---

13. For the debate on the welfare of the nineteenth-century English working class, see R. M. Hartwell et al., *The Long Debate on Poverty: Eight Essays on Industrialization and "the Condition of England"* (Old Woking, U.K., 1972).
14. Bentley 1993.

slow rate. His constituent effects include local efforts to resist or accelerate cultural conversion, and his context effects include the great waves of disease in the sixth and fourteenth centuries. The objectives of Bentley's efforts at verification are to confirm both the broad set of four periods that he proposes and the dynamics of the processes of social conversion. For both objectives, his verification is more exploratory than formal, and the narrative gives more attention to assembly than to testing of the overall argument. At the same time, Bentley's presentation of repeated cases of social conversion through social pressure serve as a sort of feedback-testing of the persistence of this pattern, especially for the expansion of Islam.

In a study of more recent times, A. G. Hopkins and colleagues begin by inquiring about the origin and nature of globalization.[15] The answer, they argue, is that globalization today is the reproduction and expansion of earlier episodes of globalization; they present a typology of the episodes and the changing character of globalization, and they argue that the earlier stages included non-Western as well as Western sources for globalization.[16] The model in which the argument is set is that of multiple systems, though they converge to become a single global system in the nineteenth century. The alternative proposed is that in which there were no substantial interregional connections before the twentieth century, and it is easily disposed of. The succeeding chapters address examples of each stage of globalization, seeking to identify the commodities and the processes of expanded interconnection in each period. In sum, this is an exploratory interpretation, in which the authors articulate an interpretive vision and document it with plausible cases. In this sense, the real point of the book is to demonstrate that the field of history provides an adequate basis for determining whether today's globalization is new or merely expanded.[17]

Jared Diamond begins *Guns, Germs, and Steel* with the question of Yali, his friend in Papua New Guinea: "Why is it that you white people developed so much cargo and brought it to New Guinea, but we black people had little cargo of our own?" Or, as he restates it, "why did human development proceed at such different rates on different continents?"[18] The answer, Diamond argues, is that the initial endowments—13,000 years ago—of various regions with edible plants, animals, and east-west space determined the size of populations and therefore the control of resources ever since. The model in which Diamond sets his analysis is that of various continental systems, and he pursues a positivist, cause-and-effect dynamic, emphasizing ultimate causes rather than proximate causes for human development. The alternative against which he sets his argument is that biological

15. Hopkins, ed., 2002. See especially the chapter by Christopher A. Bayly, "'Archaic' and 'Modern' Globalization in the Eurasian and African Arena, c. 1750–1850," in ibid., 47–73; also Hopkins 2002a.

16. The stages are archaic globalization, proto-globalization, modern globalization, and postcolonial globalization.

17. "The obligation now falls on historians to ensure that the history cited is based on evidence rather than on honorary facts, and to consider how they can apply arguments about the present to improve our understanding of the past." Hopkins 2002b: 9.

18. J. Diamond 1997: 14, 16.

differences among humans or climatic differences among the continents—or the proximate factors of guns, germs, and steel—brought about the different rates of development. His technique of verification is a feedback analysis, distinguishing ultimate from proximate causes, to argue that germs, literacy, and government developed logically out of successful agriculture. The argument includes climatic changes as context effects, and the relatively rapid evolution of germs in localized conditions appears as a constituent effect. While Diamond's test is informal, I think its results are convincing in showing the long-term importance of initial endowments of biota and space. On the other hand, when Diamond attempts in his last five chapters to show that these long-term factors also governed the shorter-term fluctuations of colonization, I think he makes a number of inappropriate simplifications.

Kevin O'Rourke and Jeffrey Williamson, in *Globalization in History*, propose two basic questions: first, what form did the "globalization boom" take, for the region they call "the Atlantic economy," from the mid-nineteenth century to the Great War? Second, what relevance does this "first globalization boom" have for the "second globalization" of the period beginning 1970?[19] Answering the first question, they argue that global convergence proceeded significantly in the first boom, bringing average wage levels of Atlantic countries closer together, especially because of the free flow of labor and resources. At the same time they argue that globalization caused inequality to rise within countries, and especially within rich countries, accounting for a "globalization backlash" that halted the free flow of commerce and migration and brought about an era of autarky from the 1920s. Second, they argue that trade can have a large though variable effect on income distribution, but that migration had a larger effect; they also argue that inequality in income distribution within nations could halt globalization again. The model of O'Rourke and Williamson is that of a single system populated by a set of analogous national units, though they also consider constituent effects (the distribution of income within nations) and context effects (the Great War).[20] The authors analyze in positivistic cause-and-effect terms; their counterfactual is a world that maintained the obstacles to free flow of resources and funds characterizing periods before and after the wave of globalization. They begin with descriptive evidence of global convergence, then turn to analysis of it. They choose wage rates rather than GDP per capita as the relevant data, because trade, migration, and capital flows acted more directly on wage rates than on per capita GDP; they also chose commodity prices and transport costs. They measure average levels and dispersion of wages and prices within nations and among nations. The basic technique of verification is linear regression—statistical analysis of the causes of inequality. The analysis is most successful in showing that migration rather than trade was the primary cause of convergence in national wage levels, and its second

---

19. O'Rourke and Williamson 2000.
20. While it has become conventional for economic historians to perform regression analysis with national units as cases, as O'Rourke and Williamson have done, the unevenness in the size of the nations and their non-random selection out of a larger economic system suggest that bias in the resulting analysis is possible.

success is in showing the income inequality within wealthy nations brought by migration. Surprisingly, however, the authors are tentative about suggesting or demonstrating that the "first" globalization had consequences for the "second" globalization: instead, they suggest that the nineteenth-century globalization brought "lessons" for the twenty-first century.

The range of these four analyses and therefore the range of my notes on verification suggest that the effort of verification is a significant and complex part of completing a study in world history. In each of my reviews, I have devoted more space to setting up the logic of verification than to reviewing the process and results of the authors' verifications themselves. But I hope that these examples show the variety of strategies and the range of accomplishments in verification, and I hope they will encourage the development of a discourse in which world historians debate their progress toward confirming the validity of their interpretations.

Methods of Presentation

Regardless of how much effort historians put into conducting their analysis and confirming their conclusions, it is the presentation of their arguments and evidence that determines whether readers understand or accept the results. Typically, an author presents the results of his or her analysis (historical or other) in a manner different from that in which the author developed the conclusion. The heuristic clarity of the successful published version may be quite different from the starts and stops of the actual process of discovery.

But if historical presentations sometimes strengthen and clarify the logic of the analysis, at other times they confound it. Sometimes authors neglect to be explicit about key assumptions in their analysis. At other times authors announce an approach or a conclusion but do not follow through to deliver it. Readers of world history textbooks, for instance, are familiar with the common difference between the preface, which announces that the work will present a truly global and fully interactive interpretation of world history, and the actual text, which too often fails into get beyond a series of area-studies chronologies in chapters that may be parallel but are rarely connected. At worst, authors make their argument simply by affirming and reaffirming their interpretive statement without going through the steps of demonstrating its consistency or its validity.

These few points are intended to emphasize that in a work of history, the method of analysis and the method of presentation are quite distinct, though of course related. A weakness in either can seriously undermine the validity of a study, and the alert reader should pay close attention to both. This focus on the distinction between methods of analysis and presentation should draw attention to the historical narrative, since it is always at the boundary of the two.

Narrative is the principal expository device of historians, a form of presentation that tells readers that they are indeed reading a work of history. Good narratives appear to the reader as simple, carrying the reader smoothly but directly to the conclusion the author has in mind. Yet, narratives were always complex, and in recent times they have become even more complex. Narratives have almost never been a simple matter of setting all of the evidence from end to end. The author weaves a tale, told on several levels at once, as a way of posing and supporting an interpretive historical proposition. The typical narrative, if dissected with care, will reveal the author's use of a variety of rhetorical devices guiding the reader to the intended conclusion. These may include disjunctures as the story line shifts from broad summary to localized details; topical shifts and chronological disorder that subtly impose the author's priorities on the reader; and the use of symbols to draw out the reader's willingness to identify with a national perspective or an individual moral quality. All of this, perhaps, in the service of conveying to readers the results of a rigorously tested historical hypothesis.

The narrative, however, may take on a life of its own, and it may do more to convey the underlying philosophical priorities of the author than to present the specifics of the historical situation under analysis. Hayden White, in his detailed analysis of nineteenth-century European historical writers, identified four basic rhetorical styles or "tropes" through which these authors presented their interpretations: metaphor, metonymy, synecdoche, and irony. He developed this argument through the cases of four philosophers of history (Hegel, Marx, Nietzsche, and Croce) and four authors of historical narratives (Michelet, Ranke, Tocqueville, and Burckhardt).

In extended analyses through which he traced the intellectual evolution of each writer in terms of critique of one trope and adoption of another, White characterized the paired philosophers and historians in terms of their principal outlook, linking their individual outlooks to the spirit of their times. For Hegel, White described his approach in terms of *synecdoche,* an integrative figure of speech using the part to symbolize an aspect of the totality; White saw in Michelet's histories of France a similar approach. In Marx, White observed the approach of *metonymy,* a reductive figure of speech in which the name of a part of a thing is substituted for the name of the whole while keeping the two conceptually separate; Ranke, though politically at odds with Marx, is classified along with him in terms of rhetoric. Nietzsche's rhetoric is read by White as *metaphor,* a representational figure of speech characterizing phenomena in terms of their similarities; he classified Tocqueville with Nietzsche. White interpreted Benedetto Croce's writings as expressing *irony,* a negational figure of speech denying on one level what is affirmed on another; he linked Jacob Burckhardt, the famed analyst of the Renaissance, to this approach.[21]

White's analysis, though wide-ranging, makes virtually no reference to the details of the revolutions, economic transformations, cultural patterns, or creative works that filled the books written by the authors he reviewed. His point, illustrated with force and depth, is that these authors had other messages that they wanted to convey, so that the historical past was the medium but not the message of the presentation. The ideological and rhetorical predispositions of these and other historians were doubtless formed by specific historical circumstances. Yet White took as his task not the reconstruction of their intellectual development but

---

21. H. White 1973.

the patterns of representing the past through a relatively fixed set of rhetorical and philosophical devices.[22]

Conclusion: Proof and Plausibility

Verifying interpretations of the past is no easy task, and verification at transregional and global levels is all the more complex. After years of excavation and debate, it has been verified that hominid species first developed several million years ago in East Africa, and later spread to much of Eurasia. We are near to verifying that *Homo sapiens* developed in Africa perhaps 200,000 years ago and then within the past 70,000 years spread to all the continents—though debate continues on many points of the interpretation. Recent work on the world economy has challenged the argument, previously accepted widely, that European merchants were the leading and expanding economic interests from 1500 forward. A counterargument, proposing East and perhaps South Asia as the center of the world economy up through the eighteenth century, has been advanced but has not yet been verified.[23]

With these examples I mean to emphasize the difference between advancing an argument and confirming it. It is important for authors and readers to be clear on whether any interpretive statement in world history is presented as a plausible interpretation or as a confirmed set of facts. World history, in its present phase, focuses mainly on identifying and documenting global patterns of change and interaction. Most studies in world history, whether presented as empirical summaries of evidence or as syntheses of historical transformation, are provisional summaries and plausible reconstructions of the past rather than verified interpretations. The emphasis on the problem of verification in this chapter has shown, I hope, how fundamental are the problems of confirming the analysis in a study of world history—even selecting a procedure of verification or deciding whether to attempt verification is a difficult choice.

Beyond the ticklish issue of verification, Hayden White's analysis reminds us of an additional set of issues: the method of presentation. Whatever the substance of the historical analysis, the mode of presenting its results can deflect the reader's response in any of several directions.[24] As readers carry out their critique of

22. Peter Gran, in informal conversation, proposes another typology of philosophical approaches to history, which he characterizes as positivism, romanticism, Marxism, and anarchism; these may be seen as approaches he has developed in *Beyond Eurocentrism*. This typology suggests that the debates among these varying approaches to history and society have continued for most of two centuries. I find this formulation plausible, and I recognize that my pairing of positivism and postmodernism seems to give more weight to postmodernist thought than it may deserve. Nonetheless, I believe that the late twentieth century brought a pervasive expansion in the understanding of changes and connections in many areas of natural and social science, so that it is appropriate to speak of a systematic contest, in the current era, of categorical and connected epistemologies. Gran 1996.

23. See pages 192–195.

24. For a manual on writing history by a world historian, see William Kelleher Storey, *Writing History: A Guide for Students* (New York, 1999).

historical works, they must distinguish between their assessment of the logic of analysis and their critique of the technique of presentation. My approach focuses mainly on how historians develop, document, and assess interpretations of world history, but White argues convincingly that in addition to the research, the author's outlook and method of presentation determine much of what the reader sees. The philosophical predispositions of world historians have been at the center of the debates on world history in earlier times, and they may return to a position of importance once an expanded literature gets beyond the early stages of developing its practices.[25]

I think that world historians need to be realistic but not pessimistic about the limitations on verifying global interpretations in history. We need a discourse in world history about what has been established, and what has not—distinguishing among provisional interpretations, interpretations that reflect a current consensus, and interpretations that seem to be confirmed beyond dispute. To this end, it is helpful for world historians to provide readers with their strongest statement of the framework of analysis and the interpretation advanced, as such statements do most to highlight the inconsistencies within the analysis and the evidence contradicting it. Such a discourse should account for varying frameworks of analysis and should identify when the debates dispute evidence, and when they dispute approaches. Meanwhile, the many plausible but unverified interpretations of the past have great value, especially because they clarify new ideas and encourage further research.

25. Costello 1993.

Chapter 18

Analyzing World History

There does exist a characteristic method for analyzing world history. As widely as studies in world history may vary in the topics and disciplines of their research, they retain a certain commonality in their underlying approach, which distinguishes them in method and not only in scope from studies at localized and specialized levels. After exploring many of the specifics in global analysis in chapters 15 through 17, I offer in this chapter a summary statement on analyzing world history. Logic, data, and carefully selected language, used in different ways, provide the materials for the six steps I have identified in preparing a global historical interpretation.

The logic of a world-historical interpretation must be evident from its top to bottom levels. At the pole of breadth and generality, the study must strike the reader as coherent. At the opposite pole of historical specificity, the analysis must show consistency with the evidence and must also fit with the logic of one or more of the established disciplines. In between these limits lies the logic of connecting the poles: the study must connect the general and the specific and must link the various disciplines, regions, perspectives, and levels of analysis.

The data provide the concrete examples through which an author expresses the logic of a world history presentation. History, although it is enriched by theory, is a fundamentally empirical and data-driven field of study. Even world historians, though seeking to escape the empiricism of just-the-facts history in order to explore broad patterns and interconnections, are really searching to enlarge and connect the containers for evidence rather than to replace data as the ultimate arbiters of historical interpretations. The data remain fundamental.

The language of a historical analysis is the medium through which the author expresses the logical consistency and factual evidence of the interpretation. Finding a language for world history is no small matter, because it is not easy for historians to agree on a common terminology. This is partly because of the breadth of historians' audience, but more fundamentally it is because most historians address a wide range of experiences and perspectives, and they are reluctant to oversimplify by selecting a uniform terminology. Specialist historians working in a given discipline (such as aviation historians and medical historians) can escape the dilemma by adopting the terminology of their discipline and limiting their studies to its confines. Historians of localities, where data are limited in quantity, can escape this dilemma by presenting the full range of their data without regard for logic, arguing that they have presented the full historical record.[1] But for world historians, the dilemma of the language of analysis is ever-present. A terminology that is too broad and general is superficial; language that is too specialized becomes inaccessible. Any terminology will have its difficulties. Civilization, society, nation, race, tribe—all these are terms that tend to reify and essentialize structures that are permeable and heterogeneous. The most common approach is for each historian to develop his or her own analytical terminology but also to use widely accepted terms.

Research Agenda: Selecting a Topic and Objective for Analysis

The first step in a world-historical analysis—selecting a topic—is at once obvious and problematic. Any topic appears too big to manage the evidence, yet too restricted to address the interpretive issue. In the face of this dilemma, my belief is that each historian should select an issue of compelling personal interest and pursue it until the broad outlines of it become clear. If the various aspects of the human experience are in fact interconnected, the selection of one topic retains access to others.[2] The study of world trade may begin as a disciplinary specialization, but in exploring trade one runs inevitably into the issues of economic production on one side, and of social organization and cultural preferences on the other side. These additional dimensions of the topic will lead the historian to reformulate and perhaps expand his or her study.

Selecting a topic leads directly to setting other priorities for the study. It involves setting an objective or posing a main question for the study. The objectives of world historical studies are generally to locate interconnections in social processes and between social and natural processes, and to show large-scale patterns over time. In addition, selecting a topic implicitly brings with it a perspective from which the analyst views the issue, one or more disciplines in which to work, a set of boundaries in time and space, and even a terminology for analysis. In setting these parameters for study, the historian is likely also to decide on what examples to emphasize and what data to collect.

---

1. For a study of plant domestication in the Fertile Crescent and surrounding regions that relies on techniques of paleobiology and archaeology see Zohary and Hopf 1993; for a cosmopolitan analysis of Los Angeles see Davis 1990. These works are limited, respectively, by discipline and space; neither is a work of world history, but both are relevant to world history.

2. I began this book with three examples of a problems for which local explanations are inadequate: the simultaneous appearance of racial segregation in many parts of the world, industrialization, and the development of agriculture in widely separated parts of the world within a few thousand years.

In this book, for instance, I began by selecting the topic of "global historiography"—the world-historical analysis of world-historical writings. My objective was to see if I could confirm the need for more research and graduate study in world history. I started with my personal perspective, that of an academic in the United States, and considered the topic as a development in intellectual history. I set the scope as a review of scholarly work of the past two or three centuries. The relevant evidence, therefore, was writings in European languages relevant to world history over that time span. This initial definition of the topic evolved, of course, as the study proceeded.

Exploratory Comparison

Having decided on a topic, but before making a definitive choice on research design, the historian should consider the topic from a maximum of different angles. He or she should make as many comparisons as possible to other topics, whether apparently related or not—comparisons ranging widely over space, time, scale, and theme—in search of links and similarities. This is the brainstorming stage of world historical analysis: at this stage, the facts may speak for themselves. The available historical data provide the main clues as to possible ways for setting up the formal analysis. In exploring the data, the historian should also explore the logic and terminology in which they are enmeshed: the exploration may reveal possible models, links, and dynamics of change.[3]

The French Revolution of 1789, often taken as a world-historical turning point, provides an illustrative example. One may begin by treating it as a national revolution and compare it with such other national cases as England in 1688, the United States in 1776, France in 1848, and Russia in 1917. Or one can consider its interactions with regions engaged in its wars: virtually all the lands of Western and Eastern Europe, plus Egypt and the Caribbean. One can consider the topic in terms of social movements, such as those of artisans (in the United States and Brazil), of peasants (such as those in South America and Russia), the antislavery movement (in Britain and the United States), movements of slaves (in Haiti and elsewhere), and political movements of republicanism and constitutional reform (in the United States, the Caribbean, and Poland). There are many more possibilities. Listing and exploring these possibilities will lead one to revise and specify priorities at the conclusion of this step.

For the example of this volume, I began by comparing works of world history to each other, then to studies in national history. Then I compared recent study in world history to earlier studies of broad scale. In addition, I compared world history to works of broad and specific scale in other social sciences, humanities, and natural sciences. I had to consider what data to collect and how the data would influence my exploratory comparison. I tried to locate all the types and levels of history and to distinguish open and closed systems of historical writing. The exploratory comparisons opened me to literatures outside European languages,

3. Tukey 1977.

to the wide range of audiences and types of world history, and to current social issues as an influence on world history. Despite my initial focus on research, the comparisons led me to include explicitly the work of teachers of world history.

Research Design: Modeling the Dynamics

Next, the historian must set the numerous aspects of a model for his or her study. Overall, this means proposing the logic of relationships and the dynamics of global interaction, to link disparate bodies of information into coherent stories. The models of historians may range from explicitly detailed and deductive theories to attractive but imprecise metaphors. Whatever model one selects, one needs to explore it to its limit in search of ideas to be tested. World history requires its analysts to consider cases, networks, systems and debates, and it relies on the art of conducting several of these activities at once.

As the first step of setting your model, you must select the discipline or disciplines in which to work. Disciplines give us topical areas of study and analytical tools with which to study them. Each discipline has one or more standard models of analysis, though the structure of these models varies greatly. Fields such as economics and demography work with a formal, deductive theory and a limited number of variables, which lend themselves to quantitative analysis. In literary analysis, the theory is typological rather than deductive, and any term refers to a whole range of variables. The disciplines of art history and music history are artisanal rather than theoretical, and they rely on developing insights through practice. Specialized historians use a single discipline and the language and logic of that discipline, be they neoclassical economic historians working on trade or medical historians working on epidemic disease. World history relies on such specialization, but is not itself specialized in that way: commonly, the approach of the discipline must be modified to fit a study of global scope. In particular, world historians commonly use two or more disciplines at a time, and they must therefore develop the logic of their own linkage of these disciplines and their models.

Once you select the disciplines of your analysis, that decision commonly entails identifying the historical dynamics on which the analysis focuses. Work in microeconomics commonly focuses on pressures within systems to reach equilibrium; work in art history centers on processes of emulation and innovation; work on biological organisms concentrates on development of steady states in open systems; some work in plant physiology and in cultural studies focuses on diffusion. You may select your principal historical dynamic as part of creating your model, or you may find that the dynamic is revealed in the analysis after you have created the model.

You must then set the scale at which to define your model in space, time, and topical breadth. For historians, whose discipline is so tied to place and who are accustomed to working within nations or civilizations, the question of scale focuses especially on region. The novelty of world history is that of exploring the overall dynamics of human society and how those are connected to regional and local dynamics. In other areas of study, the issue of scale has been addressed with considerable success. For instance, economics encompasses microeconomic analysis of firms and consumers and macroeconomic analysis of national economies; biology encompasses the analysis of cells and the behavior of whole plant and animal organisms; physics ranges from the analysis of atomic particles to the dynamics of the whole cosmos. These ranges of scale within disciplines remind us that in history, the global level is not simply the sum of the local, and long-term change is not simply an accumulation of short-term changes. Some patterns exist more at the global than the local level, and these patterns influence localities as well as the reverse.

For world historians the issue of scale is not so much that of insisting on analysis at the planetary level but rather that of developing analysis of the past at a range of scales and linking the analyses at various scales to each other.[4] The vision of human history overall is distinct from that same history as seen from Australia. By the same token, human experience as seen from Australia appears substantially different from the same experience as seen from South Asia. Yet to look within a region, one may say that human history as seen from India might look quite different to a Muslim peasant than to a Hindu prince. Analysis by region highlights the variety in human societies and the varying ecological, social, and historical circumstances that give different answers to the same questions. To the degree that analysis by region tends to emphasize exceptionalisms, one can balance these with a turn toward the planetary direction by addressing formally the systemic connections around the globe. Yet even here, the range of possible connections means that there can still be many global interpretations.

Having set the discipline, region, and time frame of your analysis, you must set other details of your analytical framework. The framework, as I have defined it, includes units of study, procedures of comparison, connection, and systemic linkage within the study, analytical strategies, and statements of the results. Comparison of experiences can take place on the basis of distinct cases, through comparison of cases able to influence each other or through emphasis on linkages among places or times. Adjusting the time frame and the geographical scope of an analysis can change the interpretation. Thus the Mongol conquest was a disaster for Iran in the perspective of the generation that experienced it, given the physical destruction and immense loss of life. Yet from the perspective of two centuries, the Mongol conquest would grow to be celebrated, even in Iran, for the economic and cultural wealth brought by the region's wider connections. Similarly, the Protestant Reformation was a disaster for Catholicism in Northern Europe, yet the Catholic response, while insufficient to reverse the loss of Northern Europe, gained new adherents in many parts of the world.

In the modeling of this study, I began with a simple model, emphasizing the gradual development of world-historical discourse among Western writers and noting the impact of warfare and rapid social change in stimulating periods of greater attention to global issues. Then I gave more thought to whom should be included as a historian, and I concluded that historians are those who assemble knowledge on the experience of the community; changes in the extent of knowledge and in the identity of the community will lead to revisions in historical interpretations. Thereafter, in giving more thought to the changes in world-historical knowledge

---

4. In particular, it remains difficult for historians to analyze patterns at the planetary level, because our data are not yet organized at that level.

and the development of various disciplines, I came up with a more elaborate model: the model of two paths to world history. In it I distinguish between the established dialogue in history and the new ideas about change over time emerging in specializations outside history. In the historical dialogue, I argue that changing perspectives, themselves stimulated by social change, are leading to more emphasis on global study. For other specializations, I argue that new discoveries about change over time are causing the frontiers of history to be expanded, as historians join those trained in other fields to explore the interactions in world history.

The dynamics of my model do not draw heavily on new theories or analytical techniques: I rely on the artisanal approaches of intellectual history and history of philosophy to explain the changes in history emerging from within the community of historians; and I rely on an eclectic survey of other fields of knowledge to trace (but not really to explain) the changes in general knowledge. Perhaps subsequent studies of global historiography will rely more explicitly on one or more disciplines to explain the dynamics of global historical analysis.

Connecting Subsystems

Modeling leads to simplification, focus, and a narrowing of the analysis. Once the model and its consistency with evidence are established, it is time to seek out broader connections. Once you have set your model and begun the analysis, you should give particular attention to connecting subsystems in your model and in your interpretation. That is, once you have defined the overall scope of your analysis, any logical, thematic, regional, temporal, or functional portions of the overall system should be considered in its interaction with others. More simply, you should identify linkages both episodic and systemic within your analysis. The connections that world historians seek out include those among events and processes in the past but also among the models and disciplines with which we explore the past. World history links both the accidental and the systemic connections in the places, times, and themes of the past to help explain the broader patterns. The effort to connect subsystems may result in identifying new and broader dynamics of historical change. In a sense, it is broadening the model.

In my analysis of global historiography, I had already observed the linkages of domestic multiculturalism in the United States, global decolonization, and the growing interest in world history; similarly, I had observed the link of area studies, new social-science theory, and world history. The link between teaching and research in world history, which I had accepted in principle, became concrete and powerful for me in the course of recent years working on secondary and college curriculum. The newest connection to me, however, was to learn of the distinctions and the links between the historians' path and the scientific-cultural path to world history.

Verifying Conclusions

Once you have pursued one or several logical threads in analysis, you can seek to complete the process by confirming the analysis, showing the consistency of evidence and model. Begin by selecting an intended level of verification among the following choices:

- Plausibility—the analyst demonstrates the plausibility or internal logic of the argument, supporting the argument with illustrative examples.
- Inspection—the analyst conducts an exploratory and usually comparative inspection of the available data. The inspection is informal in that its procedures, while perhaps detailed, are not specified systematically or compared to explicit alternatives.
- Debate—the analyst seeks to confirm the validity of his or her interpretation through critique and rejection of analyses on the same topic performed by other historians. The debate may range over any of the techniques of verification listed here.
- Hypothesis-testing—the analyst conducts a confirmatory analysis by formal hypothesis-testing. The verification requires specifying the data, assumptions, and procedures used in the test.
- Feedback-testing—the analyst conducts a confirmatory analysis by feedback-testing.

Next is carrying out the selected process of verification. Verification by inspection is still the dominant practice of historians. Within the bounds of this approach, authors should be encouraged to be as clear as possible about how they have performed their inspection of the evidence, even though the process is informal by definition. For hypothesis-testing, world historians should start up a discussion of which hypotheses on global history are testable, and how to organize data and alternative interpretations in order to conduct the tests. For feedback-testing, historians analyzing interconnected global processes should discuss how to identify the feedback among the subsystems in their objects of study and how to determine when the patterns fit their models.

Readers should pay as much attention to verification as authors. Readers should become far more alert in assessing what is verified and what is not in any historical interpretation. Going back and forth between the comparisons and connections of empirical evidence and the patterns proposed in the author's model should reveal, eventually, the weaknesses in the model, and lead to development of a stronger and more comprehensive interpretation.

In chapter 17, I have been critical of the shortcomings of historians in verifying their interpretations. In fact, I cannot claim to have achieved much more than plausibility for the presentation of the major arguments in this book. I remain convinced of my own arguments on the relative neglect of research in world historical studies, in my tracing of the main stages of the world history literature, in my interpretation of the connections of a wide range of changes in knowledge and perspective, and in my identification of main areas of debate and proposed courses of study in later chapters. Perhaps further review will establish which of the assertions in this volume are testable or verifiable according to one criterion or another, and which are simply descriptive statements.

Shifting Perspectives

The steps I have listed so far add up to a single iteration of a world-historical analysis. Having formulated the analysis, the historian should reformulate it and see whether the next iteration yields similar results. In reformulating the analysis, the key step is to shift perspectives—rethinking the analysis from inside and outside. A world historian, upon developing a perspective in the past, should evaluate it from the standpoint of one and then another person from that past time, explore it through the optic of one and then another analytical discipline, and reconsider it in short-term and in long-term perspective.[5]

The perspective from which I first analyzed global historiography was that of U.S. academics. I then reviewed the topic by seeking to adopt the perspective of academics around the world, and then of investigators of world history over a long time-perspective. Thinking about the literature from the perspective of a teacher of world history led to a revision and expansion of the manuscript. I also thought about the place of world history in popular culture, and that began a line of thinking that is not yet well organized. Another perspective from which I reconsidered my topic was the question of what groups may gain hegemony in the current wave of globalization; I also reconsidered the world history literature from the standpoint of the progress of freedom and the expansion of new sorts of inequality. Exploring each of these new perspectives encouraged me to rewrite a few sentences at least; in other cases the result was changing the chapter organization and rewriting extensively.

I think that all six steps of the guidelines for interpreting world history are necessary. Three of them, however, are really distinctive for world history—exploratory comparison at early stages of the work, connecting subsystems in later stages of the work, and shifting perspectives to launch a new phase of study. These are the analytical steps that distinguish the broad and interconnected approach of world historians from the work of those specializing in smaller arenas of history. Of the six steps, these are the three that broaden the analysis; the other three steps work in the opposite direction to focus and specify the study.

Presenting the Results

Having advanced and verified, insofar as possible, the various analyses, one now faces the question of how to assemble them into an overall interpretation. Historians are divided on the degree to which one may formally combine various analyses—for instance, social class, imperial policy, religious faith, and environmental change—into an encompassing statement. Fred Spier, an insightful proponent of Big History, maintains the hope that the framework of Big History is the basis for a "unified method" of world history. In his view, the notion of "regimes" in various areas of the human and natural world links to the more

5. How many new perspectives should the researcher explore? As long as the last new perspective has added significant insights to the study, I think it is worth trying one more.

specific arenas of social, environmental, and other analyses that fit systematically into the larger framework.[6] These regimes are patterns of interaction and change in the various systems of the social, biological, and physical world.

Spier proposes, for regimes of human society, that there exist social regimes, ecological regimes, and individual regimes. These regimes define the dynamics and the limits in each of these areas of human activity. Spier defines a regime in general as "a more or less regular but ultimately unstable pattern that has a certain temporal permanence."[7] In implementing this definition historically, he seeks out the sources of both stability and instability of each regime, and then traces those times of "regime transformation" during which the regime undergoes major reorganization. One of the main instances he explores is that of the human ecological regime and its transformations. The key developments, in this perspective, are the controlled use of fire (about 400,000 years ago), the development of agriculture (about 10,000 years ago), and the development of industry in the past 200 years.[8]

The idea of developing a single logic for analyzing history at all scales has its attractions: the historian carries out an analogous exercise at each level of temporal, spatial, and thematic breadth, fitting the specifics of the data into the general pattern of regimes and regime transformations. In contrast, I would say that we do not yet have the knowledge to create an adequate model for all of history and that there are particular benefits to replicating analyses from different standpoints rather than seeking one unified approach.

Spier has found his language for broad patterns: a terminology that enables him to identify the parallels in patterns of change at various scales in history. There remains a significant difference, however, between what Spier has called a "unified method" for history and what I have called "tasks" for the study of world history.[9] In my opinion, historians and social theorists do not yet know how to weight or balance the disparate though overlapping analyses of art, politics, and agriculture that they conduct. Rather than seek to aggregate or assemble the various perspectives into a comprehensive interpretation, I think historians should focus for the present on alternating among various views of world history.[10] For instance, in the case of Spier's regime transformations in human ecological history, I would say that the development of language, somewhere about 200,000 years ago, is an important transformation that was left out. Spier, in turn, might respond that language represented a regime transformation at the individual or perhaps the social level and was thus not excluded from his analysis. A practical way of determining whether one of these approaches to world-historical studies has advantages over the other would be to conduct a number of world-historical analyses, some focusing on a unified method and regimes, others concentrating on a more eclectic set of analyses from various standpoints, and see whether one

6. Spier 1996. This search for a "general paradigm" in history parallels the search for "unification" in the theory of physics.

7. Ibid., 14.

8. Ibid., 45–80.

9. See chapter 23.

10. See pages 253–254.

produces results that are at once more detailed and stronger in establishing over-all patterns. In both Spier's view and mine, however, the analyst's alternation among various frames of analysis and effort to establish an interpretation encom-passing them are a priority in world history.

How is one to analyze cases for which significant data are missing? How are we to work from situations for which we have sufficient data to broader situations, for which data are in shorter supply? It may be helpful to consider the circum-stance of archaeologists, who study single sites through laborious and expensive effort, yet who find ways to propose generalizations applicable to broad regions. The links between empirical case studies and global interpretations include inter-polating, extrapolating, projecting, and generalizing from known data to related circumstances for which data are unknown. All these are types of speculation, but with care they can become speculation that is guided and limited by data and theory.[11] To put it in terms of hypotheses, one may say that this sort of analysis involves developing verified interpretations out of theory and documentation on the past and then posing hypotheses on wider times and spaces that might be ana-lyzed with future data.

In the preceding passages, analysis has been presented as the task of solving a problem posed in a historical question. This much of analysis is true to the posi-tivist tradition of disaggregating the world into small pieces and studying each in isolation. Analysis in history, and especially in world history, requires us to look beyond the specific cases or networks under study. It asks us to pose additional problems and to identify dynamics that operated in other situations.

The dilemma in overall interpretation is how to assemble the various perspec-tives one has developed in the course of the study. To provide a decisive analysis, the author must summarize the historical processes under study at various levels, from local to global and from short-term to long-term. Where is the nexus of change?

Two recent efforts in world history reveal the benefits of alternating perspec-tives in deepening the analysis of global topics. Eric Martin undertook a study with the intent to portray twentieth-century globalization as a global movement. His device for this portrayal was to explore the careers of three major leaders—Mohandas K. Gandhi, Kwame Nkrumah, and Ernesto Guevara.[12] In his initial design for the study, he focused on the debate of anti-colonialists with the defend-ers of colonialism, taking newspapers, radio, and the construction of history as fields of discourse. With time, however, Martin shifted from the perspective of political and social history to consider this same topic as intellectual history. There, after exploring the work of Wilhelm Dilthey and his successors on the

---

11. The role of speculation in reconstruction of the history of Africa is often discussed by specialists in the field. When I mentioned speculation to David William Cohen, he responded, "It's oxygen," thus conveying, I think, that the speculative element of his-torical reconstruction is not the majority of what we do but that it is essential to the success of the interpretive exercise.

12. Eric L. Martin, "Decolonization in the Twentieth Century" (Ph.D. dissertation, Northeastern University, 2001).

concept of worldview, he focused on the development of the world views of the young Gandhi, Nkrumah, and Guevara. In each case he found that a cosmopolitan life combined with the facing of great personal difficulties turned each of these men into a person wishing to devote his life to challenging the existing system of colonial dominance. Each of them wrote a major interpretative statement on the world around them at this turning point in life, and these statements provide a clear guide to their policies and actions in later years.

In another study, Pamela Brooks explored parallels in the participation of women in bus and pass boycotts during the 1950s in Montgomery, Alabama, and in Johannesburg, South Africa.[13] In exploring the antecedents of women's consciousness, she located parallels over time, and indicated the global dynamics at work in racial discrimination, industrialization, and urbanization. The women of Johannesburg were in very little direct contact with those of Montgomery, but each group found itself in conditions after World War II in which conservative political elites sought to turn back the clock on incremental improvements in the status of blacks. Brooks's initial analysis focused on the consciousness of the women activists and their decision to become active in the demonstrations against racial restrictions. In reforming the analysis, however, she found a nexus of change in the reproduction of a culture of resistance to discrimination in the black families of Alabama and the Transvaal. Her initially short-term investigation was transformed into an eighty-year review tracing the accumulation of generations of self-development that enabled the women of the 1950s to commit themselves to social confrontations. The coincidence of those distant confrontations in turn reinforced their impact as visible and effective challenges to racial discrimination worldwide.

Both authors selected historical sub-disciplines—intellectual history and descriptive social history—and followed fairly closely the canons of those approaches. Each invoked historical dynamics from the local level to the planetary but centered on forces of change within the reach of individuals in the twentieth century.

Conclusion: Research Design and Execution

All of these steps combine into the larger categories of design and execution of a research project, whether it becomes a book-length project, the preparation of materials for a day's class session, or anywhere in between.

The chapters in part IV have moved from the descriptive mode to the prescriptive. This chapter thus serves as a bridge to the next and concluding section of the book, in which I suggest ways for those involving themselves in world history to prepare themselves in study, research, writing, and teaching.

13. Pamela E. Brooks, "Buses, Boycotts and Passes: Black Women's Resistance in Montgomery, Alabama and Johannesburg, South Africa from Colonization to 1960" (Ph.D. dissertation, Northeastern University, 2000).

Part V

Study and Research in World History

N avigating world history requires well-developed habits of study. The world historian needs to be clear in selecting and pursuing well-chosen objectives but also alert to discoveries along the way that may call for a change in course. The basic habits of the world historian, at levels from the beginning to the advanced, provide the most consistent support for an enjoyable and productive exploration of world history. Those habits include attention to patterns in history, exploration of linkages and comparisons among historical patterns, concern for details that reveal historical connections, and readiness to shift perspective in assessing the past.

Part V presents guidelines for study of world history. Principally, it focuses on formal graduate study for researchers and teachers of world history. In these chapters, I argue that the field of world history is sufficiently complex and distinctive that a program of formal study presents substantial advantages over informal, self-directed study. For researchers, I focus primarily on doctoral programs of coursework, preparation for teaching, and dissertation research. In addition, I suggest programs in research for graduate students who will not do a doctoral specialization in world history.

For teachers, I present programs of study at three levels of intensity. I believe formal study at the graduate level to be as important for teachers—working at elementary, secondary, and college levels—as it is for researchers. At the same time, there are many who, while wishing to learn more about world history, find programs of graduate study to be either unavailable or inconvenient. So I have also developed a parallel set of suggestions for independent study in world history. These suggestions, which occupy a portion of each chapter in part V, are directed toward general readers or scholars in fields outside history who wish to expand systematically their understanding of world history. More especially, I offer these suggestions for independent study to those many teachers, at all levels of the educational system, who are thrust into the responsibility of teaching world history to young people but who have little or no formal preparation in the field.

In addition to my review of the various types of graduate program for study of world history that have developed in recent decades, successive chapters propose programs for graduate study in world history at three different levels of intensity, describe details of five categories of resources in world history, and propose models of research projects from microprojects to book-length studies. Throughout, I discuss the priorities for faculty members and graduate students in programs addressing world history.

Throughout the chapters of this section, I caution the student of world history to be alert to the inescapable dilemmas of the field: the need to be comprehensive yet selective, to show connections without tying history into jumbled knots, and to recall important detail of the past without rote memorization. Faced with such dilemmas, world historians need to be decisive but open to revising their decisions.

The concluding chapter recapitulates what I believe to be some of the main items on the world historian's daily checklist. The keynotes for study and research in world history include points of emphasis in reading, analysis, coursework, research, writing, and oral and multimedia presentations.

Chapter 19

Programs and Priorities in Graduate Education

G raduate programs in world history now have enough of a history to warrant a brief survey of their development. After sketching such an overview, I turn to commentary on the main decisions within such programs: choices for the organizers of the programs, the participating faculty members, and the students entering the programs. The most basic choice in program organization has been whether to take an area-studies approach or a global-studies approach to world history. Once that choice is made, universities and faculty members face choices in their level of commitment to world history, the nature of their course offerings, and their approach to mentoring of students. Prospective students must decide on the type of program they wish to enter and their level of involvement in world history. In this and subsequent chapters, I discuss these choices within the categories of world history as a research field, a teaching field, and a minor field.

World History Programs: Growth and Constraint

The University of Wisconsin–Madison program in Comparative Tropical History, created under the leadership of Philip Curtin in 1959, was arguably the first formal program of graduate training in world history. By the end of the 1960s the program had been renamed Comparative World History. This program trained a large number of active and successful scholars who have since become very prominent in the fields of African, Latin American, Southeast Asian, and South Asian history. To encourage global research, Curtin organized a 1975 summer workshop in comparative history at which a dozen young scholars, mostly with recent Ph.D.s from the University of Wisconsin, wrote and exchanged studies in comparative history.[1] These students did not receive degrees in world history, nor

---

1. Curtin, "Graduate Teaching in World History," *Journal of World History* 2 (1991), 81–89. The summer 1975 workshop, held in Madison, was supported by the Carnegie

did they initially publish in world history. Yet the experience of comparative work during graduate study encouraged them, in their second books and thereafter, to begin contributing actively to study field of world history. In an outstanding instance, Ross Dunn, who became president of the World History Association in 1984, completed his Ph.D. in 1970 with a study of anti-colonial resistance in Morocco, which he published in 1977. He went on to write a biography of world traveler Ibn Battuta and turned next to a high school text in world history. A second outstanding figure was Michael Adas, who published a study of economic development in Burma, then in 1977 published his comparative study of millennarian movements, the first volume in a series on comparative world history. Adas sustained this emphasis on comparative world history in directing graduate students at Rutgers.[2]

As Curtin moved to the Johns Hopkins University in 1975, the global dimension of historical studies languished at Wisconsin. At Hopkins, however, several of Curtin's graduate students took up global historical topics. This was no longer a formal program in world history, because of the particular mix of collegial and individualistic approach of the Hopkins department, but doctoral students in history were able to gain a cosmopolitan and interdisciplinary program of study through participation in the Atlantic Seminar.[3]

Beginning in 1977, the doctoral program in sociology at Binghamton University trained historical sociologists in association with the Fernand Braudel Center.[4] In the 1980s, programs of graduate study in world history developed at the University of Hawaii, Ohio State University, University of California–Santa Cruz, and the University of Minnesota. At the University of Hawaii, Jerry Bentley, who had completed his doctoral work at Minnesota in Renaissance and early modern history, began teaching a graduate seminar in world history and gained approval from his department for world history as a secondary field in Ph.D. examinations. The students in that seminar gained further experience as teaching

Continued

Foundation. Among the participants were Ross Dunn and Michael Adas. I participated in a similar group the previous year, in which a dozen young scholars researched, wrote, and exchanged manuscripts on African economic history: included were Anthony G. Hopkins, Sara Berry, Ralph Austen, James Spiegler, Paul Lovejoy, Jean Hay, E. J. Alagoa, and Babatunde Agiri.

2. Ross E. Dunn, *Resistance in the Desert: Moroccan Responses to French Imperialism 1881–1912* (Madison, 1977); R. Dunn 1989; Ross Dunn, *Links Across Time and Place: A World History* (New York, 1990); R. Dunn 2000; Michael Adas, *The Burma Delta: Economic Development and Social Change on an Asian Rice Frontier, 1852–1941* (Madison, 1974); Adas 1979; Adas 1989; Adas 1993.

3. Among the students were William Storey, Helen Wheatley, Lauren Benton, and David Gutelius.

4. William G. Martin completed his degree in the program and is now on the faculty at Binghamton and a member of the board of the Fernand Braudel Center. William G. Martin and Michael O. West, eds., *Out of One, Many Africas: Reconstructing the Study and Meaning of Africa* (Urbana, 1999); Martin, ed., *Semiperipheral States in the World-Economy* (New York, 1990).

assistants in an undergraduate survey course in world history, and the totality of these activities provided the basis for Bentley's launching of the *Journal of World History* in 1990. At Ohio State University, Carter V. Findley and John Rothney began teaching an undergraduate course in twentieth-century world history in the early 1980s and wrote a successful textbook to go with it. Findley and fellow Middle East specialist Jane Hathaway then created a graduate seminar in world history, and the history department approved world history as a secondary field in doctoral examinations. At Santa Cruz, Edmund Burke III, David Sweet, and Dilip Basu created a graduate concentration in world history during the 1980s. It was discontinued after a couple of years and then reinstituted in the late 1990s. At the University of Minnesota, a doctoral concentration on early modern world history developed in the 1980s, through collaboration of Carla Rahn Phillips, William D. Phillips, Stuart Schwartz, James D. Tracy, and others. Rutgers University, through the activities of Michael Adas and Allen Howard, developed a teaching specialization in world history for its graduates. Most of the graduates of these programs were employed as regional specialists, but some have become well-known as world historians: David Chappell, who completed his degree at the University of Hawaii and now teaches there, is one example.[5]

Master's-level programs began including specializations in world history in the 1980s, notably at the University of Manitoba. The two processes were combined at Northeastern University where, in 1990, the Department of History resolved to add a significant world history component to its M.A. program and to create a Ph.D. program with degrees in World, U.S., and European history. The new Ph.D. program admitted its first students in 1994 and began awarding degrees in 1999.[6] By 1998 world history became the only field available for the Northeastern Ph.D. With the doctoral program came creation of a World History Center focusing on linkage of research, curriculum development, and teacher preparation. The Center was soon involved in multimedia production, with production of *Migration in Modern World History*, a CD-ROM, for which work began in 1995 and publication came in 2000. In 1996 the program began work in preparing secondary teachers of world history, and in 1998 it began an active campaign of teacher workshops.[7]

In the late 1990s a larger number of institutions began creation of programs in global historical studies. Additional graduate programs in world history began at Georgia State University and at the University of California campuses at Riverside and Irvine; doctoral programs in Atlantic history opened at Florida International University and University of Texas–Arlington. At the University of Pennsylvania, a graduate certificate program in world history opened.

This development of graduate study in world history took place in an era of severe constraints.[8] The profession of history was in demographic decline during most of this time among students and faculty members. As a result, while the

---

5. These notes on early programs come from my discussions with the persons noted.
6. Manning 1999.
7. Manning et al. 2000. Teacher institutes are listed in the "Workshop" section of the World History Center website, www.worldhistorycenter.org.
8. See chapter 7.

research was developing and changing dramatically, the institutions for the study and the teaching of history changed very little. Changes in graduate education took place in a few institutions, as I have mentioned, but with little reinforcement from colleagues in those same institutions or from history departments at other institutions.

Each of the programs has faced its ups and downs. Especially for public institutions, the fluctuations in state finances led to cycles of boom and bust. The states of Hawaii and Ohio each underwent severe budget cuts in the 1990s that led to restrictions in history programs; one result was the cancellation of a promising Ph.D. program in world history at Miami University of Ohio before it could even begin. Private institutions also had their problems: in 1997 the Northeastern University administration proposed closing down the new History Ph.D. program.[9]

Beyond the boundaries of the United States, studies of world history have grown more slowly and according to different dynamics. In few cases is there a substantial expansion in teaching world history at secondary or college levels, though India and parts of Africa may present examples of such a trend. In Australia, Netherlands, Canada, China, Turkey, Brazil, Singapore, and some other countries, significant numbers of individual scholars have taken up activity in global historical projects, though rarely do they have institutional support. In Britain a national directorate of world historical studies has been organized under the leadership of Patrick K. O'Brien, a noted economic historian, at the London School of Economics. Thus, while scholarly study of global issues is heavily focused on the United States, university communities survive throughout the world, and internet connections make possible a shared discourse, and shared resources.

Dilemma in Focus: Regional or Global Approaches

What should be the balance between area-studies training and global training in the preparation of world historians of the future? This question was addressed explicitly in a meeting of many leading world historians at the University of Texas in February 2000.[10] Philip Curtin of the Johns Hopkins University began the

9. With a new president in office, in 1997 the Northeastern University administration proposed closing down the new History Ph.D. program, along with other social science doctoral programs. The History Department campaigned vigorously for continuation of the program, receiving wide support from world historians. The administration ultimately reconfirmed the program, but changed its definition from a Ph.D. in U.S., European, and world history to a Ph.D. in world history, and provided no new resources.

10. This meeting, organized by Philip White as part of World 2000, a larger conference on world history and geography, brought sixteen leading scholars together to discuss directions in graduate study of world history, to assist the University of Texas history department and administration decide whether and how to undertake graduate study in world history. Philip L. White, ed., "Doctoral Training in World History: What Changes in Ph.D. Programs Will it Require?" *World History Bulletin* 17 (Spring 2002), 8–17.

discussion by emphasizing his belief that area-studies training should be the primary emphasis in doctoral training and that global studies should be a field of study within that rubric. Jerry Bentley of the University of Hawaii, Carter Findley of Ohio State University, Peter Stearns of George Mason University, and others supported this judgment, arguing that an attempt to place global training first would risk creating historians insufficiently knowledgeable of the specifics of the historical situation on which they worked. Of all the historians to speak, I was surprised to find myself the only one to argue in favor of a Ph.D. in world history, with a dissertation in world history and with area-studies training subordinate to global preparation.[11] The program in which I participate at Northeastern was then the only one to grant a Ph.D. in world history. My argument was that at least some programs should give primary focus to identifying and analyzing global processes, to provide those graduate students an opportunity to focus their entire career on global patterns rather than take up global connections as a second priority somewhat later in life. It was agreed that the actual courses of study and the actual dissertations completed in the two types of program had many similarities. The two positions were laid out, and one must now wait several years to see the results emerging from the various programs of doctoral study to see how the two policies have fared.

Curtin's recommendations for graduate study in world history restated the approach he had expressed in a 1991 article.[12] Both his practice and his article present a program for the comparison of large subunits of human society—culture areas, nations, and continents. His approach emphasizes broad reading and interdisciplinary analysis, but it also emphasizes that Ph.D. candidates should have strong grounding in the empirical data and the literature on one or two world regions. Curtin presents a brief for comparative study in world history at the graduate level and for the production of dissertations that are monographs in world history. Implicitly, Curtin's comparative approach tends to assume that the various regional cases have been independent of one another and that one can elucidate patterns and social science principles through comparing them. There remains a lot of autonomy in the affairs of the world's regions, and this sort of research design is best for analyzing the generality of autonomous experiences. His sensible approach should help to fill the growing need for solid research in world history. But if the literature as a whole is to be a symbiosis of monographic and synthetic work at the global level, graduate study in world history must also address the issue of synthesis.

Graduate study in world history should pursue global and interactive approaches as well as the comparative method. I have no interest in rejecting the comparative, area-studies approach to world history. But I do reject the idea that

---

11. Immanuel Wallerstein described the Binghamton doctoral program in global sociology; William McNeill expressed skepticism as to whether the dissertations I described were sufficiently global. Ibid.

12. Curtin 1991. See also Jerry H. Bentley, "Graduate Education and Research in World History," *World History Bulletin* 3 (Fall/Winter 1985–86), 3–7 (reprinted in R. Dunn 2000: 526–33); and Curtin 1986.

graduate study in world history should be limited to that approach. Unless we have some strong programs giving primacy to global approaches to world history, the development of ideas and information and world history will be slowed down. I have been working since the early 1990s to develop a more systematically global approach to graduate study.[13] The world is rife with interactions among regions and societies, and our research frameworks need to account explicitly for the global as well as the regional level of experience. The rise and decline of racist ideology, as I argued in chapter 1, took place not so much in any given country as in the interaction of many regions. To try to make sense of it by comparison of a set of national studies is to avoid addressing directly the issues of the international development of racism and the international campaigns against it.

Developing doctoral programs at the global level may require changes in the structure of graduate education. The inherited system of doctoral study is characterized by students working with mentors in a single department and their specializing in studies of a chosen region, working with a chosen methodology. According to this model, the way in which broad and transregional studies are developed is for the historian to do an apprenticeship in local studies and broaden out to consider wider areas at mid-career and beyond. This model of graduate study in history seems limited in its usefulness for preparing graduate students either to participate, during their graduate study, in work of this breadth or later on to direct such work.

The area-studies monograph becomes, in effect, the "first language" of such historians, and they take the benefits and the limits of such imprinting and social-ization with them through to the end of their careers. When they undertake global studies linking regions or exploring planetary patterns, they are working in a "sec-ond language," in which they often find themselves thinking in area-studies terms and have to translate their thinking into the language of global interactions. With rehearsal they can be skilled in their second language, but it never reaches the instinctive level of their first language.

Graduate school is a unique and formative period in academic life. The habits and framework of the local study tend to stay fixed in the mind. The perceptions of global interactions, if studied late in the career of a single mind for no more than ten or fifteen years, never get explored in the depth that could occur after forty years of rethinking the issue. Further, the student's formal training is only in the initial region and the initial methodology addressed—the later study, going beyond these initial limits, risks being less concentrated, more informal, and more superficial. In addition, if the student's exploration of successive regions and methods takes place one by one, the result is more likely to lead to good compar-isons among distinct regions and methods than it is to develop global insights into the global and systemic interactions spanning the larger field of study.

We need at least a solid faction of historians whose "first language" is world history. These will begin as graduate students who focus from the beginning of their graduate studies on global patterns and global interactions, on working in multiple disciplines, and on addressing the interplay of multiple themes. Their

13. Manning 1992; Manning 1999.

initial studies with this complex set of connected factors should not be expected to be triumphs of innovation. But if these young scholars make the comprehension of global historical patterns and connections their top academic priority, and if they continue to study and reformulate their interpretations for a long lifetime, there is hope that in the later years of their careers, some genuinely innovative and insightful observations and interpretations will emerge from their studies.

In practice, I think that a comparison of dissertations completed in the Northeastern program with those completed at Johns Hopkins, Rutgers, Hawaii, and Ohio State will reveal no drastic difference in their scope, conceptualization, or handling of documents. But there will surely be a difference, over a lifetime of research and teaching, between those trained in global programs and those trained in area-studies programs with a global emphasis. I think that the best investment of overall energies, in the establishment of doctoral programs in world history, is to find an appropriate mix of programs privileging comparative work and programs privileging global and interactive work. In each case the result of graduate study would be students skilled in performing and writing up monographic research in world history.

In fact the difference between the area-studies approach and the global-studies approach is not as drastic as an either-or selection. The distinction can be as small as the order in which students take their courses. This has become clear to me in observing graduate students at Northeastern. Students who began their graduate study at Northeastern (rather than entering after a Master's program) took an initial year or two of courses focusing on world history and then selected a region (and theme) of concentration as they began to develop their dissertation proposal. At this point they undertook energetic study of their chosen region in area-studies terms. Yet what became clear was that their approach to the area-studies literature was distinctive because of their previous global-studies training. They learned details on their chosen region as well as area-studies specialists, but they found themselves critical of the essentialist approaches of the specialists.[14]

For University Faculties: Creating Programs and Mentoring Students

Creating a graduate program in world history is a big step, because it requires significant innovation in the curriculum, dependable interaction among faculty members, and successful selection, mentoring, and placement of students. It involves substantial responsibilities for the faculty members leading the world history program but also for the entire History Department, cooperating departments, and the university as a whole. The commitment must be taken seriously if

14. Several of the students in the Northeastern program studied world history first and took up area-studies specializations thereafter. This simple alternation of the order of study brought significant difference in the historical questions they sought to investigate. Others among the Northeastern students had began with an area-studies concentration and took their global courses subsequently. Still others focused on global approaches systematically. I enjoyed working with students taking each approach, and I found benefits to each.

the program is to succeed. It must be as firm, I believe, for programs of teacher preparation as for programs of doctoral research.

Commitment. To begin with, in my experience it is very difficult to build a successful graduate program in world history unless at least three members of the faculty are actively committed to its tasks and unless they have clear support from the majority of their department. The tasks of this group include developing a curriculum and scheduling individual courses, recruiting and mentoring students, locating funding for students, finding placements for students (notably in apprentice teaching), and building library and other resources.

Typically, it has been easier for small and medium-sized departments than for large departments of history to launch programs of world history, because personal and professional communications are better in the smaller departments. In the major research departments, historians are focused on individual research projects and on subfields of history rather than on collaboration and connection among fields. That is how it came to pass that at UCLA, for a time in the late 1980s, a department with eighty faculty members could find no one to teach world history.

Preparation of researchers and college teachers of world history can take place within departments of history. But preparation of high school and middle school teachers generally requires cooperation between departments of history and of education. These programs can succeed only if there is a firm commitment to world history in each department as well as a firm commitment to work together. If the work of linking the two departments is left to one or two individuals, one may safely predict a few years of activity followed by decline.

Scope. If a commitment to world history is the fundamental choice, the next basic choice is in the approach of the program to the field of world history and to the education of graduate students. The doctoral program at Binghamton was defined to focus on the historical sociology of world-systems. The doctoral program at Northeastern was defined to be open to comparative, interactive, and synthetic approaches to world history, though its curriculum gave particular attention to historical interactions. The point is that each program should be clear in the academic orientation of its approach. Will it focus on research or on teaching? What time frame and regional emphasis will it adopt? How will faculty members express their involvement?

Structure. The first issue in the structure and organization of a graduate program is the identification and scheduling of course requirements to be completed by the students. The courses fall into the categories of courses in historical methods, courses in world history, and courses in history that are specialized by region, time period, or theme. In addition, students may take courses in fields other than history (in social sciences, cultural studies, natural sciences, or computer techniques). Identifying and planning courses is one part of the process; arranging for them to be taught at times that meet the needs of students is another.

A second set of issues is that of non-course requirements. These include preparation for teaching, requirements for language study, and requirements in methodology. For doctoral students, the program must set procedures for selection of major and minor fields and a dissertation committee.

A third set of issues for doctoral study is the linkage of dissertation proposal and comprehensive exams. In most history Ph.D. programs, students complete comprehensive exams in their major and minor fields of study before submitting their dissertation proposal for approval. In my experience, it is more effective for students and for the program to reverse the order and require that students gain approval of their dissertation proposal before taking their comprehensive exams. In the Northeastern program, we have systematically urged students to complete their dissertation proposal before taking their exams. The result tends to speed completion of the dissertation in a field in which dissertation research and writing often continues for several years. While this benefit of the proposal-first approach is easily recognized, members of our department were asked in a recent external review if it was not the case that asking students to develop a dissertation proposal within two or three years was moving too fast, and whether faculty members were simply assigning topics to students or whether students were adopting topics for which they were not ready.[15] In practice, I believe, the proposal-first approach is an effective way of making the dissertation the centerpiece of the doctoral program, so that courses, exams, and other activities are centered on designing and executing an effective piece of original research. In particular, if the comprehensive exams are taken before a dissertation proposal is approved, the student is more likely to postpone the exams out of fear that he or she has not mastered the literature sufficiently. In any case, the exams are a stressful experience, and it is almost inevitable that the student will have a period of rest (and, not uncommonly, depression) after the exams. If the dissertation project is already designed and approved, it is much easier for the student to begin serious work on it.

The structure of the comprehensive exams is another matter for careful choice. It is possible to organize exams by standard fields (e.g., world history in general, area-studies fields such as East Asia or Europe, or temporal fields such as the twentieth century). It is also possible to tailor the exams to the specific concentrations of the students (e.g., social and economic history of the Atlantic world, history of interactions among major religious traditions). I have found it beneficial to require all world history students to take a general exam on the overall literature in world history, an approach developed earlier at the University of Hawaii and Ohio State University.

Recruitment. What are to be the admission standards for graduate students in world history? How much background in history or in world history do they need before admission? Should students be admitted who already have Master's degrees in history or other fields, or should programs limit admissions to students who

---

15. These questions came in an October 2001 site visit to the Northeastern doctoral program in world history by members of the AHA Committee on Graduate Education.

have just completed their B.A. degree? My experience is that the best students in world history have been those who expressed the greatest interest in the subject. This criterion, in my opinion, has been a better measure of student achievement than student grades, scores on Graduate Record Exams, or the reputation of their previous schools. Of course these other measures are valuable, and students who do not rank at the top on the latter criteria may need extra study to produce work of excellence. A further criterion that I have found of importance is the number of history courses the applicant has taken. Graduate students who are new to history, however well they score in other fields, generally do not do well until they have done a substantial amount of reading in history.

Students who are admitted to a graduate program in world history should receive a clear message as to the content and organization of the program. I have observed cases of students who entered the Northeastern program without a clear knowledge of its focus in world history and who therefore had to undergo a difficult (but sometimes successful) adjustment. Also important, especially for students hoping to undertake research in world history, is that there be a clear connection of the student and a faculty member from the start. Students who enter a doctoral program without a faculty advisor are likely to miss a great deal in their first year of work, for lack of effective mentoring. Of course, students may find after a time that they wish to shift to working primarily with a different faculty member, and this sort of change should be facilitated.

Mentoring. Faculty members in graduate programs of world history need to work actively as mentors. This mentoring is to coach students in their understanding of the field of history at a global scale, in developing a methodological specialization, in becoming effective teachers, in putting a high priority on language study, in research, and in styles of collaboration. Since a graduate program in world history addresses such a wide range of issues, most students will require multiple mentors. The student must then learn to navigate the different approaches of various mentors and may escape the firm direction that would come from a single mentor. Faculty members must then work in committees and keep up with a larger number of students. This collaborative approach will commonly require working with students and faculty members at other departments and other institutions, so that differing schedules and institutional requirements will add complexity to the program of training. This broad approach to graduate training thus requires institutional change as well as individual initiative. Advisors must develop more flexible systems of providing and supervising coursework; students must gain practice in maintaining ties with other researchers through correspondence and (thanks to technical advances) the internet.

Collaboration. World history is too big a subject to be addressed effectively by historians working on their own. Individual study is of course the core of the historical profession, but for world historians, it must be supplemented with collaborative work. This includes collaboration with colleagues at the same institution and at other institutions, between faculty and students and among students. It includes meeting regularly with students and seeking out colleagues who will assist in arranging for students to take courses with other departments and other institutions.

There do already exist some institutions for collaborative research. Some history departments have seminars at which results are regularly reported and discussed: I am familiar with those at the Johns Hopkins University. Perhaps, based on this model, one could set up some transinstitutional seminar on global history, collect relevant theoretical literature, list bibliography, or summarize current research. This might take the form of a website. The National Endowment for the Humanities program in Collaborative Research supported a large project on Louisiana in which I was active, and it might be prepared to support collaborative research that included an explicit component for graduate training.

But who will lead in defining and executing large and collaborative projects? Will there be staff for coordinating the work of faculty members and graduate students who may be at several institutions? Again, there are precedents if we know how to look for them. A fine series of studies on the colonial Chesapeake, appearing mostly in the 1980s, benefited from informal cooperation among doctoral students and faculty members based at different universities.[16]

If there existed large projects requiring collaborative work by several researchers, either within a given history department or spanning various departments, doctoral students could participate in them either as an apprenticeship or as part of their own research. From this experience, students could learn how to draw on the full range of collaborative work yet mark their own individual contribution; they would encounter such issues as the standardization of citations and research techniques.

Collaboration among historians also has its limits as well as its benefits. If project leadership is insufficiently strong, or if there are too many changes in personnel in the research and administration of the project, or if publishers are put off by multiple authorship, these difficulties can interfere with the process of bringing collaborative research to completion and to dissemination. Publication on the web is an option, but one must also seek out new types of relationships with print publishers. A further danger is that collaboration may be easiest for those based at the most privileged institutions, so that the voices of scholars in Latin America or Southeast Asia may not be heard in large projects—that is, their work is implicitly treated as local studies rather than of cosmopolitan import. One can work to address this imbalance through attention to the internet, journals based in these regions, non-English publications, theses and dissertations in African and Latin American universities, and agencies through which these scholars work, including UNESCO and, for African scholars, CODESRIA.

16. Allan Kulikoff, *Tobacco and Slaves: The Development of Southern Cultures in the Chesapeake, 1680–1800* (Chapel Hill, 1986); Russell R. Menard, *Economy and Society in Early Colonial Maryland* (New York, 1985); Lois Green Carr, *County Government in Maryland, 1689–1709* (New York, 1987).

Teaching and research. Graduate programs must provide solid preparation in both teaching and research beyond that given in formal courses. For teaching, this includes informal mentoring, scheduling of internships and preparation of curriculum, and supervised teaching at either school or college level. For research, it includes small research projects to develop basic skills and then a master's thesis or dissertation as a major project.

To turn to the nature of the graduate experience in global research: I have favored requiring students to complete a master's thesis that is an individual piece of study, in which the student learns the discipline of defining and executing a single sizable piece of research and writing. In effect, it is a rehearsal for the doctoral dissertation. Research at the doctoral level must meet more complex requirements: it needs to develop habits of collaborative work that will serve the student well during a career of interconnected study, but it must also enable the student to write a monographic dissertation that can provide the basis for employment and tenure in a field that remains dominated by individual work styles.

Research support. While one point of having students conduct major research projects is for them to learn to fly on their own, the program in which the students are enrolled must do more than nudge them out of the nest. The history department and the university should devise techniques of support for students conducting their research, beginning with direct financial assistance where possible but also including access to university resources and connections to other institutions. Departmental and university support can be very effective in assisting students in making effective applications for foundation support of their research.

For Graduate Students: Options in Study of World History

Graduate school is at once a time for selection of a concentration and for intensive study within that concentration. While some students are certain of their interests and never waver in their course, others find disappointments in some areas of study and new excitement in others and shift their program of study as a result. While graduate programs in world history should emphasize depth of study, they should also allow students to shift their programs of study in response to changing interests. World history, as an interdisciplinary field, should be expected to lead its specialists from topic to topic in the course of a career. Following are the main frameworks for graduate study in world history as I conceive of them.

World history as a research field. This is the option for professional concentration in world history. The program of study includes thorough review of the literature, and selection of concentrations in temporal and regional emphasis and in methodology. This is a research degree, but it also provides the basis for advanced work in curriculum development and preparation for teaching at secondary, undergraduate, and graduate levels. The program includes three years of coursework, plus research and writing of a dissertation.

Recent trends in the job market for historians makes clear that there will be faculty positions in world history. But because of the recent development of world history as a professional field, it is not yet clear how those positions will be filled. That is, we do not yet know whether departments will hire world historians to be specialists in world history or area-studies historians with an emphasis on world history or thematic historians with an emphasis on world history. With the expansion of area-studies appointments from the 1960s, history departments became confirmed in making appointments strictly on a regional basis. Then with the development of studies in gender and ecological history, some departments have begun to create thematic as well as regional appointments. Whether this trend expands will be important for the nature of employment of world historians.[17]

World history as a teaching field. For those wishing to focus on the teaching of world history, it is possible to gain thorough preparation in doctoral programs, master's programs, and professional development programs. Work at this level involves courses in world history and specialized courses in areas of the student's interest, small research projects, and specific preparation for teaching.

I strongly encourage teachers of world history to earn a master's degree with a specialization in world history, and I believe that at least two graduate courses in world history are the minimum requirement for fully qualified teachers of world history at all levels of instruction: middle school, high school, college undergraduate, and graduate school.

Why does good teaching of world history require graduate study? If an undergraduate curriculum in math or science or U.S. history is adequate to prepare a new teacher for the classroom in those fields (when supplemented by necessary courses in pedagogy and educational psychology), why should an undergraduate major not be sufficient for teachers of world history? For graduate-level instruction, I do not think faculty members should teach graduate courses in world history if they have not taken graduate courses in world history. Similarly, I believe it is insufficient for a prospective teacher to take a freshman-level college world history survey course, then use that single, general course as the basis for teaching world history to middle-school, secondary-school, or college-level students.

While it is possible that, over time, undergraduate history curricula could develop a comprehensive focus on interactions and global patterns in history, such a curriculum would be far different from the present circumstance in which courses become steadily more specialized as one approaches the senior year of college. Only at the graduate level (and only at a few institutions) do there now exist

---

17. For the six Northeastern students who completed doctorates by 2001, four had tenure-track appointments, one continued as a tenured community-college professor, and one had a succession of one-year jobs teaching world history. Of the tenure-track appointments, two were defined as world history, one was defined as African diaspora history, and one was defined as U.S. history.

courses ranging across the globe, across long periods of time, and across varying themes and thus able to provide preparation for those who must lead secondary students through courses of this complexity.

The research to be conducted by secondary teachers will rarely lead to monographic publications, though it may often lead teachers to publish their creations and commentaries in curriculum.[18] In order to keep track of new developments in this rapidly changing field, and in order to select materials to teach a course that is coherent yet sufficiently wide ranging, teachers need the experience and the intellectual challenge that is to be found in graduate courses in world history.

The potential result, if high school world-history classes are led by teachers with strong graduate courses under their belt, is that the classes will break past the limits of textbook-driven caution and memorization and will reveal world history to the students as an exciting, challenging, creative, and rewarding—though still frustrating—area of study.

World history as a minor field. This is a framework of study for those who want some experience in world history but do not wish to make a major commitment either to research or teaching in the field. Such graduate students may include researchers in other fields of history or in fields other than history, teachers focusing in other areas but who may occasionally teach world history, and general readers hoping to gain substantial understanding of world history. Study in this model program centers on the literature and methods in world history. According to preferences, individuals taking world history as a minor field can do additional study in the teaching of world history or in research.

Conclusion: Prioritizing Priorities

The single most important unmet need in the establishment of a strong field of world history is programs of graduate training. Graduate study in history is the most dependable way to achieve readiness for teaching in depth and for conducting research at the global level. Both teachers and researchers will benefit immensely from programs of formal study in world history. The relative absence of graduate programs, with their intensive and specialized study, means that world history has yet to benefit from the fresh thinking that can emerge from concentrated and detailed study.

To put it in more positive terms, the establishment of full-scale programs of graduate study in world history will enliven world history and enable world

18. The World History Resource Center at Northeastern University, founded in 1998, has accumulated several dozen lessons and multilesson units in world history, created by teachers in graduate-level courses and workshops. These are available in print or, in some cases, on the web; all are listed in the online catalogue at www.worldhistorycenter.org. Julie Gauthier served as the initial director of the Resource Center; she rapidly gathered and catalogued a wide range of textbooks, teaching activities, monographs, and multimedia.

historians to face the challenge of an immense and complex field of study. For researchers, such programs will provide training for a lifetime of specialization in identifying and analyzing connections and large processes in human history. For college and secondary teachers, programs of graduate study will provide them with expertise in the literature and debates on global historical interactions, so that they may inspire students to pose the questions and interpretations appropriate to the next generation.

What I found from a week with fourteen leading teachers and researchers in world history, gathered for a program of intensive curriculum development, made me think that world history may soon develop many new issues and results.[19] At the end of the week's work and discussion, the result was not a consensus in which all agreed on the main lines of world history, or a debate in which the participants lined up on two or three sides for some defining issue, but rather a blossoming of new ideas and new perspectives brought about by the interaction of individuals who usually have to work alone. I expect that the same is true for world history generally: there are really a great deal more new ideas and patterns waiting to be enunciated.

Yet the benefits of further study of world history cannot be achieved without development of solid programs, and without careful choices by the directors, faculty, and students in those programs. Faculty members and institutions offering graduate study in world history, or those considering doing so, need to select their areas of emphasis and identify ways of supporting students through to completion of their studies. Students of world history need to select their approach and level of involvement. Researchers need to know key debates and research techniques. Teachers need a narrative plus resources and techniques for teaching. Non-historians need an introduction to historical methods, global thinking, and a sampling of issues. To announce that graduate study in world history will open is one step, and to schedule and conduct classes is a second step. Even then, many more steps remain to be taken before such a program reaches a high level of effectiveness.

---

19. These fourteen were gathered in July 2001 for work in preparing curriculum for the AP World History Course.

Chapter 20

Courses of Study

I have had the challenging though rewarding experience of teaching world history in a wide range of graduate courses and teacher institutes. While teaching such a range of courses took me to and beyond the limits of my own expertise, it has at least given me a glimpse into the possibilities for programs of global historical study.[1] It convinced me that no general-purpose graduate course is sufficient to prepare students for either research or teaching in this vast and variegated field.

Here are what I believe to be the curricular elements of a model program of graduate training in world history. These ideas have developed out of my experience at Northeastern University. I think that the range of courses designed and taught during that time added up to a good first approximation to a graduate curriculum in world history, though the small number of participating faculty and students meant that the courses could rarely be presented to students as a coherent program. As I will indicate here, the courses and activities apply in different ways to those pursuing secondary teacher credentials, M.A. degrees, and doctoral degrees.

For a research field in world history, I suggest a model program with a total of twelve courses in three years plus language study, to provide preparation for advanced research in global issues. After comprehensive exams and the research and writing of a doctoral dissertation, this program would lead to a Ph.D. in world history.

For a teaching specialization in world history, I suggest two approaches, one for those entering the field of teaching and one for practicing teachers who are beginning or upgrading their teaching of world history. For those entering the teaching of history, I propose a two-year graduate program including eight courses plus language study, providing preparation for advanced teaching and introductory research in world history. This program would be appropriate for Ph.D. candidates who will be teaching but not researching in world history, community college teachers, and advanced secondary and middle-school teachers of world history.

---

1. Syllabi for most of these courses and descriptions of the workshops are on-line at www.worldhistorycenter.org/manning.

For practicing teachers who are already qualified in history and seek to prepare themselves as teachers of world history, I propose a professional development program of four graduate courses or major workshops to provide the requisite background.

For world history as a minor field, I suggest that a one-year graduate program including four courses can offer preparation for teaching world history at a basic level or an introduction to major research issues. An active program of self-directed study or a set of professional development workshops could provide the informal equivalent to this program of study.

In this chapter, I describe each of these programs in some detail. I begin with the Ph.D. program, as it is the most comprehensive and contains all the elements in the other programs. Those specially interested in the teaching or minor-field programs may wish to go first to them, then locate the fuller discussion of each category of courses and activities as presented in the section on the Ph.D. program.

Preparation for Research: The Ph.D. in World History

The doctoral program in world history, as described here, includes nine categories of courses and nine categories of non-course activities. I describe the content and objectives of each course or group of courses and suggest the order in which courses should be taken.[2]

Graduate training in history should begin with a methods course (to be taken in year one of the program). This methods course should review the organization of the historical profession, techniques of historical research and writing, introduction

2. Here is a list of the courses and non-course activities to be described in the following paragraphs:

Courses for a Ph.D. Program in World History
1. Historical methods
2. Global historiography
3. Survey or narrative of world history
4. Teaching world history
5. Global research (one or more courses)
6. Disciplines in history
7. Regional courses (one or more courses)
8. Thematic courses (one or more courses)
9. Advanced methods (one or more courses)

Non-course Activities for a Ph.D. Program in World History
A. Prepare curriculum unit
B. Supervised teaching
C. Research project
D. Electronic and other resources
E. Language for reading
F. Language for research
G. Dissertation proposal
H. Comprehensive exam
I. Dissertation research and writing

to the various fields of study in history, and contending perspectives in historical analysis, including modernist and postmodernist critique. The course should introduce students to case studies, comparative studies, and studies that are global and interactive. In it I have taken classes to visit archives and have assigned papers on book reviews, annotated bibliographies, and research designs.

A global historiography course (year one) reviews the literature on world history. The most basic assignment for graduate students in world history is to read widely in the field. This course lays out a list of major contributions to the literature and provides a forum for discussing historiography and the patterns of history represented in the literature. In it students build their strength in analyzing and debating history at the global level.

A narrative of world history (year one) is a course surveying the events and processes of world history. Such a course must move very rapidly in order to address, within a single term, the full range of time, regions, and themes in world history. This course could be taught as a human history course, and begin with human evolution or the Neolithic era, or it could be taught as a Big History course. Readings in the course may be a world history textbook and sourcebook, and several pieces of literature or short interpretive works in history to draw attention to a few key issues. The course provides preparation for teaching, but it also emphasizes for students the breadth of world history and provides practice in making linkages and comparisons.

Taking a course on the teaching of world history (year one) is quite a different matter from taking a survey course. This course presents the problems in exploring not just what the teachers will say but what the students will hear and how they will process it and fit it into their view of the world. In this case those taking the course explore the assigned teaching materials and then construct their own.[3] They grapple with the dilemma of covering the material yet providing students with time to reflect on it.

Global research seminars (one or more courses, taken in years one and two) in world history provide an opportunity for supervised research in which students address the practical problems of locating and assembling materials for global historical analysis. I have offered research seminars focusing on global political, social, and cultural history, and I would hope to see seminars in global economic and ecological history. While students in these seminars must often rely on secondary works, since they are faced with the need to define and execute a research project within a single short term, it is equally important that they gain experience in locating and evaluating primary documents.[4]

History and the disciplines (year one) introduces students in detail to the disciplinary sub-fields in history, enabling them to establish basic literacy in each of

3. I have taught this not as a graduate course but in the form of intensive teacher workshops (with graduate credit available), along with co-director Deborah Smith Johnston. Materials for these workshops are on file at the World History Resource Center, www.worldhistorycenter.org.
4. For further details on approaches and materials appropriate for research seminars, see chapter 22 on research.

the major sub-fields and to develop substantial strength in sub-fields of their choice. As I have taught this course, individual students take responsibility for presenting political, social, economic, ecological, diplomatic, gender, and cultural history by selecting articles and book chapters conveying both the methods and controversies in each sub-field and leading a class discussion on those readings. Then students write substantial papers articulating a methodology of their choice and indicating how it will be applied to a historical problem. Generally, I found that on this assignment students went beyond emphasizing a single field, such as social history, and preferred to link two or even more sub-disciplines into a methodology for world-historical analysis. One student chose to link diplomatic history and game theory; another linked cultural anthropology and political economy. Looking to the longer run, world historians require formal training in selected disciplines and continued practice in discussing the connections, parallels, and differences among disciplines. There is no hope of learning every discipline, but there is hope of learning one or two well and studying to read and understand summary publications in most disciplines. It is certain that the career of a world historian will require learning one or several new disciplines during the course of a long career.

Regional and topical courses (two or more courses, years two and three), led by faculty members focusing on their specialty, are necessary to ensure that the global approaches developed here are tied to specific historical situations. For many students, this will mean an area-studies specialization with two or three courses on that area. For other students, it will mean a topical specialization. I have taught a course on the African diaspora, but other courses taught in our program included environmental history, gender and colonialism, U.S. maritime history, urban history, and courses on the Soviet Union and Eastern Europe and Caribbean history.[5]

In advanced methods (one or more courses, year three), I have favored a formal methodological specialization by doctoral students, including coursework taken in adjoining disciplines to make sure that students become skilled in their method of choice. Two difficulties have kept the formal methodological training of our graduate students at a minimum. First and most important is that the methodological interests of graduate students today are complex and overlapping, so that no one course or neighboring department can satisfy their needs. Second is the difficulty of matching the relevant course with the student. In practice, graduate students are sometimes left with the need to train themselves in the fine points of their methods. In other cases, students have been able to benefit from courses in literary theory, sociology, demography, and social statistics.

In addition to such courses in a world history doctoral program, there are many other activities required for completion of the degree, of which I have identified nine. The first two of these focus on preparation for teaching.

---

5. The range of courses can never meet the range of students' interests. In the Northeastern program, as commonly in graduate study, students earned credit for "independent study" or "directed reading" by meeting as individuals or in small groups with faculty members for reading and discussion on selected issues.

Along with graduate survey courses in world history, students should have experience in developing curriculum units (year one) of their own, whether at college or high school level, in association with practicing teachers at either level or both. This work in curriculum development not only addresses the shortage of good class materials in world history but also presents the student with further details of interpretation and presentation in global connections.[6]

Having completed their curriculum units, students should next undertake supervised teaching (year one or two). They should direct classes (with lectures or group activities), and participate in creating assignments or exams and grading student work in association with their faculty supervisor. With this preparation, students should be ready to teach their own college classes. In our program, students have begun teaching their own classes once they have passed their comprehensive exams.

A third non-course requirement is that students become skilled in electronic media and technology. This requirement includes several dimensions. Students must certainly be skilled in word processing and data bases. They must be able to do web searches, and should be able to set up their own web pages. They should develop skills in conducting bibliographic searches through electronic media. They should also use the internet to develop their professional identity. I have required graduate students to subscribe to electronic and other resources, including H-WORLD (year one and thereafter) and to submit postings to the list on a topic relevant to a world-history course in which they are enrolled. This discussion list (and similar lists in other fields) enables graduate students to participate as equals in the scholarly community. The practice of joining in the scholarly give-and-take is very effective in advancing the interpretive skills and the expressive powers of graduate students.

I believe it is advantageous for doctoral students to conduct at least one sizable research project (year two) at a global level before taking on a doctoral dissertation. The requirement of a master's thesis is perhaps excessive, in that it may add as much as a year to the student's doctoral program. I think, however, that the experience of defining and executing a project in world history will help protect students against errors in judgment on a large project—misjudgments that could cost them a year or more of time before completion.

Language study must become central to programs of world history at all levels. Although English is a convenient language, having become the language of choice for today's international communication, it would be a pernicious development for world history to become an English-only field of study and to undermine its own emphasis on multiple perspectives by leaving out of the discussion those expressing themselves in languages other than English, in past and present. Doctoral students in world history should begin work on an additional language for reading (year one and thereafter). The purpose of this requirement is that

---

6. At Northeastern, the curriculum development is conducted through the World History Resource Center, often by editing and completing curriculum units drafted by practicing teachers.

students should be able to read secondary literature in languages other than English. I have proposed French and Spanish as the languages that all world historians should know, and I have tried in recent years to include assignments in both languages in all my graduate courses. For students who already have a second language, it is important to begin work on another language: Portuguese, German, and Dutch are very useful in world history, but any number of other languages presents materials of importance for world history. This requirement is not a matter of passing an exam but of actually using writings in multiple languages as part of graduate study.

At a somewhat more stringent level is the requirement of language study for research (year two and thereafter). This is preparation for reading primary sources. Programs in world history require support for language study and an atmosphere that urges students on with the time-consuming task of learning one or more new languages. The range of languages relevant to world history presents students with a daunting choice. Beyond the Romance and Germanic languages, there are vast literatures in Chinese, Arabic, Japanese, and Russian. Beyond those, the list of relevant languages continues at length, depending on the area of the world in which a student expects to work. No one can learn them all, but world historians should work at expanding their language skills.

The dissertation proposal (years two and three) is a crucial step in graduate study, incorporating coursework and general reading into a plan for the major research project that will set each new scholar's direction. I believe each dissertation project should come from the student who will research and write it rather than from the advisor. But I also believe that students should be urged to begin sketching out possible dissertations from the very beginning of graduate study. For some students the project becomes clear early on; for others it emerges only in the third year of study, or only after a few projects have been developed and then discarded. The advantage of carrying on a discourse about the dissertation throughout graduate school is that it makes clear the centrality of the dissertation to the whole of graduate school. As I explained in chapter 19, I believe there are great advantages to requiring that the dissertation proposal be approved before each student takes the comprehensive examination.

The form of dissertation proposals is relatively standard, especially since the competition for dissertation fellowships sets standards that are widely understood. The proposal should set forth the topic, make clear its significance, review the literature on the topic, describe the details of the proposed research, indicate the time frame and the resources required for the research, and present the expected results and significance of the research.

Comprehensive exams (year three) typically involve three examinations. At Northeastern, one exam has been on world history as a whole. The other two exams are designed according to the specialization of the student. They might be on two regional literatures, or on a region and a disciplinary methodology, or a region and a theme. Thus one might have exams on South Asia and gender, or South Asia and the Pacific, or South Asia and economic history. In each case, the student is expected to be able to summarize and debate the major issues in substantial fields of study.

A comprehensive examination on the literature in world history, while a daunting exercise, is proving to be an effective way to unify a world history graduate program and is steadily developing a shared discourse on world history among students and faculty. The list of books addressed on the Northeastern exam in world history has grown at this moment to about seventy. No such list can be either definitive or sufficiently comprehensive, but the debates over what to add to the list, what to remove from it, and how to organize the list can lead to a broad and shared understanding of the debates in the field.

Research and writing on the dissertation require at least two years and commonly continue for a longer period. If the dissertation proposal is well thought out, research may begin without much delay after the comprehensive exams. Depending on the topic, the location of relevant libraries, archives, and other sources, and the availability of funds for travel and research, the writer of the dissertation may spend substantial time away from home. While dissertation writing is in some ways a lonely task, regular contact with advisors and with other students in world history can assist in bringing the dissertation to completion.

Preparation for Teaching

A two-year program of eight courses can prepare graduate students for advanced work in teaching and introductory research in world history. Such a program is most likely to be followed by a pre-service teacher preparing to teach advanced middle-school or high-school courses in world history or by a Ph.D. candidate who will write a dissertation that is regional rather than global.[7]

For the first year, this program includes the general graduate-level introduction to historical methods, a course on global historiography, a course providing a narrative of world history, and a course on the practice of teaching world history. (Each of these courses is described in more detail in the previous section on the doctoral program of study.) In the second year, students take one or two research seminars in world history and two or three regional and thematic courses in areas that the student finds to be of special interest.

With these courses, the student also completes several non-course requirements. These include language study for reading secondary sources in French or Spanish beginning in year one of the program, the preparation of a curriculum unit during year one, a supervised teaching experience during year two, and an independent research project during year two.[8]

While this program gives primary experience to developing the teaching experience of its students, it also gives substantial emphasis to research. Teachers need

---

7. Pre-service teachers at Northeastern completed their training either in the MAT (Master of Arts in Teaching) program, in association with the Education School, or in the M.A. program in History. Among the graduates of these two programs who went on to successful careers teaching world history in Massachusetts high schools were Brian Carr, Christopher Cook, Julie Gauthier, Mark Johnson, and Molly Duffy.
8. For materials on the preparation of curriculum units, see the World History Resource Center at www.worldhistorycenter.org.

to keep up with current research, and they need to discover the techniques for learning recent research results. Teachers also need to complete small research projects in order to prepare for class, and they need especially to encourage and assist their students in doing small research projects.

In professional development in world history, for teachers who are prepared in history and experienced in the classroom, a somewhat different program of study may be equally satisfactory. Such a program could include teacher institutes and workshops adding up to the equivalent of four graduate courses in world history. At Northeastern, teachers completing outstanding work in professional development received certificates of excellence.[9] Assuming that teachers already have studied historical methods, the workshops should address the teaching of world history, a narrative of world history, global historiography (with an emphasis on new research), and a workshop with a regional or thematic emphasis. In addition to or as part of these workshops, the participants should prepare a curriculum unit and focus on reading world history in a second language.

The professional development workshops should be very helpful to teachers with a strong background in history. The problem is that, unfortunately, many persons assigned to teach history (including world history) had majors that were not in history but in other social sciences or in other fields entirely. Of course experience, talent, and dedication to the mission of teaching have enabled many people to become good teachers of history even though they lacked formal training. However, there is a definite advantage to formal enrollment in graduate courses, in addition to workshops and self-directed study for these teachers to be able to do their best in the world history classroom.

World History as a Minor Field

World history as a minor field is a set of courses intended for those who wish to have a solid introduction to world history without making it the principal emphasis of their study. This program of four courses provides preparation for teaching introductory courses in world history. The courses include the graduate introduction to historical method, the graduate survey of world history, global historiography, and the course in teaching world history. The non-course requirements of this program are study for reading world history in a second language, preparation of a curriculum unit in world history, and supervised teaching.

For those seeking teaching credentials in world history, I believe that the minimum set of courses in history should include one term of method, global historiography, a graduate survey of world history, and completion of a curriculum project. In addition, this program could also stand as a second field or a teaching field for a Ph.D. candidate working primarily in national, area studies, or thematic history.

Parallel to this formal minor program in world history, one can imagine a program of self-directed study. Such a program can be followed by practicing

9. Recipients of the World History Center's Certificate of Excellence in Teaching World History in 2000 were Abigail Cox, Julie Gauthier, Kristin Palmer, Lori Shaller, and Rebecca Vizulis.

teachers seeking to improve their background, scholars in other fields reading about world history on their own, or general readers who are simply interested in the field.

For these readers, it is possible to gain a good introduction to historical methods through a set of standard works on the subject. For global historiography, one can do the readings suggested in syllabi for graduate courses, or one can read a selection of the works cited in this book. For a narrative of world history, one can read one or more of the many college texts in world history, along with documentary readers in world history. For the teaching of world history, numerous hints are available on the web, as indicated in chapter 21. For regional histories and thematic histories, it is helpful to visit the websites of the various regional and thematic historical organizations. And for those doing reading on their own, as for others, it is important to cross the language barrier and study writings on world history in languages other than English.

Conclusion: Practice in Global Thinking

Proficiency in world history requires practice: world historians should read widely, discuss readings with others, and follow the threads in their study that lead across boundaries in space, time, and subject matter. Travel over both short and long distances is helpful in developing world historical insights, but those unable to travel can make up for the gap with more reading.

To review, the formal coursework for graduate study in world history for either teachers or researchers should include:

- *Survey.* An empirical and interpretive survey course, providing an introduction to the full sweep of world history.
- *Global historiography.* Critical analysis of the interpretive literature in world history.
- *Historical methodology.* Coursework on the basic methodology of history—documentary research and analysis, bibliography, historiography, and research design.
- *Disciplines.* Thoughtful introduction to many of the disciplines (in social science, the arts, and environmental studies) with which history now interacts.
- *Regional and topical courses.* These courses, encouraging the pursuit of individual interests, might include national units, such as ones on France or Burma. But they could just as well address an economic region such as the Black Sea, a social network such as the Chinese diaspora, or a global topic such as forestry.
- *Language.* Study of historical documents and interpretations in more than one language.
- *Writing and discussion.* Students should write a substantial study in world history, which might include research or creation of curriculum units for teaching at college or secondary level. They should get involved in discussions at conferences or participate in H-WORLD or other electronic discussion groups.

In addition, this advanced study for teachers and researchers should give them experience in collaborative research and curriculum development. Historians have worked for eons as individuals, but it is time for us to learn to work in groups, in both research and teaching. That means sharing data, sharing insights, and sharing authorship. It means working with colleagues in the same library or classroom, and also sharing ideas with colleagues by mail or over the internet. In addition to working as individuals and in small groups, it is logical to expect that some historians will work on large-scale projects to collect and organize data at the global level.

Those planning to earn a Ph.D. and conduct research in world history need to deepen their general preparation in several ways. First, of course, is a continued program of active and wide reading. Second is formal training in selected disciplines and continued practice in discussing the connections, parallels, and differences among them. There is no hope of learning every discipline, but there is hope of learning one or two of them in depth, and learning enough to read and understand summary publications in most disciplines. It is certain that a successful world historian will need to learn one or several new disciplines during the course of a long career. Third is additional language training: learning languages should become a habit, like learning disciplines. Fourth is training in data collection, retrieval, and analysis; and fifth is field research, preferably in more than one region of the world.

Teachers of world history, whether new to the profession or experienced in the classroom and in the teaching of history, need a steady program of professional development in order to provide students with access to the best understanding of world history. This means a continuing program of reading, travel, contacts across language barriers, and study of recent research and new curriculum ideas.

Chapter 21

Resources for Graduate Study

This brief listing cannot in any way survey the vast resources available for study of world history. Yet by offering some of the general categories in which these resources are to be found and some examples within each category, I hope to encourage students and practitioners in world history to pursue in detail a few more of the many questions they may have about the global past.

The distinctiveness of world history is that as a field only recently organized, it lacks the great compendia of archives and finding aids that have been developed patiently by scholars in more established fields. The task of world historians is in part to create new such compendia at a global level. In the meantime, much can be done with intelligent use of the resources already pulled together under other frameworks. World historians should identify connections among documents and interpretations already assembled in regional context and add new materials developed in a global framework.

Libraries

The Library of Congress in Washington, D.C., built up systematically since Thomas Jefferson contributed the majority of his personal library to its founding, is now the most extensive library in existence.[1] Its collection, built through gifts from heads of state and private donors, through deposit of all books published in the United States, and through purchases by the library, may be reviewed through its online catalogue. In recent years a substantial effort has gone into cataloguing and displaying the collection in area-studies categories. The library is of immense value to world historians simply through the size of its holdings, but one could imagine a project to develop finding aids aimed at tracing historical connections

---

1. The remainder of Jefferson's library became the basis of the collection at the Massachusetts Historical Society.

through the collection and thereby facilitating the documentation of global patterns and connections.[2]

Other great national libraries are the British Library in London, the Bibliothèque Nationale in Paris, and national libraries in Spain, the Netherlands, Sweden, Denmark, Mexico, and other long-standing national powers. Additional major libraries are those of major cities, especially the public libraries of New York and Boston. Universities are next in maintaining major libraries. University libraries in the United States have become the largest and most up-to-date research libraries—notably those at Harvard, Yale, Columbia, and the University of California at Berkeley. The university libraries at Oxford and Cambridge in England and other European university libraries are large and modern, but their particular strength is holdings of early books and manuscripts.

The strength of libraries is not only in books but in periodicals, manuscripts, and access to other materials.[3] Further, it is not only general-purpose libraries that are useful for world historians. Sometimes more specialized libraries have particular strengths, such as the U.S. National Agricultural Library, located in Maryland (www.nalusda.gov). In the United States and Canada, the system of interlibrary loan makes it possible to borrow many books from anywhere within the two countries after allowance for time to locate and send the book to the reader. Reference sections in libraries include the many finding aids that have been developed to assist the researcher, along with librarians, some of them with remarkable skills, to help locate relevant materials. An example of a finding aid of interest to historical researchers is the *AHA Guide to the Historical Literature,* which has an excellent section on the literature in world history.[4]

Archives

The archives of national governments are the central store of documentation on political events, often reaching back several hundred years and occasionally further. In addition, records on administration, taxation, trade are sometimes systematic, sometimes scattered. Reports of local officials and official travelers, submitted to central governments, sometimes have wide and eclectic selections of information.

Among the principal national archives that will be of interest to world historians are those of Britain, the Netherlands, Portugal, Spain, France, China, Japan, Russia, Turkey, the United States, Mexico, and many other countries. The Vatican

2. I wish to express my appreciation to Carolyn Brown, Assistant Librarian, and to Prosser Gifford for meeting with me to suggest ways in which the Library of Congress can facilitate research in world history.
3. Further, it is not only general-purpose libraries that are useful for world historians. Sometimes more specialized libraries have particular strengths, such as the U.S. National Agricultural Library, located in Maryland (www.nalusda.gov).
4. Mary Beth Norton and Pamela Gerardi, eds., *The American Historical Association's Guide to the Historical Literature,* 2 vols. (New York, 1995). The section on world history, edited by Kevin Reilly and Lynda N. Shaffer, is on pages 42–76 of volume 1.

and orders of the Catholic Church have preserved extensive archives, many of them centralized in Rome. Archives of provincial and local governments, more erratically preserved, may hold uniquely valuable documents. Official documents were commonly copied to various levels of government, so that documents missing in one archive may exist in another. When Brazil abolished slavery in 1888, the federal government destroyed all central records on slavery, ostensibly to protect the identity of the former slaves. Researchers on slavery have been able to reproduce some of the missing information through study of local archives.

Archives created recently and of immense importance for modern world history are those of the wide range of international organizations. Most obviously these include such paragovernmental organizations as the League of Nations, the United Nations, the International Labor Office, the World Bank, and the World Court. In addition they include non-governmental organizations such as the Red Cross, Amnesty International, and the World Wildlife Federation. Further, and not least, are the multinational or transnational corporations, such as Royal Dutch Shell, Nestlé, Standard Oil, ITT, and Sumitomo Bank.

Families and small firms occasionally set up their own archives. In countries such as the United States, where local historical societies are well established, firms and families may donate materials for holding in the archives of local historical societies.

Thus, since the archives of the Massachusetts Historical Society are a few blocks from my university, I was able to send students there with the assignment of designing a global history project based on a local archive. They were to explore the many individual and family archives, and try to locate stories that showed global linkages. While a number of students found good global connections, perhaps the best was of a family of Boston merchants who sent younger sons to China and elsewhere in Asia to carry on the family's mercantile affairs. Another was a shipbuilder in Duxbury, Massachusetts, who for a time in the mid-nineteenth century was the largest and most successful shipbuilder in the United States. Yet another global connection was that of a leader in Boston's African American community, who in her earlier life was the pivotal person in the migration from Barbados to the United States and England of numerous members of her family.

Electronic Resources

World history is the first major field of history to develop in the era of electronic communication. It is therefore reliant on—and able to benefit from—electronic resources to a degree unprecedented in other fields of history.

The World Wide Web Virtual Library, a remarkable scholarly resource maintained at the University of Kansas by historian Lynn Nelson and a dedicated set of volunteers, lists journals, archives, discussion lists, bibliographies and more, in multiple languages and for all regions of the world.[5] The Library of Congress website includes an online catalogue of its holdings. It also provides access to the

---

5. World Wide Web Virtual Library, www.ku.edu/history/VL/index.html.

American Memory Project, a massive collection of historical data online, and to the smaller but promising collection of online data based on the Spanish and Latin American materials held at the library.[6] In addition, the websites of the national archives and libraries of most nations with sufficient resources include historical documents relevant to global issues. It is usually easy to locate these websites through standard search engines.

Many newspapers around the world are available online.[7] While the newspapers are online for earlier years in only a small number of cases, the access to current information for a given city or publication can be very helpful in locating newspapers for earlier times.

Finding aids and catalogues are increasingly available online and on CD-ROM. Lexus-Nexus, WorldCat, and UnCover enable searches for articles and books. The British Parliamentary Papers, an extremely important collection of data for world historians, have been excellently indexed on a CD-ROM published by Chadwick-Healy. The combination of the Chadwick-Healy finding aid and the microfiche version of the original Parliamentary Papers (available in more libraries than the original folio volumes) greatly eases research for many issues in the nineteenth and twentieth centuries. Guides to other major collections of published and archival materials will be appearing in years to come.

Electronic discussion groups have permitted, since the early 1990s, an unprecedented expansion in the flow of information and debate among scholars. Historians have been among the most energetic users of discussion lists, and the largest number of discussion lists is affiliated with H-Net (Humanities On-Line), with its list of over 130 discussion lists, most of them focusing on fields of history, administered at Michigan State University. H-WORLD, founded in 1994, is one of the H-Net lists; its daily postings are available to all subscribers, and the archive of all its past postings is available on the web.[8] H-WORLD reached 1,500 subscribers by 2001, with somewhat over two-thirds of its subscribers based in the United States. Several dozen subscribers were based each in Canada, the Netherlands, the United Kingdom, and Australia, and additional voices from Turkey, Hong Kong, Japan, South Africa, Brazil, Mexico, Italy, Russia, India, and elsewhere enliven the discussion. Among other discussion lists of relevance to world history, the World System Network supports discussion on world-system research, H-TEACH and H-High-S are H-Net lists focusing on the teaching of history, and ap-world is a list for discussion of the AP World History course supported by the College Board.[9]

For researchers, teachers, and especially for students, the question of how to evaluate electronic resources remains serious. Centuries of publication and review have allowed the development of reviews and standards that assist the reader in selecting and assessing books and articles. For materials online, the process of evaluating is just beginning. Some teachers have developed useful exercises to

6. "American Memory"—www.memory.loc.gov; "Spain, the U.S., and the American Frontier: Histórias Paralelas"—www.international.loc.gov/intldl/eshtml/eshome.html.
7. For a comprehensive source, see Online Newspapers: www.onlinenewspapers.com.
8. www2.h-net.msu.edu/~world.
9. For H-TEACH and H-HS, see the H-Net home page at www.h-net.msu.edu. For the ap-world list, see the College Board site at www.collegeboard.com.

encourage students to develop critical skills in assessing websites for history. Style sheets, citation guides, and suggestions for users are available on a number of major websites.[10] As this book is being completed, the World History Center has begun work on a World History Network intended to address these issues and to promote accessibility of all sorts of historical information to researchers, teachers, and students.[11] It is also intended to present a critical review of the resources identified and to help users of the internet develop their own skills for critical assessment of the resources in world history they locate.

Individual websites come and go, and those that disappear leave fewer traces than is the case for a book that has lost its initial audience yet still sits on a shelf. Nevertheless, the steady advances in technology are matched by growing scholarly inventiveness and by the steady input of energy into electronic datasets. One can be sure that, one way or another, the electronic resources for world history will improve with time.

Teaching Resources

Of print materials for teaching world history, the most elaborate are textbooks at college, high-school, and middle-school levels, many of which come with substantial packages of additional resources such as maps, images, and test banks. The number of new textbooks seems to be increasing far more rapidly than the number of students.[12] A second category of print resources is documentary readers, usually consisting of primary sources with introductions and questions.[13] These are the materials actually assigned to students, and they are undergoing many small improvements. It is widely recognized, however, that these texts (and, to a lesser degree, the readers) are too voluminous and too encyclopedic to encourage active and critical learning by the students who read them. Textbooks are a useful crutch, but they are not, in my opinion, the solution to the problem of teaching world history.

The spread of this realization puts more pressure on the development of additional guidelines and materials that teachers in middle school, high school, and college can use to improve the experience and expand the learning of students in world history survey courses. Several leaders in the support of world history teachers have published books including methods and examples of world history teaching techniques: notable are *Bring History Alive!* and Heidi Roupp's *Jump Start Manual.*[14] Published teaching units have been circulated by a number of

---

10. Internet citation guide: www2.h-net.msu.edu/about/citation/. The World Wide Web Virtual Library has a detailed guide for users.
11. www.worldhistorynetwork.org.
12. In 2001 Adam McKeown prepared a current list of high school and college texts in world history. It is available in the Resource Center section of the World History Center website, www.worldhistorycenter.org.
13. For instance Andrea and Overfield 1990 (now available in a 4th edition).
14. Ross Dunn and David Vigilante, eds., *Bring History Alive! A Sourcebook for Teaching World History* (Los Angeles, 1996); Heidi Roupp, ed., *Jump Start Manual for Teaching World History* (Aspen, Colo., 2000).

groups. Most prominent among them have been the National Center for History in the Schools, based in Los Angeles at UCLA, which includes detailed lessons on a variety of world history issues. The Teachers Curriculum Institute in Palo Alto produces social-studies materials that are extremely effective as interactive curriculum, though they tend to take a local rather than global approach to instruction. The College Board has published teaching units for the AP World History course.[15] In addition, textbook publishers make available a variety of published lessons.

Another type of teaching resource is provided by outreach centers, which are libraries maintained by organizations supportive of teaching. To begin with, all of the university-level area-studies centers supported by Title VI funds from the U.S. government have outreach centers providing information to teachers and communities nearby.[16] The National Geographic Society, in association with some local and state associations, supports global-studies outreach centers focusing especially on curriculum in geography. Various other organizations maintain outreach centers relevant to world history, including World Affairs Council offices in major cities and Facing History and Ourselves, which provides a curriculum on Holocaust awareness and other issues in social justice.[17] Regional, statewide, and national meetings of the National Council of Social Studies offer workshops and display resources prepared for teaching world history.

For explicit concentration on world history, not many resource centers have yet developed. One example is the World History Resource Center, founded in 1998 as a branch of the Northeastern University World History Center. It established a pilot program with a wide range of activities in professional development for in-service teachers, curriculum development, building a resource collection for teachers of world history, and preparation of new teachers in world history. Beginning with intensive summer workshops for teachers, supported by the Massachusetts Department of Education, the Resource Center led nearly thirty workshops for teachers (most of them multisession workshops) in its first five years.[18]

In a further effort to broaden the resources and workshops available for teachers, Deborah Smith Johnston led in organizing the World History Symposium, an annual workshop beginning May 1999, in which participants from as many as

15. The National Center for History in the Schools had published twenty-nine teaching units in world history by 2003, including Daniel Berman and Robert Rittner, *The Industrial Revolution: A Global Event* (Los Angeles, 1998), and Jean Elliott Johnson and Donald James Johnson, *Emperor Ashoka of India: What Makes a Ruler Legitimate?* (Los Angeles, 1999). The College Board has in press, as of this writing, fifteen teaching units in world history prepared by the World History Center. In addition, the World History Center has published units including Kristin Palmer, *Mecca: Islam's Mosque* (Boston, 1999); and Jessica Goonan, *Africa: Cultural and Geographic Diversity* (Boston, 1999). Further, the Teachers' Curriculum Institute in Palo Alto publishes interactive curriculum materials applicable to world history.
16. U.S. Department of Education: www.ed.gov.
17. Facing History and Ourselves: www.facing.org.
18. "Workshop" section of the World History Resource Center: www.worldhistorycenter.org.

twenty outreach organizations joined to present workshop sessions on various aspects of teaching world history, from primary grades through high school. The strength of the symposium was its gathering of area-studies resources to cooperate in support of the teaching of world history. At its most successful, the two-day symposium in May 2000 attracted 150 teachers to its twenty sessions.[19] Participating organizations in the symposium included six area-studies outreach centers, five global-studies centers supported by the Massachusetts Geographic Alliance, the Boston Children's Museum, the World Affairs Council, and leading independent curricular organizations such as Primary Source and Facing History and Ourselves. James Diskant coordinated the third and fourth Symposia, in May and October of 2001, and obtained a grant from the Massachusetts Foundation for the Humanities to support the May 2001 meeting. The nexus of collaboration among the participating teacher-resource organizations was not the least of the benefits of this approach.

Based on its record in professional development work, the World History Resource Center was awarded two projects by the College Board to support the development of the AP World History course, for which the first exam was given in May 2002. In the summer of 2000, the Resource Center led a week-long workshop to train thirty-six nationally selected teachers and professors to become workshop leaders, who would in turn serve as College Board consultants, to direct programs introducing the details of the new course to teachers. In the summer of 2001, the Resource Center led another week-long workshop, this one including fourteen leading teachers and professors of world history to create a set of teaching units, a web guide, and a "best practices" guide for AP World History to be made available on the AP Central website of the College Board.

Funding

However bright the long-term prospects for study of world history, they can be achieved only step by step. In chapter 9, I discussed the relative scarcity of funds for world historical study and the relative invisibility of world history in the calculations and allocations of major funding agencies. For the present, we are caught between the growing intellectual appeal of global historical studies and the continuing belief, at the level of governments and funding agencies, that such studies and teaching work can be completed without new resources.

While there is so far no example of ongoing funding or endowment of programs for research or teaching in world history, the number of grants made for individual projects is impressive. The Woodrow Wilson Foundation supported three years of workshops for teachers of world history in the early 1990s, and The Annenberg-CPB Project supported development of a CD-ROM on migration in world history. The National Endowment for the Humanities supported a two-year project at Northeastern from 1994 to 1996 and has since supported two years of summer institutes for teachers of world history in association with the World

19. Ibid.

History Association and later the World History Center. The NEH has provided support for a total of over one dozen educational projects in world history, including a collaborative pre-service teaching project directed by Heidi Roupp, a curriculum-development project based at San Diego State, a world history website based at Northeastern University, a program for documents in teaching world history based at George Mason University, and a CD-ROM project on the West African epic of Sundiata based at Tufts University.

Conclusion: Gathering and Creating Resources

The quantity of resources relevant to research and teaching in world history is vast. The quantity of materials prepared explicitly for research and teaching in world history is quite modest by comparison. This puts the world historian in the position of searching through materials prepared for other purposes and keeping an eye out for global connections. Studies in world history will always present the teacher and the researcher with a surplus of materials, but in time it should become clearer which materials are about the world in general and which are about connections in the world—and world historians will focus on the latter. Further, it is in the process of locating and documenting connections in the world that researchers will create new resources for the study of world history.

Chapter 22

Researching World History

The greatest need in the field of world history is that for original research. Whether carried out on a large scale or small, whether based on local library research or on extended field research, new research in world history will help the field get past the current logjam in which it is bound, in which sequences of local studies are made to stand in for analysis of connections across space and time. The highest level of world-historical research will require teams of experts, including broadly based and cosmopolitan scholars. But as in every other field of intellectual endeavor, the work of the most sophisticated analysts advances only in intimate interconnection with scholars and observers working in less specialized fashion or with smaller amounts of resources. Research in world history takes place at all levels. In one sense it is just more historical research. In another sense, crossing all those boundaries makes it different.

This chapter presents a few pointers on research at several different levels to indicate both the similarities in problems at all levels and the specific issues to be addressed in research on large-scale or small-scale levels.

Research Seminars in World History

I begin with formal graduate courses in world-historical research because they provide structured examples. These are summaries of research courses that I have taught, presented in narrative form because it is the most direct way to summarize the range of experiences. From 1994 forth, I have offered roughly one global historical research seminar every second year, with the emphasis shifting among political, social, and cultural history. All of these courses have included a mix of M.A. and Ph.D. candidates except the first one, given before the arrival of doctoral students.

I began with a graduate research seminar in global social history, taught in the spring of 1994. The Northeastern doctoral program had not opened yet, but two M.A. candidates, both new to world history, completed the seminar. In the first few weeks they read selections in world history, social history, and migration.[1] They then had to define and execute (within a ten-week term!) a transnational project in social history, using resources available at the university library. One compared baby booms after World War II in the United States, Italy, and Japan, and the other explored twentieth-century changes in family structure in industrial countries.

Next I taught a seminar on global political history in the fall of 1995. Four students (a mix of M.A. and Ph.D. candidates in this and in subsequent seminars) read five major studies on the politics of nations and empires from the eighteenth to the twentieth centuries and then developed research topics, choosing between empires and international organizations.[2] Three of the papers worked out quite well: they focused respectively on contraband trade in the seventeenth and eighteenth centuries (especially in the Caribbean), and its effect on the Spanish and English empires; on British negotiations with the Catholic Church following the seizure of Quebec in 1759 and Trinidad in 1797; and the creation and experience of the World Court from the turn of the twentieth century to the 1960s. All these papers relied on published sources, but they addressed global issues solidly.

I completed the first cycle with a seminar on global cultural history, taught in the spring of 1998 to six students. This group had first to address the range of cultural history, then to identify global dimensions of cultural history. To begin with, they read four substantial but differing volumes on cultural history.[3] I asked them to select topics and develop annotated bibliographies of twenty items each on the global history of their topic, and then to select the best of these for the class to read and discuss. Thereafter they were to define, complete, and present a piece of research in cultural history at the global or transregional level. The best work of this group was in selecting and discussing a good set of readings.[4] Because of the time it took to develop an orientation to cultural history on a global scale, the research projects themselves were hurried. The research projects addressed the connections of Hindi and Hollywood films, textiles and fashion in Europe, coffeehouses worldwide, rock music in the United States and Britain, and transformations of Hula dance in Hawaii.

1. The selections in social history were Kathleen Canning, "Gender and the Politics of Class Formation: Rethinking German Labor History," *American Historical Review* 97 (1992), 736–768; Tamara K. Haraven, "The History of the Family and the Complexity of Social Change," *American Historical Review* 96 (1991), 95–124; Sabean 1984; Quale 1988; and John K. Thornton, "African Dimensions of the Stono Rebellion," *American Historical Review* 96 (1991), 1,101–13.
2. Liss 1983; Gellner 1983; B. Anderson 1983; Samuel Huntington, *The Third Wave: Democratization in the Late Twentieth Century* (Norman, 1991); Carter V. Findley and John Alexander Murray Rothney, *Twentieth-Century World*, 3rd ed. (Boston, 1994).
3. The initial readings were Kuper 1988; Morris-Suzuki 1998; R. Williams 1983; and Prakash 1995.
4. Among the readings located by students for their bibliographies were: Wolfgang Schivelbusch, *Tastes of Paradise: A Social History of Spices, Stimulants, and Intoxicants* (New York, 1992); Geoffrey Parrinder, *Sexual Morality in the World's Religions* (Rockport, Mass., 1996); Rex M. Nettleford, *Inward Stretch, Outward Reach* (London, 1963); and James Laver, *Costume and Fashion: A Concise History* (New York, 1995).

While the students were doing well in absorbing and critiquing the literature and in defining research projects, this approach was not giving them much depth in research experience, and in particular it was not focusing them on primary documents or on archival research. For the global social history seminar of fall 2000, therefore, I concentrated (as firmly as one can in a ten-week term) on primary and archival research. I defined the research to focus on migration, demography, and family. The five students were first to read some major publications relevant to these topics.[5] Then they were instructed to complete an individual project on the global history of the family and a collaborative project on the global history of migration. The former research was to be completed out of family archives at the Massachusetts Historical Society or other local archives; the latter was to be completed out of the British Parliamentary Papers, of which a full set was available in a nearby library.

The individual papers, based entirely on archival research, explored a Boston family in the China trade, Scottish migration to North America, and Barbados-Boston migration in the twentieth century. Of the collaborative papers, one paper prepared by three persons focused on plague epidemics from 1894 to 1897 and their impact on trade and migration in Hong Kong, Bombay, and environs; the other two students analyzed labor migration from China and British Central Africa to South Africa from 1890 to 1907. The research experiences took the students through the experience of formulating and reformulating their projects and yielded well-documented, well-argued papers. That was a good start, but the experience demonstrated to me that a graduate program in world history must put energy into locating archival collections that enable students to define and complete small research projects at a global or transnational level.

For the seminar on global political history in the winter of 2002, I returned to reliance on secondary sources in order to give attention to a conceptual issue. My general argument was that empires have not been analyzed and theorized as much as nations, and they deserve to be so theorized; the syllabus identified twenty-five empires in the modern era. The nine students began their work by reading a series of books on empires in various times and situations.[6] Their initial assignment was to write a conceptual paper on an aspect of empire, including at least three empires in their articulating of the concept. Their second assignment was to write an interpretive paper developing a view of a single empire in global perspective. The conceptual papers, each addressing at least three empires, addressed such topics as bureaucracy, citizenship, incorporation of subject peoples, unfree labor, and the behavior of elites as empires began to decline. Interpretive papers addressed such issues as imperial food supply, decolonization, the choice between trade and domination, the Belgian empire, and Spanish incorporation of imperial subjects.

---

5. Readings included Cohen 1997; Daniel Scott Smith, "Recent Change and the Periodization of American Family History," *Journal of Family History* 20 (1995), 329–46; excerpts from Wrigley and Schofield 1981; and Peter Laslett, *Household and Family in Past Time* (Cambridge, 1972).
6. Readings included Abernethy 2000; Wong 1997; Gore Vidal, *The Decline and Fall of the American Empire* (Monroe, Maine, 2000); Colin Wells, *The Roman Empire*, 2nd ed. ([1984] Cambridge, Mass., 1992); Huntington 1996.

These five research seminars reflect a modest but sustained effort to use the university classroom as a forum for teaching research design and research techniques in world history. No one of these courses measured up to the dream of a rigorous course in world history research. But I believe the results show that a structured course in research on a global topic, giving the opportunity to share notes with fellow students, gives students an initial experience on which they may build. With this basis in supervised research, supplemented by other coursework in world history, one should be able to design and execute research projects at any of the levels considered here.

Microprojects: Preparing for Class

Microlevel research projects are quick studies of global issues to answer such short-term needs as preparing materials for a class. In this approach, the researcher makes the best use of readily available materials and pulls together a report with easily communicable results. Encyclopedias, standard works in print, textbooks, major journals, and internet search engines are the sorts of resources one consults in order to get an answer to the question under study within a day or so.

The microproject approach often works well in pulling together information on subjects and for comparisons. It is fairly easy to learn what areas the Mongols had conquered by 1160 or what empires were in existence in 1600. For processes and connections (how did iron technology spread to Southeast Asia?), it is more difficult to get information in a hurry.

Suppose I take as my topic the impact of the Mongol Empire on Russia. If I begin with college textbooks (such as those by Bentley, Spodek, or Stearns), I will learn about the Mongol conquest and the subsequent formation of the Golden Horde or Kipchak Khanate, but not much more. Bertold Spüler's classic little book on the Mongols reveals more detail.[7] Entering "Mongols" in web search engines yields sites with information on Chinghiz Khan and the initial conquests. A search on "Golden Horde" yields more detail: narratives of Golden Horde rule in Kazan and the rise of Muscovy as a vassal state, images of Kipchak coins, and other artifacts and art works at the Museum of the Hermitage (Moscow) and Kazan State University. A further look at a library catalogue (or Amazon.com) reveals the recent book by Hartog on Russia and the Mongols.[8] With these materials I can prepare a class of an hour's length. I can choose to wrap the story up in 1462, when Ivan IV ("the Terrible") halted tribute payments to Kazan, or a century later when Russia conquered Kazan.

---

7. Spüler [1969]; Jerry H. Bentley and Herbert F. Ziegler, *Traditions and Encounters: A Global Perspective on the Past* (Boston, 1999); Peter N. Stearns, Michael Adas, Stuart B. Schwartz, and Marc Jason Gilbert, *World Civilizations, the Global Experience*, 3rd ed. (New York, 2000); Richard W. Bulliet, Pamela Kyle Crossley, Daniel R. Headrick, Steven W. Hirsch, Lyman L. Johnson, and David Northrup, *The Earth and Its Peoples: A Global History*, 2nd ed. (Boston, 2001); Howard Spodek, *The World's History*, 2nd ed. (Upper Saddle River, N.J., 2001).

8. Leo de Hartog, *Russia and the Mongol Yoke: The History of the Russian Principalities and the Golden Horde, 1221–1502* (London, 1996).

Small Projects: Master's Theses and Articles

A small project involves defining and resolving a substantial but limited historical issue. The researcher should locate and scrutinize primary documents, using them to make an original evaluation of the issue but also to review the relevant secondary literature and link the original research to published studies.

The content and mode of preparation for an article or a master's thesis are well enough known, and I have only two major points to emphasize. The first is that a master's thesis or a first article can indeed focus on world history, if the author has sufficient clarity in selecting a topic, identifying relevant resources, and constructing an appropriate research design.[9]

My second point is that a small research project provides an excellent opportunity for graduate students to develop their system of taking and filing notes. Once they have tested their system for at least a year, they are ready to try it on a large research project. When one is meeting the deadlines of finishing a paper for class, it is often easiest to scribble notes on pads of paper or enter text in a single large computer file. But if one wishes to refer to notes after several months or years, then a more precise and dependable system is required.

I remember my major professor showing his graduate students a system of five-by-eight-inch cards, typed and labeled at the top, and filed neatly in metal drawers. I too started a system of five-by-eight cards, and kept it going for several years. Those are still my most dependable notes. There was a time in the 1970s when stationers made it almost impossible to buy five-by-eight cards, as they were pushing four-by-six cards. I got past that crisis, but I should have known that it was the harbinger of things to come.

With electronic text files, the short-term benefit of convenience and searchability was offset in large measure by periodic changes in platform. I now have notes in CPM, DOS, Windows, and Macintosh operating systems and the prospect of more changes to come. In addition to this was the difference among Word Star, Word Perfect, and Word, not to mention the various bibliographical database programs. Since we can be sure that platforms and programs will be changing in the foreseeable future, each graduate student must design not only a good note-taking system, but a system for updating and transforming it with each change in platform and program.

Large Projects: Dissertations and Books

The numerous recent studies cited in part III of this book contain a wealth of information on the design and execution of large studies in world history. Most often, however, these are second or later books by scholars whose first book was a

---

9. A recent example is Athanasios S. Michaels, "Masculinity and Imperialism in England: A Patriotic Construct from the Indian Mutiny to the South African War, 1857–1902" (M.A. thesis, Northeastern University, August 2000).

localized study. In this section, I want to emphasize doctoral dissertations as large projects in world history that are also the scholar's first major research effort. Such a project requires addressing the issues of research design, locating resources, language preparation, and financial support for travel and study—and doing it for the first time and dealing with a supervising committee. The time required for a dissertation is about a year for developing the proposal, one or two years for research, and a year for writing. This means that one may hope for completion of a history dissertation two or three years after the completion of comprehensive exams; the work often takes longer to complete, but the delays are commonly associated with taking breaks from research and writing, as for teaching.

Doctoral dissertations in world history may be divided into those that are comparative, interactive, and global in scope. For studies that are explicitly comparative, the initial research design is likely to remain largely unchanged throughout the study: the researcher will document the two or more cases and the comparisons among them. For studies intended to be interactive or global in scope, the research design and the interpretation are more likely to change in the course of the study. That is, the initial design of the study is a comparison of cases, and only with time does the analyst reformulate it to emphasize interactions or global patterns for the materials under study.

This evolution of research leads to what I have chosen to call "McKeown's dilemma," after Adam McKeown, my former colleague at Northeastern, who has posed the difficulty clearly in discussion with doctoral students. He tells the tale based on his own experience: in his dissertation, he presented the story of early twentieth-century migration from South China to Chicago, Peru, and Hawaii. The dissertation included many global insights, but was written in the form of comparative chapters. McKeown found that in revising it into a book that gave greater emphasis to global patterns, relying on the concept of a network linking migrants all across the Pacific, the earlier organization left patterns of regional organization that were difficult to break out of.[10] This, then, is the dilemma: whether to complete a dissertation as a comparative study of localities and hope to be able to articulate the global connections later on—or to spend more time on the dissertation, restructuring it until the global historical patterns become clear.

Yinghong Cheng's dissertation, while explicitly comparative, also elucidates global tensions and transformations. It provides an example of McKeown's dilemma. The principal chapters of this study compare the Revolutionary Offensive of Cuba to the Cultural Revolution of China, with a focus on the issue of the "new man," the idea of changing human nature to direct people toward social welfare rather than individual advance. The analysis shows that although the Cuban and Chinese regimes were in very little contact, their movements of social mobilization stemmed from similar problems, they expanded and

10. Adam McKeown, "Chinese Migrants among Ghosts: Chicago, Peru, and Hawaii in the Early Twentieth Century" (Ph.D. dissertation, University of Chicago, 1997); McKeown 2001.

contracted according to remarkably similar dynamics, and they influenced other movements of their time.[11]

Deborah Smith Johnston completed a dissertation on world-history curriculum, reviewing past and current practice and emphasizing the advantages of a thematic approach to teaching world history to secondary students.[12] The analysis emphasizes the tensions among competing needs to organize course material by theme, time period, region, discipline, and pedagogical approach. The data for the study include numerous interviews with teachers and professors working in the field of world history, as well as the monographic and secondary resources used in teaching world history. The analysis traces the limitations of courses based on political narratives in area-studies context but also demonstrates that establishing thematic approaches to the teaching of world history will require setting criteria for and definitions of world historical themes.

Global Research Projects

One of the great limitations on the study of world history is that historical data are not organized in global terms. Historical data, in their original form, are commonly found at the local level. It has taken over a century of historians' work in compilation and transformation to develop our present collection of historical data at the national level. At one level, it is a simple matter of aggregating national data to develop a global picture—this is the approach taken in some studies of population and economic history. At another level, the political units of today are not necessarily the appropriate geographical units to study the history of the last several millennia.

I favor the organization of collaborative efforts to develop and analyze historical data at the global level. One approach is the creation of a formal or informal think tank of research-oriented world historians. It should nurture ties among specialists in world history and build dependable links to other scholars and funding agencies. It should articulate major items on the research agenda for world history and present them in a well-reasoned fashion to those private and public agencies that support research in history, the humanities, and social sciences. The think tank should pursue links to the International Congress of Historical Sciences, to the American Historical Association, the Organization of American Historians, to the Social Science History Association, historians of science, and other groups within the history profession. It should also seek out ties with area-studies associations, to develop the strength of cross-disciplinary, interregional

11. Yinghong Cheng, "Creating the New Man—Communist Experiments in China and Cuba: A World History Perspective" (Ph.D. dissertation, Northeastern University, 2001). In revising his dissertation for publication, however, Cheng has articulated a global issue: he reinterprets twentieth-century communist movements to emphasize that they were movements for spiritual renewal addressing major religious issues, not just a sociopolitical response to capitalism.

12. Johnston 2003.

study. Similarly, the think tank should pursue links to disciplinary organizations in sociology, economics, anthropology, geography, literature, political science, art history, and so forth. Finally, this group of research activists should also establish firm ties to groups of teachers from primary school through university, both to convey the latest research results and to learn of the priorities and the outcomes of classroom work and the guidelines that they offer for further research.

As an example of a global research project I have defined, with colleagues at Northeastern and elsewhere, a World History Databank as a project to take a next step in organizing global historical data.[13] The project is to collect—for the period of the last four centuries—four sorts of quantifiable data related to each other and to current debates in world history. The current debates address the temporal and cross-national dimensions of such key questions as growth, inequality, dominance and systemic behavior in the world economy; the transformation of world regions through migratory movements, and current and past changes in family and social life. Specifically, the data to be collected are: (a) volumes of trade and prices for such commodities as precious metals, textiles, grains, and for wage rates; (b) systematic estimates of migration flows within and among regions, as compared with population sizes; (c) data on family size and family structure; and (d) weights and measures, currency values, places, geographic regions, and demographic units to permit the conversion and standardization of other data. Collection and processing of the data will require extensive collaboration.

In addition to collecting the data and displaying the unprocessed data on a website, the key step of the project is conducting transformations to make the data comparable. For prices and quantities of goods, this means finding equivalent currency values and weights and measures. For population and migration estimates, it means defining relatively standard and equivalent regions throughout the world over the time period to which the data apply. For data on family size and family structure, comparability requires some system for translating descriptions of families into terms comparable from one historical situation to another.

This project, which is still at the planning stage and has yet to receive funding, is an example of the sort of work that will have to be done to develop advanced analyses of world history. World historians still work primarily as individual researchers relying on their own energies and insights, which are sometimes formidable. Putting scholars in contact with one another so that they can share results and commentaries extends further the capabilities of these individual scholars.[14] But it is clear that world history, if it is ever to be analyzed systematically, requires collaborative efforts and major institutional support to create and edit collections of data enabling systematic study of the whole world over long periods of time.

13. World History Databank: www.worldhistorydatabank.net.
14. Examples of connections among world historians include the discussions on H-WORLD and other electronic lists, the presentations at Northeastern University's World History Seminar, and the compilation and publication of slave trade data in the Du Bois Institute database. See Eltis et al. 1999.

Conclusion: Creating Global Knowledge

The task of advancing knowledge about world history can be accomplished far more readily if the basic principles of world-historical study are applied broadly and soundly. Some of the principles, it seems to me, are quite straightforward. First, world historians should travel. Second, world historians should work collaboratively. It is not only the collection of data that is more efficient when the work is shared, but also the decisions over which data to collect and how to identify and explain the results. Third, research projects in world history require strong research design. It is important to be clear on the big question in world history under study, on the specific data to be collected, and on the relationship between the data and the question. Fourth, the data used in world historical studies need to be assembled carefully and modified for consistency. Working across boundaries means that data sets will rarely if ever provide neatly consistent information, and research will therefore include a laborious but important task of linking different data sets, making them consistent for analysis across boundaries.[15]

Historians are numerous enough to divide their energies between those who focus on small problems and on big problems and between those who focus on localized issues and those who follow the connections from issue to issue. For those who pursue the connections, there is the certain frustration of repeatedly selecting which branch to follow and of never having mastery of the terrain one traverses. Yet these armchair navigators of the past have the satisfaction of glimpsing the larger patterns and of demonstrating that historians, too, can venture as far and wide as have their subjects, those endlessly curious humans.

15. To demonstrate this issue, I assembled commercial and tax data from Cameroon under German, French, British, and independent rule. The results provided an estimate of economic continuity and change for the whole region of modern Cameroon 1890–1985. Manning 1990–1991.

Chapter 23

Conclusion: Tasks in World History

The field of world history continues its spontaneous and somewhat disorderly expansion.[1] Should this laissez-faire approach of scholars continue, it will probably allow more such incremental growth and a gradual clarification of the specific contributions that can come from a global approach to history. My experience, however, is that a purposeful campaign of developing world-historical insights can speed up the process greatly. Much of my argument centers on a call for the establishment of comprehensive institutions for research, study, and teaching in world history, to facilitate accelerated learning.

In this concluding review, I summarize the materials in this book as a set of tasks: five areas of practice in world history. Attention to these tasks will enable world historians to develop and maintain an acquaintance with the five areas of knowledge and skills emphasized in this book. At the same time, most of the work in research and teaching history is done at an individual level, whether in a supportive institution or on one's own. What I offer in this section is a set of tasks whose accomplishment should strengthen individuals and small groups in their programs of study in world history. These five categories of reminders and priorities, in my opinion, will assist students of history at all levels in using the knowledge at hand to create useful interpretations of human history.[2]

1. The significance of world history in current research is indicated in the following figures, developed from articles published in the American Historical Review. For the year 2000, out of sixteen articles, two were structured in global terms and two were comparisons; out of five review essays, two were on books taking a global approach. For the year 2001, out of twelve articles, three were global and two were comparative; out of five review essays, one was on a global book and one on comparison. Ten years earlier, the global and comparative works were far less common.
2. Each set of task for navigating world history is the subject of a set of chapters in this volume: a review of global historiography (part I), an exploration of approaches to world history (part II), a focus on themes and debates in world history (part III), a rehearsal of the logic of world history (part IV), and a discussion of graduate education (part V).

Past Accomplishments

One task of world historians today is to sustain contact with the lessons already learned. World history records many old and important debates and restates many valuable old truths. The new knowledge is exciting and gains our attention, but the previously accumulated knowledge retains its importance. To put it somewhat differently: if the material world is now greatly different than before, the human mind is much the same as ever.

For at least three centuries, a thin line of prodigious readers and deep thinkers has sought to comprehend the general outlines of human development, and in so doing has elaborated schemas of world history. In chronological order, I have emphasized Guicchiardini, Bodin, Bossuet, Fontenelle, Vico, Voltaire, Gibbon, Herder, Condorcet, Hegel, Comte, Ranke, Marx, Weber, Spengler, Toynbee, and McNeill.[3] The treatises of these thinkers often mixed history with philosophy, religion, social advocacy, and economic analysis; still, the complexity of these mixtures has not made them any less influential on our understanding of history. And those who have written explicitly on world history have done so voluminously. Toynbee's ten-volume world history, written in the twentieth century, was exceeded in size by several nineteenth-century German compendia on world history; the English *Universal History* of the eighteenth century reached sixty-five volumes.[4] The issues debated in these interpretations have included the causes of change in material, social, intellectual and spiritual arenas, the nature and achievement of dominance, the long-term patterns of culture and politics, and the roles of individuals in historical change.

By the end of the eighteenth century, some key lessons had been learned, and other major debates had been set. The debate over polygenesis versus monogenesis in the human species had largely been resolved in favor of monogenesis, though that did not prevent acceptance of a doctrine of racial distinctiveness. The notion of evolution and development in human society—the doctrine of progress—had been articulated and documented. The idea of distinct civilizations had been codified, though along with it came the notion of European (or "Western") preeminence. Long-term dynamics of the rise, transformation, and decline of empires were documented.

As the historical profession took form in nineteenth-century universities, in a process not unrelated to the broader crystallization of nation-states and national education systems, those who assumed the formal charge of preserving the past gradually divorced themselves from the discourse on world history. Professional historians, nurturing their reputations for objectivity in the well-fertilized gardens of each national tradition, expressed disdain for the speculative, philosophical and overgeneralized aspects of broader studies in human history. Still, the debate continued—albeit in public squares rather than ivory towers.

---

3. Gibbon 1776–1778; Hegel [1830]; Ranke [1833]; Ranke 1883–1887. In addition, as historians read more broadly, the importance of contributions by writers outside Europe will be recognized increasingly.
4. Toynbee 1933–1961; Oncken 1879–1890; Sale et al. 1736–1765.

By the mid-twentieth century, several hypotheses were under debate on the dynamics of human development. From Marx came the view that economic structures, themselves emerging out of conflicting influences, governed the social and cultural superstructure of society. From Weber came a vision of commerce and bureaucracy as powerful yet contending social forces. Spengler traced a recurrent civilizational life-cycle, Toynbee envisioned a long-term pattern of challenge and response, and Freud emphasized the recurrent human repression of animal urges. Mumford traced the impact of the city in history, while Lattimore focused on the dynamic of the frontier.[5] For all of these thinkers, the concentration on connections in history began to be enunciated with some clarity.

The founding texts of the professional study of world history include William McNeill's 1963 *Rise of the West* and the works published in subsequent years by Alfred Crosby, Philip Curtin, Andre Gunder Frank, and Immanuel Wallerstein. But in another sense, these volumes translated and updated discussions that had been initiated decades and even centuries earlier. Wallerstein's 1974 *Modern World-System*, asserting that the sixteenth-century expansion of commerce was the essential change in the modern world economy, took the side of Sweezy in the Dobb-Sweezy debate of the 1940s, supported Weber's essays in the early twentieth century disputing Marx, and echoed the writings of Herder in the 1780s.[6] One side in this debate saw the sixteenth-century expansion of capitalist commerce as the key change in the modern economy; the other saw the transformation as centered on the rise of industrial production and wage labor in the eighteenth century. The debate continues.

Developing Disciplines

A second task of world historians is to be alert to changes and advances in the growing number of disciplines that are relevant to study of the past. The disciplines range from social sciences to humanities, arts, and natural sciences. They include regional and topical specializations. Within each specialization, one should keep track of advances and debates in theory and improved techniques of analysis. The current rate of expansion of knowledge is so great that new disciplines constantly emerge and existing disciplines combine. The result is to provide the world historian with new knowledge and new types of knowledge.

World historians naturally renounce the possibility of learning all the relevant approaches and evidence, but they just as naturally develop and trust their instincts in selecting which disciplines to study in detail. Historians are not simply consumers of the work of other specialists. Instead, the broad and interactive perspective of world historians makes it possible to connect and combine the disciplines on which they draw and to develop new applications. In this way, historians can be critics and creators of theory as well as an audience for the interpretations of other scholars.

5. Lenin [1899]; Lenin [1917]; Freud 1930; Polanyi 1944.
6. Wallerstein 1974; Brenner 1975; Dobb 1946; Sweezy et al. 1976; Weber [1904]; Marx [1867–1894]; Herder [1784–1791].

Old and New Debates

A third main task of world historians is to locate key debates—new debates and old debates, whether resolved or continuing. The new scholarly debates in world history center on new empirical evidence and on the theories and analytical techniques producing new evidence. The continuing debates in world history stem from the philosophical dilemmas inherited from the ages.

One great current debate focuses on the world economy from the sixteenth to the nineteenth centuries. Phrased in terms of dominance, the question is whether East Asia or Western Europe led the world in output, productivity, and profit. The same question, phrased in terms of connection, is what global trade in silver reveals about the mutual dependence of East Asia, Western Europe, and other regions in a global system of production.[7] The debate is not only about the answers but about which question is most useful for developing an interpretation.

Another great debate centers on nationalism and nationhood. In terms of dominance, the question is whether nationhood was a system of politics developed in the North Atlantic that was later exported to the rest of the world. In terms of connection, the question is how nationhood became the political organization of everyone.[8]

The debates on earlier times are more likely to be interdisciplinary than these economic and political debates on recent times. Scholars in linguistics, archaeology, and other fields have been tracing the early movements of Indo-European speakers across the Eurasian heartland up to ten thousand years ago, debating their points of origin, their concentrations of population, and their cultural exchanges with each other and with other groups. Similarly, scholars have focused on the interplay among religious traditions in this same Eurasian space. While connections among Judaism, Christianity, and Islam are well known, recent work examines commercial, cultural, and philosophical links among Buddhism, Hinduism, Zoroastrian religion, and other traditions.[9]

Other debates have yet to develop in detail, but we can certain that they will emerge once new evidence is located. What is the place of gender in world history? What have been the global patterns in the structure and behavior of the family? The list of topics ripe for debate is a long one.[10]

In these debates about the patterns of historical change, certain underlying analytical or philosophical choices reappear at every turn. Are the similarities between different regions to be accounted for by independent invention or by diffusion? Should we emphasize the uniqueness of each historical situation or the commonality linking them all? Should we give more emphasis to continuity or to change? Did the changes in material life bring about new ideas, or did the changes in human ideas bring about changes in the material world? Are we to think of the

---

7. Pomeranz 2000; Wong 1997; Frank 1998; Landes 1998.
8. B. Anderson 1983; Huntington 1992.
9. Mallory and Mair 2000; Liu 2001.
10. Under the editorship of Bonnie Smith, the American Historical Association launched a series of publications entitled "Women's and Gender History in Global Perspective."

changes over the long term of human history as a progression, or have there been significant cycles of change (be they economic, cultural, or philosophical)? Are the developments of the past to be seen as determinate and having some meaning, or were they accidents for which we can only trace the consequences? In analyzing the past, do we seek to identify unique factors that propelled the changes of history, or do we explain change through the correlation and interpenetration of many influences? For world history, as with history of families, localities, or nations, the experience of each generation leads to different emphases on these perennial questions. We can be certain, then, that each generation will have to write its own world history.

Global Logic

A fourth task of world historians is to explore world history logically. The logic of world history is both general and specific, and it requires attention to the range of evidence on the experiences in human life. World history, as may be seen from its origins, is an array of approaches to the past rather than a single formula for explaining our history. It is an umbrella of histoi cal themes and methods, unified by the focus on connections across boundaries ut allowing for diverse and even conflicting approaches and interpretations. It is not a single approach or a single insight, or the outlook of a single interest group. It is rather a terrain of convergence and interaction of many historical interests. The methods and rules of world history are therefore the rules for combining and analyzing data of different sorts.[11] Rather than serving as a precise recipe for creating a preordained product, the analytical rules of world history provide a set of methodological and philosophical priorities.

Here is my statement of the general steps for creating or evaluating an interpretation of issues in world history. At each stage, these interpretive steps focus on logical consistency, empirical documentation, and the identification of global dynamics. The steps alternate in widening and narrowing the analysis, in order to ensure that the result is at once broadly connected and grounded in specifics.

Select a topic and purpose for study. World history faces one with too much to cover, so one must develop a readiness to select specific topics for study. Happily, many—perhaps most—topics in history are susceptible to global analysis. Each historian may take responsibility for selecting what part of the past to analyze, rather than accept someone else's definition of the problem, debate, or dilemma. He or she should be able to defend the logic of that choice, and should accept the interpretive consequences of the selection.

Exploratory comparison. Having selected a topic, one should explore comparisons with a wide range of related or parallel topics, considering any possible

11. Locating this range of approaches under an umbrella does not prevent them from being arranged in a hierarchy or in competing hierarchies. I am indebted to Stacy Tweedy for this point, which indicates an opening for further study.

connections, similarities, and contrasts among topics. This sort of brainstorming is essential to guaranteeing breadth and comprehensiveness in world historical analysis, and it may reveal unsuspected dimensions of the topic or suggest patterns and interactions to be documented.[12]

Modeling the dynamics. World historians, in seeking to link disparate bodies of information into coherent stories, must formalize their logic rather than wait for the facts to speak for themselves. The models of historians may range from explicitly detailed and deductive theories to attractive but imprecise metaphors. In any case, the model needs to be explored to its limit in search of ideas to be tested, and it must highlight the dynamics of global change. Modeling world history requires that analysts consider cases, networks, systems, and debates and develop the art of conducting several of these activities at once.

Identify connections. World historians seek out connections among events and processes in the past and also among the models and disciplines with which we explore the past. In particular, since the models may refer to one or several areas under study, it is important to seek out linkages among the subsystems within the topic under study. World history links both the accidental and the systemic connections in the places, times, and themes of the past to help explain the broader patterns.

Verify the conclusions. Having developed an argument about the past, the world historian must next seek to verify it. In general, this means analyzing historical data to measure one model against another for sections of the analysis and for the study overall. This task, even when only partially completed, addresses the questions "How do you know?" and "Compared to what?" It thereby takes the analysis from insight to confirmation.

Shift perspective. A world historian, having completed the steps so far and developed a perspective on the past, will find that there is always another relevant way to look at the issue. The next step, then, is rethinking the analysis from inside and outside and reiterating the earlier steps from a new perspective. One should evaluate the interpretation from the standpoint of one and then another person from past time, explore the interpretation through the optic of one and then another analytical discipline, and reconsider it in short-term and in long-term perspective.

Overall interpretation. Having carried out all the preceding steps at least twice, one may offer an overall interpretation that is not necessarily a synthesis of all

12. Exploratory comparison presumes that the analyst begins with a framework and categories of information within which to compare. See chapter 16 for a discussion of the formalities of a world-historical framework, and for such categories of information as cases, networks, comparisons, and connections.

available information and probably not a definitive conclusion but more likely a provisional summary. Such a summary, if presented forcefully, can prove insightful and can stimulate further discussion and research.

Within each of these seven steps lies a great range of details and possible sub-categories. This summary of the logic of world history provides a framework for analysis of the past that gives systematic emphasis at once to broad patterns and to specific links in historical experience.

Comprehensive Study Habits

The fifth task of world historians today is to follow a rigorous program of study. Proficiency in world history requires practice: world historians should read widely, discuss readings with others, and follow the threads in their study that lead across boundaries in space, time, and subject matter. Graduate study in history is the most dependable way to achieve readiness for teaching in depth or for conducting research at the global level. The single most important unmet need, in the establishment of a strong field of world history, is programs of graduate training. Both teachers and researchers will benefit immensely from programs of formal study in world history. The relative absence of graduate programs means that world history has yet to benefit from the fresh thinking that can emerge from concentrated and detailed study.

To put it in more positive terms, the establishment of full-scale programs of graduate study will enliven world history and enable world historians to face the challenge of an immense and complex field of study. For researchers, such programs will provide training for a lifetime of specialization in identifying and analyzing connections and large processes in human history. Programs of graduate study will provide college and secondary teachers with expertise in the literature and debates on global historical interactions, so that they may inspire students to pose the questions and interpretations appropriate to the next generation. With or without formal graduate study, world historians need to keep their study habits sharp, so they can retrieve and apply the most significant new information and new approaches to understanding the global past.

Conclusion

Historians are usually ill-advised to claim uniqueness for their topic. Yet it is difficult to avoid suggesting that we live at a unique moment in our perception of history. Large numbers of well-informed people have recently learned to see global interconnections as a major topic for study. Similarly, one cannot avoid noting the specificity of the world history movement in the United States, which surely stems from its place as the global hegemon, a center of the cosmopolitan connections that were once called a melting pot, and through the peculiarities of its public education system. At the same time, a few days' travel is sufficient to verify the commonality of global perspectives and insights at every crossroads—Port-au-Prince, Singapore, Kinshasa, Turin, or Vladivostok.

At the beginning of this book, I sought to suspend the whole enterprise of world history from the term *connection*. Is this single word sufficient to characterize world history? The term *connection* works only if it brings several other descriptors along in its wake. I have added *selection, comparison, modeling, shifting perspectives,* and more. But *connection* conveys the character of world-historical analysis better than any other term. It acknowledges locality and uniqueness, yet also invokes broad patterns.

The study of world history is an exercise in complex thinking. As such it presents a thought-provoking challenge not only for mature adults but also for the educational priorities of the United States and every other nation. School policy everywhere emphasizes math and grammar: the emphasis is on learning the rules of grammar and developing sophisticated levels of skill in solving problems that involve a small number of variables. World history, in contrast, presents students with large numbers of variables and requires that they select a problem and define its limits as well as try to solve it. World history provides practice in selecting issues, handling multiple variables, adjusting scope, developing an analysis, then shifting perspectives to review the same issues. The study of world history is a good rehearsal for life itself.

Bibliography

Entries in this bibliography are organized by date of first publication to help indicate the development of the world-history literature. The categories parallel the time frame of chapters 2 through 5. That is, the four categories are works first published before 1900, between 1900 and 1964, between 1965 and 1989, and since 1990. Works are organized alphabetically by author within these categories, and works published in multiple editions are classified according to the first edition.

A. Works First Published before 1900

Astley, Thomas (Pub.), 1745–1747. *A New General Collection of Voyages and Travels*. 4 vols. London.

Augustine, Saint. [426] 1957. *City of God against the Pagans*. 7 vols. Trans. George E. McCracken. Cambridge, Mass.

Bancroft, George. 1873–1874. *History of the United States of America from the Discovery of the Continent*. 10 vols. Boston.

Bodin, Jean. [1566] 1960. *Method for the Easy Comprehension of History*. Trans. Beatrice Reynolds. New York.

———. 1577. *De la vicissitude ou variété des choses on l'univers*.

Bossuet, Jacques-Benigne. [1681] 1976. *Discourse on Universal History*. Trans. Elborg Forster, ed. Orest Ranum. Chicago.

Burckhardt, Jacob. [1860] 1958. *The Civilization of the Renaissance in Italy*. Trans. S. Middlemore. New York.

Carneiro, Robert L., ed., 1967. *The Evolution of Society: Selections from Herbert Spencer's Principles of Sociology*. Chicago.

Comte, Auguste. [1830–1842] 1974. *The Positive Philosophy*. Trans. [1855] Harriet Martineau, introduction by Abraham S. Blumburg. New York.

———. 1830–1842. *Cours de philosophie positive*. 6 vols. Paris.

———. 1851–1854. *Système de philosophie positive*. 4 vols. Paris.

Condorcet, Marie Jean Antoine Nicolas Carstat, Marquis de. [1795] 1900. *Tableau historique des progrès de l'esprit humain*. Paris.

Crone, G. R. trans. and ed. 1937. *The Voyages of Cadamosto and other Documents on Western Africa in the Second Half of the Fifteenth Century*. London.

Dapper, Olfert. [1670] 1967. *Beschreibung von Afrika*. New York.

Darwin, Charles. 1856. *The Origin of the Species by means of natural selection*. London.

Diderot, Denis, ed. 1751–1765. *Encyclopédie, ou dictionnaire raisonné des sciences, des arts et des métiers*. 17 vols. Paris.

Engels, Friedrich. [1884] 1968. *The Origin of the Family, Private Property, and the State*, in Marx and Engels, *Selected Works*, pp. 455–593.

Eusebius. [325] 1953. *The Ecclesiastical History*. Trans. Kirsopp Lake. Cambridge, Mass.

Gibbon, Edward. 1776–1778. *The Decline and Fall of the Roman Empire*. 6 vols. London.

Hegel, G. W. F. [1830] 1975. *Lectures on the Philosophy of World History.* Trans. H. B. Nisbet. Cambridge.

Helmolt, H. G., ed. [1899] 1901–1907. *The History of the World.* 8 vols. New York.

Herder, Johann Gottfried von. [1784–1791] 1997. *On World History.* Eds. Hans Adler and Ernest A. Menze, trans. Ernest A. Menze and Michael Palma. Armonk, N.Y.

Herodotus. [ca. 450 B.C.E.] 1987. *History.* Trans. David Grene. Chicago.

Khaldun, Ibn. [1377] 1967. *An Introduction to History: The Muqaddimah.* Trans. Franz Rosenthal, ed. N. J. Dawood. London.

Labat, J.-P. 1728. *Voyage du Chevalier des Marchais.* 3 vols. Paris.

Las Casas, Bartolomé. [1566] 1986. *História de las Indias.* 3 vols. Caracas.

Lenin, V. I. [1899] 1964. *The Development of Capitalism in Russia.* Lenin, *Collected Works,* Vol. 3 (Moscow).

Lenzer, Gertrud, ed., 1975. *Auguste Comte and Positivism: The Essential Writings.* Chicago.

Leo, Heinrich. 1835–1844. *Lehrbuch der Universalgeschichte.* 6 vols. Halle.

Locke, John. [1690] 1988. *Two Treatises of Government.* Ed. Peter Laslett. Cambridge.

Ma Huan. [1433] 1970. *The Overall Survey of the Ocean's Shore.* Ed. Feng Ch'en Chun, trans. J. V. G. Mills. Cambridge.

Macaulay, Thomas Babington. [1849–1861] 1953. *History of England from the Accession of James II.* 4 vols. London.

Maine, Henry. 1861. *Ancient Law: Its Connection with the Early History of Society and its Relation to Modern Times.* London.

Malthus, Thomas. [1803] 1992. *Essay on the Principle of Population,* 2nd ed. Cambridge.

Marees, Pieter de. [1602] 1987. *Description and Historical Account of the Gold Kingdom of Guinea.* Trans. and ed. Albert van Dantzig and Adam Jones. Oxford.

Marx, Karl. [1844] 1964. *Economic and Philosophic Manuscripts of 1844.* Ed. Dirk J. Struik, trans. Martin Milligan. New York.

———. [1852] 1968. *The Eighteenth Brumaire of Louis Napoleon.* Marx and Engels, *Selected Works,* 95–180.

———. [1857] 1973. *Grundrisse: Introduction to the Critique of Political Economy.* Trans. Martin Nicolaus. London.

———. [1867, 1885, 1894] 1971. *Capital.* 3 vols. Moscow. Vol. 1, trans. Samuel Moore and Edward Aveling, ed. Friedrich Engels (London, 1887; first published 1867); Vol. 2, ed. Friedrich Engels (first published 1885); Vol. 3, ed. Engels (first published 1894).

Marx, Karl, and Friedrich Engels. [1848] 1968. *Manifesto of the Communist Party.* Karl Marx and Frederick Engels, *Selected Works* (New York), 35–63.

———. 1968. *Selected Works.* New York.

Michelet, Jules. [1847–1853] 1967. *History of the French Revolution.* Trans. Charles Cocks, ed. Gordon Wright. Chicago.

Mill, John Stuart. [1865] 1965. *Auguste Comte and Positivism.* Ann Arbor.

Montesquieu, Charles de Secondat, baron de. [ca. 1721] 1946. *Lettres persanes.* Ed. Gonzague Truc. Paris.

———. [1748] 1949. *The Spirit of the Laws.* Trans. Thomas Nugent. New York.

Morgan, Lewis Henry. 1877. *Ancient Society; or, Researches in the Lines of Human Progress from Savagery, through Barbarism to Civilization.* New York.

Oncken, Wilhelm, ed., 1879–1890. *Allgemeine Geschichte in Einzeldarstellungen.* 32 vols. in 4 series. Berlin.

Pliny. [ca. 70 C.E.] 1949. *Natural History, with an English Translation.* 10 vols. Ed. and trans. H. Rackham. Cambridge, Mass.

Ploetz, Carl. 1880. *Auszug aus der alten, mittleren und neueren Geschichte [Epitome of Universal History].* Berlin.

Ranke, Leopold von. [1833] 1981. "The Great Powers." Ranke, *The Secret of World History: Selected Writings on the Art and Science of History.* Trans. and ed. Roger Wines. New York, 1981.

———. 1883–1887. *Weltgeschichte.* 8 vols. Leipzig.

Raynal, Guillaume-Thomas-François. 1781. *Histoire philosophique et politique des établissemens et du commerce des Européens dans les deux Indes,* 3rd ed. Amsterdam.

Ridpath, John Clark. 1894–1897. History of the World. 4 vols. New York.

Sale, George, George Psalmanzar, Archivald Bower, and others, publishers. 1736–1765. *Universal History.* 65 vols. London.

Smith, Adam. [1776] 1976. *An Inquiry into the Nature and Causes of the Wealth of Nations.* Eds. R. H. Campbell and A. S. Skinner. Oxford.

Spencer, Herbert. 1876, 1882, 1896. *Principles of Sociology.* 3 vols. London.

Ssu-ma Chien [Sima Qian]. [ca. 100 B.C.E.] 1961. *Records of the grand historian of China.* Trans. Burton Watson. 2 vols. Oxford.

Tabari, Abu Ja'far Muhammad b. Jarir al-. [ca. 915] 1987–1998. *History of Prophets and Kings.* 39 vols., Numerous translators. Albany.

Thucydides. [ca. 420 B.C.E.] 1989. *History of the Peloponnesian War.* Trans. Thomas Hobbes. Chicago.

Tönnies, Ferdinand. [1887] 1957. *Community and Society.* Trans. Charles P. Loomis. East Lansing.

Vico, Giambattista. [1725] 1984. *The New Science of Giambattista Vico.* Trans. Thomas Goddard Bergin and Max Harold Fisch. Ithaca.

Voltaire. [1753–1754] 1969. *La Philosophie de l'histoire.* Ed. J. H. Brumfitt. Toronto.

———. [1756] 1991. *Candide, or Optimism.* Ed. and trans. Robert M. Adams. New York.

———. [1754–1757] 1901–1904. *The General History [Essai sur les Moeurs et l'Esprit des Nations].* Trans. William Fleming. Akron.

B. Works First Published from 1900 to 1964

Almond, Gabriel A., and James S. Coleman. 1960. *The Politics of Developing Areas.* Princeton.

Almond, Gabriel A., and Sidney Verba. 1963. *The Civic Culture: Political Attitudes and Democracy in Five Nations.* Princeton.

Anstey, Vera. 1929. *The Economic Development of India.* London.

Apter, David E. 1961. *The Political Kingdom in Uganda: A Study in Bureacratic Nationalism.* Princeton.

Baran, Paul. 1957. *The Political Economy of Growth.* New York.

Barnes, Herry Elmer, ed. 1925. *Ploetz's Manual of Universal History.* Trans. William H. Tillinghast. Boston.

Beard, Charles. 1913. *An Economic Interpretation of the Constitution of the United States.* New York.

Bendix, Reinhard. 1960. *Max Weber, an Intellectual Portrait.* New York.

Berlin, Isaiah. 1939. *Karl Marx: His Life and Environment,* New York.

Bloch, Marc. 1953. *The Historian's Craft.* Trans. Peter Putnam. New York.

Boulding, Kenneth. 1953. *The Organizational Revolution: A Study in the Ethics of Economic Organization.* New York.

Boxer, Charles R. 1951. *The Christian Century in Japan.* London.

———. 1952. *Salvador de Sá.* London.

———. 1957. *The Dutch in Brazil.* London.

Boxer, Charles R. 1962. *The Golden Age of Brazil.* London.

Braudel, Fernand. [1949] 1995. *The Mediterranean and the Mediterranean World in the Era of Philip II.* 2 vols. Trans. Sian Reynolds. Berkeley.

Bury, J. B. 1932. *The Idea of Progress: An Inquiry into its Growth and Origin.* New York.

Butterfield, Herbert, 1951. *The Whig Interpretation of History.* New York.

Carr, Edward H. 1961. *What is History?* New York.

Childe, V. Gordon. 1925. *The Dawn of European Civilization.* London.

——. 1934. *New Light on the Most Ancient East: The Oriental Prelude to European Prehistory.* London.

——. 1936. *Man Makes Himself.* London.

——. 1942. *What Happened in History.* Harmondsworth.

Coleman, James Smoot, 1958. *Nigeria: Background to Nationalism.* Berkeley.

Collingwood, R. G. [1946] 1993. *The Idea of History.* Ed. Jan van der Duesen, revised ed. Oxford.

Conrad, Alfred H., and John R. Meyer. 1958. "The Economics of Slavery in the Ante Bellum South." *Journal of Political Economy* 66, 95–130.

Curtin, Philip D. 1955. *Two Jamaicas: The Role of Ideas in a Tropical Colony 1830–1865.* Cambridge, Mass.

——. 1964. *African History.* New York.

——. 1964. *The Image of Africa: British Ideas and Action, 1780–1850.* Madison.

Dawson, Christopher. 1956. *The Dynamics of World History.* Ed. John J. Mulloy. New York.

Delbrück, Hans. 1931. *Weltgeschichte.* 5 vols. Berlin.

Dobb, Maurice. 1946. *Studies in the Development of Capitalism.* Oxford.

Du Bois, W. E. B. [1915] 2002. *The Negro.* Amherst, N.Y.

Elias, Norbert. [1939] 1978–1982. *The Civilizing Process.* 2 vols. Trans. Edmund Jephcott. New York.

Elkins, Stanley M. 1959. *Slavery: A Problem in American Institutional and Intellectual Life.* Chicago.

Emerson, Rupert. 1960. *From Empire to Nation: The Rise of Self-Assertion of Asian and African Peoples.* Cambridge, Mass.

Fairbank, John K. 1953. *Trade and Diplomacy on the China Coast: The Opening of the Treaty Ports, 1842–1854.* Cambridge, Mass.

Fanon, Frantz. [1961] 1968. *The Wretched of the Earth.* Trans. Constance Farrington. New York.

Fogel, Robert William. 1964. *Railroads and American Economic Growth: Essays in Econometric History.* Baltimore.

Freud, Sigmund. [1930] 1961. *Civilization and Its Discontents.* Trans. and ed. James Strachey. New York.

Freyre, Gilberto. [1933] 1946. *The Masters and the Slaves: A Study in the Development of Brazilian Civilization.* Trans. Samuel Putnam. New York.

Gerth, H. H., and C. Wright Mills, ed. and trans. 1946. *From Max Weber: Essays in Sociology.* New York.

Gramsci, Antonio. [1975] 1996. *Prison Notebooks.* 2 vols. Trans. and ed. Joseph A. Buttigieg. New York.

Halpern, Manfred. 1963. *The Politics of Social Change in the Middle East and North Africa.* Princeton.

Hancock, W. Keith. 1937–1942. *Survey of British Commonwealth Affairs.* 2 vols. London.

Hartwell, R. M. et al. 1972. *The Long Debate on Poverty: Eight Essays on Industrialization and "the Condition of England."* Old Woking, U.K.

Hitti, Philip K. 1937. *History of the Arabs.* London.

Hobsbawm, Eric J. 1962. *The Age of Revolution: Europe, 1789–1848.* New York.

Hobson, J. A. 1902. *Imperialism: A Study.* London.

Jahn, Jahnheinz. [1958] 1961. *Muntu, The New African Culture.* Trans. Marjorie Grene. New York.

Jaspers, Karl. [1947] 1953. *The Origin and Goal of History.* Trans. Michael Bullock. New Haven.

Johnston, H. H. 1900. *The Discovery and Colonization of Africa by Alien Races.* Cambridge.

Johnston, H. H. 1910. *The Negro in the New World.* London.

Kedourie, Elie. 1960. *Nationalism.* Oxford.

Kliuchevskii, Vasilii. [1904–1922] 1911–1931. *A History of Russia.* 5 vols. Trans. C. J. Hogarth. London.

Knowles, Lilian C. A. 1924. *The Economic Development of the British Overseas Empire.* London.

Kohn, Hans. 1944. *The Idea of Nationalism: A Study in its Origins and Background.* New York.

———. 1962. *The Age of Nationalism: The First Era of Global History.* New York.

Kroeber, A. L. 1963. *An Anthropologist Looks at History.* Berkeley.

Kroeber, A. L., and Clyde Kluckhohn. 1952. *Culture: A Critical Review of Concepts and Definitions.* New York.

Kuczynski, Robert Rene. 1948–1949. *The Population of the British Colonial Empire.* 2 vols. London.

Kuhn, Thomas. 1962. *The Structure of Scientific Revolution,* Chicago.

Kuznets, Simon. 1941. *National Income and Its Composition, 1919–1938.* Vol. 1. New York.

Lamprecht, Karl. [1904] 1905. *What is History? Five Lectures on the Modern Science of History.* Trans. E. A. Andrews. New York.

Langer, William L., ed. 1940. *An Encyclopedia of World History, Ancient, Medieval, and Modern, Chronologically Arranged.* Boston.

Lattimore, Owen. 1940. *Inner Asian Frontier of China.* Boston.

Lenin, V. I. [1917] 1967. *Imperialism, the Highest Stage of Capitalism: A Popular Outline.* Trans. Yuri Sbodnikov, ed. George Hanna. V. I. Lenin, *Selected Works.* Vol. 1 (New York), 673–777.

Link, Arthur S. 1947. *Wilson.* 5 vols. Princeton.

McNeill, William H. 1963. *The Rise of the West: A History of the Human Community.* Chicago.

Mitchell, B. R., ed. 1962. *Abstract of British Historican Statistics.* Cambridge.

Morgenthau, Ruth Schachter. 1964. *Political Parties in French-Speaking West Africa.* Oxford.

Mumford, Lewis. 1934. *Technics and Civilization.* New York.

———. 1938. *The Culture of Cities.* New York.

———. 1944. *The Condition of Man.* New York.

———. 1961. *The City in History.* New York.

Murdock, George Peter. 1959. *Africa: Its Peoples and Their Culture History.* New York.

Nehru, Jawaharlal. 1934. *Glimpses of World History.* London.

Nettleford, Rex M. 1963. *Inward Stretch, Outward Reach.* London.

Niane, D. T. [1960] 1965. *Sundiata, An Epic of Old Mali.* Trans. G. D. Pickett. London.

Ortega y Gasset, José. [1949] 1973. *An Interpretation of Universal History.* Trans. Mildred Adams. New York.

Palmer, Robert R. 1950. *A History of the Modern World.* New York.

Park, Robert Ezra. 1950. *Race and Culture.* New York.

Parsons, Talcott. 1937. *The Structure of Social Action: A Study in Social Theory with Special Reference to a Group of Recent European Writers.* New York.

Parsons, Talcott, and Neil Smelser. 1956. *Economy and Society: A Study in the Integration of Economic and Social Theory*. Glencoe, Ill.

Pipes, Richard. 1954. *The Formation of the Soviet Union: Communism and Nationalism, 1917–1923*. Cambridge, Mass.

Pirenne, Jacques. 1948. *Les Grands Courants de l'Histoire*. Brussels.

———. 1962. *The Tides of History*. 2 vols. Trans. Lovett Edwards. New York.

Polanyi, Karl. 1944. *The Great Transformation: The Political and Economic Origins of Our Time*. New York.

Polanyi, Karl, Conrad Arensberg, and Harry W. Pearson. 1957. *Trade and Market in the Early Empires*. New York.

Reed, John. [1919] 1967. *Ten Days that Shook the World*. New York.

Robinson, James Harvey. 1912. *The New History*. New York.

Rostow, W. W. 1960. *The Stages of Economic Growth: A Non-Communist Manifesto*. Cambridge.

Rudé, George. 1964. *The Crowd in History: A Study of Popular Disturbances in France and England, 1730–1848*. New York.

Seligman, C. G. 1930. *The Races of Africa*. New York.

Skinner, G. William. 1964–1965. *Marketing and Social Structure in China*. Tucson.

Smith, Huston. 1958. *The Religions of Man*. New York.

Sorokin, Pitirim. 1937. *Social and Cultural Dynamics*. 4 vols. New York.

Spengler, Oswald. [1918–1922] 1926–1928. *The Decline of the West*. 2 vols. Trans. Charles Francis Atkinson. London.

———. 1932. *The Decline of the West*. Abridged by Helmut Werner, English abridged edition prepared by Arthur Helps. Oxford.

Tannenbaum, Frank. 1946. *Slave and Citizen: The Negro in the Americas*. New York.

Tarbell, Ida M., 1925. *The History of the Standard Oil Company*. New York.

Tawney, R. H. [1926] 1962. *Religion and the Rise of Capitalism*. Gloucester, Mass.

Teggart, Frederick J. 1939. *Rome and China: A Study of Correlations in Historical Events*. Berkeley.

Teng, Ssu-yu, and John K. Fairbank. 1954. *China's Response to the West: A Documentary Survey, 1839–1923*. Cambridge, Mass.

Thompson, E. P. 1964. *The Making of the English Working Class*. Harmondsworth.

Thomas, Brinley. 1954. *Migration and Economic Growth*. Cambridge.

Tilly, Charles. 1964. *The Vendée*. Cambridge, Mass.

Toynbee, Arnold J. 1933–1961. *A Study of History*. 12 vols. Oxford.

———. 1946. *A Study of History: Abridgement of Volumes I–VI*, by D. C. Somervell. Oxford.

———. 1957. *A Study of History: Abridgement of Volumes VII–IX*, by D. C. Somervell. Oxford.

———. 1948. *Civilization on Trial*. New York.

UNESCO. 1953. *Interrelations of Cultures: Their Contribution to International Understanding*. Paris.

———. 1963. *History of Mankind: Cultural and Scientific Developments. Vol. 1: Prehistory and the Beginnings of Civilization,* by Jacquetta Hawkes and Sir Leonard Woolley. New York.

Van Dalen, Deobold B., Elmer D. Mitchell, and Bruce L. Bennett. 1953. *A World History of Physical Education*. New York.

Van Loon, Hendrik Willem. 1921. *The Story of Mankind*. New York.

Vansina, Jan. [1961] 1965. *Oral Tradition: A Study in Historical Methodology*. Trans. H. M. Wright. Chicago.

Von Neumann, John. 1951. "The General and Logical Theory of Automata." L. A. Jeffries, ed., *Cerebral Mechanisms in Behavior* (New York), 1–31.

Wallerstein, Immanuel. 1961. *Africa, the Politics of Independence: An Interpretation of Modern African History.* New York.

Webb, Walter Prescott. 1952. *The Great Frontier.* Boston.

Weber, Max. [1904] 1958. *The Protestant Ethic and the Spirit of Capitalism.* Trans. Talcott Parsons. New York.

———. [1916] 1968. *The Religion of China. Confucianism and Taoism.* Trans. Hans H. Gerth. New York.

———. [1916–1917] 1958. *The Religion of India. The Sociology of Hinduism and Buddhism.* Trans. Hans H. Gerth and Don Martindale. Glencoe, IL.

———. [1917–1919] 1952. *Ancient Judaism.* Trans. Hand H. Gerth and Don Martindale. Glencoe. IL.

———. 1956. *Wirtschaft und Gesellschaft, Grundriss der verstehenden Soziologie,* 4th ed. Ed. Johan Winckelmann. Tübingen.

———. [1956] 1968. *Economy and Society: An Outline of Interpretive Sociology.* 3 vols. Ed. Guenther Ross and Claus Wittich, trans. E. Fischoff et al. New York.

Wegener, Alfred. 1915. *Die Entstehung der Kontinente und Ozeane.* Braunschweig.

Wells, H. G. 1920. *The Outline of History, Being a Plain History of Life and Mankind.* London.

Williams, Eric. [1944] 1994. *Capitalism and Slavery.* Chapel Hill.

C. Works First Published from 1965 to 1989

Abu-Lughod, Janet 1989. *Before European Hegemony: The World System A.D. 1250–1350.* New York.

Adas, Michael. 1974. *The Burma Delta: Economic Development and Social Change on an Asian Rice Frontier, 1852–1941.* Madison.

Adas, Michael. 1979. *Prophets of Rebellion: Millenarian Protest Movements against the European Colonial Order.* Chapel Hill.

———. 1989. *Machines as the Measure of Men: Science, Technology, and Ideologies of Western Dominance.* Ithaca.

Allardyce, Gilbert. 1982. "The Rise and Fall of the Western Civilization Course." *American Historical Review* 87, 695–725.

Amin Samir. [1970] 1974. *Accumulation on a World Scale: A Critique of the Theory of Underdevelopment.* Trans. Brian Pearce. New York.

———. [1973] 1976. *Unequal Development: An Essay on the Social Formations of Peripheral Capitalism.* Trans. Brian Pearce. New York.

———. [1988] 1989. *Eurocentrism.* Trans. Russell Moore. New York.

Anderson, Benedict R. O'G. 1983. *Imagined Communities: Reflections on the Origin and Spread of Nationalism.* London.

Anderson, Michael. 1980. *Approaches to the History of the Western Family, 1500–1914.* London.

Anderson, Perry. 1974. *Passages from Antiquity to Feudalism.* London.

———. 1974. *Lineages of the Absolutist State.* London.

Anstey, Roger. 1975. *The Atlantic Slave Trade and British Abolition, 1760–1810.* Atlantic Highlands, N.J.

Apter, David E. 1965. *The Politics of Modernization.* Chicago.

Asiwaju, A. I. 1976. *Western Yorubaland Under European Rule, 1889–1945.* London.

Ayandele, E. A. 1966. *The Missionary Impact on Modern Nigeria, 1842–1914.* London.

Bairoch, Paul. 1971. *Le Tiers-monde dans l'impasse.* Paris.

Baldick, Julian. 1989. *Mystical Islam: An Introduction to Sufism.* New York.

Baran, Paul, and Paul Sweezy. 1966. *Monopoly Capital: An Essay on the American Economic and Social Order.* New York.

Barfield, Thomas J. 1989. *The Perilous Frontier: Nomadic Empires and China, 221 B.C. to A.D. 1757.* Cambridge, Mass.

Barth, Fredrik. 1969. *Ethnic Groups and Boundaries: The Social Organization of Culture Difference.* Oslo.

Bayly, Christopher A. 1988. *Indian Society and the Making of the British Empire.* Cambridge.

———. 1989. *Imperial Meridian: The British Empire and the World, 1780–1830.* London.

Beckles, Hilary McD. 1989. *Natural Rebels: A Social History of Enslaved Black Women in Barbados.* London.

Beckwith, Christopher I. 1987. *The Tibetan Empire in Central Asia: A History of the Struggle for Great Power among Tibetans, Turks, Arabs and Chinese during the Early Middle Ages.* Princeton.

Ben-Amos, Paula. 1977. "Pidgin Languages and Tourist Arts." *Studies in the Anthropology of Visual Communication* 4, 128–39.

Bentley, Jerry H. 1985–1986. "Graduate Education and Research in World History." *World History Bulletin* 3 (Fall/Winter), 3–7; reprinted in Dunn, *New World History*, 526–33.

Berlin, Isaiah. 1976. *Vico and Herder: Two Studies in the History of Ideas.* London.

Bernal, Martin. 1987–1991. *Black Athena: The Afroasiatic Roots of Classical Civilization,* 2 vols. New Brunswick, N.J.

Blassingame, John. 1972. *The Slave Community.* New York.

Bodnar, John. 1985. *The Transplanted: A History of Immigrants in Urban America.* Bloomington.

Bongaarts, John, Thomas K. Burch, and Kenneth W. Wachter. 1987. *Family Demography: Methods and Their Application.* Oxford.

Boserup, Ester. 1965. *The Conditions of Agricultural Growth.* Chicago.

———. 1970. *Woman's Role in Economic Development.* New York.

———. 1981. *Population and Technological Change: A Study of Long-term Trends.* Chicago.

Boxer, Charles R. 1965. *The Dutch Seaborne Empire.* London.

———. 1969. *The Portuguese Seaborne Empire.* London.

Braudel, Fernand. [1967] 1973. *Capitalism and Material Life.* Trans. Miriam Kochan. New York.

———. [1979] 1981. *Civilization and Capitalism, 15th–18th Century.* 3 vols. Trans. Sian Reynolds. New York.

Breisach, Ernst. 1983. *Historiography: Ancient, Medieval, and Modern.* Chicago.

Brenner, Robert. 1976. "Agrarian Class Structure and Economic Development in Pre-Industrial Europe." *Past and Present* 70, 30–74.

Bridenthal, Renate, and Claudia Koonz, eds. 1977. *Becoming Visible: Women in European History.* Boston.

Buisseret, David, and Steven G. Reinhardt, eds. 2000. *Creolization in the Americas.* College Station, TX.

Burgière, André, Christine Klapisch-Zuber, Martine Segalen, and Françoise Zonabend, eds. [1986] 1996. *A History of the Family.* 2 vols. Trans. Sarah Hanbury Tenison, Rosemary Morris, and Andrew Wilson. Cambridge, Mass.

Burke III, Edmund, ed. 1988. *Global Crises and Social Movement: Artisans, Peasants, Populists, and the World Economy.* Boulder.

Burke, Peter. 1980. *Sociology and History.* London.

———. 1985. *Vico.* New York.

Butterfield, Herbert. 1981. *The Origins of History.* New York.

Cameron, Rondo. 1989. *A Concise Economic History of the World from Paleolithic Times to the Present.* New York.

Campbell, Mary B. 1988. *The Witness and the Other World: Exotic European Travel Writing, 400–1600.* Ithaca.

Carr, Lois Green. 1987. *County Government in Maryland, 1689–1709.* New York.

Cell, John. 1982. *The Highest Stage of White Supremacy: The Oreigins of Segregation in South Africa and the American South.* New York.

Chandler, Tertius. 1987. *Four Thousand Years of Urban Growth.* Lewiston, N.Y.

Coale, Ansley J., and Susan Cotts Watkins, eds. 1986. *The Decline of Fertility in Europe.* Princeton.

Cohen, David, and Jack P. Greene, eds. 1972. *Neither Slave nor Free: The Freedmen of African Descent in the Slave Societies of the New World.* Baltimore.

Comaroff, Jean. 1985. *Body of Power, Spirit of Resistance: The Culture and History of a South African People.* Chicago.

Cook, Noble David. 1981. *Demographic Collapse: Indian Peru, 1520–1620.* Cambridge.

Cooper, Frederick. 1980. *From Slaves to Squatters: Plantation Labor and Agriculture in Zanzibar and Coastal Kenya, 1890–1925.* New Haven.

Cordell, Dennis D., and Joel Gregory, eds. 1987. *African Population and Capitalism: Historical Studies.* Boulder.

Crabtree, Charlotte, 1989. *Lessons from History.* Los Angeles.

Craton, Michael. 1978. *Searching for the Invisible Man: Slaves and Plantation Life in Jamaica.* Cambridge, Mass.

Crawford, M., and D. Marsh. 1989. *The Driving Force: Food in Evolution and the Future.* London.

Cronon, William. 1983. *Changes in the Land: Indians, Colonists, and the Ecology of New England.* New York.

Crosby, Alfred W. 1965. *America, Russia, Hemp, and Napoleon: American Trade with Russia and the Baltic, 1783–1812.* Columbus, OH.

———. 1972. *The Columbian Exchange: Biological and Cultural Consequences of 1492.* Wesport.

———. 1976. *Epidemic and Peace, 1918.* Westport. Reissued as *America's Forgotten Pandemic: The Influenza of 1918* (Cambridge, 1989).

———. 1986. *Ecological Imperialism: The Biological Expansion of Europe, 900–1900.* Cambridge.

Curtin, Philip D. 1968. "Epidemiology and the Slave Trade." *Political Science Quarterly* 83 (June), 190–216.

———. 1969. *The Atlantic Slave Trade: A Census.* Madison.

———. 1974. *Precolonial African History.* Washington.

———. 1975. *Economic Change in Precolonial Africa: Senegambia in the Era of the Slave Trade.* 1 vol. and supplement. Madison.

———. 1976. "Measuring the Atlantic Slave Trade Once Again." *Journal of African History* 17, 595–605.

———. 1981. "Recent Trends in African Historiography and their Contribution to History in General." UNESCO, *General History of Africa*, vol. 1 (London, 1981), 54–71.

———. 1984a. "Depth, Span, and Relevance." *American Historical Review* 89, 1–9.

———. 1984b. *Cross-Cultural Trade in World History.* Cambridge.

———. 1986. "World Historical Studies in a Crowded World." *Perspectives* 24 (January), 19–21.

———. 1989. *Death by Migration: Europe's Encounter with the Tropical World in the Nineteenth Century.* Cambridge.

Curtin, Philip D., ed. 1967. *Africa Remembered: Narratives by West Africans from the Era of the Slave Trade.* Madison.

Daget, Serge, ed., 1988. *De la traite à l'esclavage.* 2 vols. Nantes.

Davis, David Brion. 1975. *The Problem of Slavery in the Age of Revolution, 1770–1823.* Ithaca.

De Vries, Jan. 1976. *Economy of Europe in an Age of Crisis, 1600–1750.* Cambridge.

Degler, Carl. 1971. *Neither Black nor White: Slavery and Race Relations in Brazil and the United States.* New York.

Derrida, Jacques. [1967] 1976. *Of Grammatology.* Trans. Gayatri Charavorty Spivak. Baltimore.

Dollar, Charles M., and Richard J. Jensen. 1971. *Historian's Guide to Statistics: Quantitative Analysis and Historical Research.* New York.

Dunn, Ross E. 1977. *Resistance in the Desert: Moroccan Responses to French Imperialism 1881–1912.* Madison.

———. 1985. "The Challenge of Hemispheric History." *The History Teacher* 18, 329–38.

———. 1989. *The Adventures of Ibn Battuta: A Muslim Traveler of the Fourteenth Century.* Berkeley.

DuPlessis, Robert S. 1987. "The Partial Transition to World-Systems Analysis in Early Modern European History." *Radical History Review* 39, 11–27.

Eagleton, Terry. 1983. *Literary Theory: An Introduction.* Oxford.

Eltis, David. 1987. *Economic Growth and the Ending of the Transatlantic Slave Trade.* New York.

Embree, Ainslie T., ed. 1988. *Sources of Indian Tradition.* 2 vols., 2nd ed. New York.

Emerson, Rupert. 1960. *From Empire to Nation: The Rise to Self-Assertion of Asian and African Peoples.* Boston.

Evans, Peter, Dietrich Rueshemeyer, and Theda Skocpol. 1985. *Bringing the State Back in.* Cambridge.

Fabian, Johannes. 1986. *Language and Colonial Power: The Appropriation of Swahili in the Former Belgian Congo 1880–1938.* Cambridge.

———. 1983. *Time and the Other: How Anthropology Makes its Object.* New York.

Fairbank, John K., and Edwin O. Reischauer. 1978. *China: Tradition and Transformation.* Boston.

Finley, Moses I. 1980. *Ancient Slavery and Modern Ideology.* New York.

Finney, Ben. 1979. *Hokulea: The Way to Tahiti.* New York.

Firestone, Shulamith. 1970. *The Dialectic of Sex: The Case for Feminist Revolution.* New York.

Fishlow, Albert. 1965. *American Railroads and the Transformation of the Antebellum Economy.* Cambridge, Mass.

Flinn, M. 1966. *The Origins of the Industrial Revolution.* New York.

Floud, Roderick. 1973. *An Introduction to Quantitataive Methods for Historians.* Princeton.

Fogel, Robert W., and Stanley L. Engerman. 1974. *Time on the Cross: The Economics of American Negro Slavery.* 1 vol. and supplement. Boston.

Formisano, Ronald P. 1971. *The Birth of Mass Political Parties, Michigan, 1827–1861.* Princeton.

Foucault, Michel. [1954] 1976. *Mental Illness and Psychology.* Trans. Alan Sheridan. New York.

———. [1963] 1973. *The Birth of the Clinic: An Archaeology of Medical Perception*, trans. A. M. Sheridan Smith. New York.

———. [1966] 1970. *The Order of Things: An Archaeology of the Human Sciences.* Trans. anon. London.

———. [1976–1984] 1978–1986. *History of Sexuality.* 3 vols. Trans. R. Hurley. Harmondsworth.

Fox-Genovese, Elizabeth, and Eugene D. Genovese. 1983. *Fruits of Merchant Capital: Slavery and Bourgeois Property in the Rise and Expansion of Capitalism.* New York.

Frank, Andre Gunder. 1966. "The Development of Underdevelopment." *Monthly Review* 18, 17–31.

———. 1967. *Capitalism and Underdevelopment in Latin America.* New York.

———. 1969. *Latin America: Underdevelopment or Revolution.* New York.

———. 1978. *World Accumulation 1492–1789.* London.

———. 1980. *Crisis: In the World Economy.* New York.

Frederickson, George. 1981. *White Supremacy: A Comparative Study in American and South African History.* New York.

Gans, Herbert J. 1974. *Popular Culture and High Culture: An Analysis and Evaluation of Taste.* New York.

Garraty, John A., and Peter Gay, eds. 1972. *The Columbia History of the World.* New York.

Gaur, Albertine. 1985. *A History of Writing.* New York.

Geertz, Clifford. 1973. *The Interpretation of Cultures: Selected Essays.* New York.

Geiss, Imanuel. [1968] 1974. *The Pan-African Movement: A History of Pan-Africanism in America, Europe, and Africa.* Trans. Ann Keep. New York.

Gellner, Ernest. 1983. *Nations and Nationalism.* Ithaca.

Genovese, Eugene D. 1976. *Roll, Jordan, Roll: The World the Slaves Made.* New York.

———. 1979. *From Rebellion to Revolution: Afro-American Slave Revolts in the Making of the Modern World.* Baton Rouge.

Geyer, Michael, and Charles Bright. 1987. "For a Unified History of the World in the Twentieth Century." *Radical History Review,* No. 39, 69–91.

Giddens, Anthony. 1981. *A Contemporary Critique of Historical Materialism.* Berkeley.

———. 1979. *Central Problems in Social Theory.* London

Gordon, David C. 1989. *Images of the West.* New York.

Gordon, Michael, ed. 1973. *The American Family in Social-Historical Perspective.* New York.

Gould, Stephen Jay. 1980. *The Panda's Thumb.* New York.

———. 1987. *Time's Arrow, Time's Cycle: Myth and Metaphor in the Discovery of Geological Time.* Cambridge.

Gran, Peter. 1979. *Islamic Roots of Capitalism: Egypt, 1760–1840.* Austin.

Green, Harvey. 1983. *The Light of the Home: An Intimate View of the Lives of Women in Victorian America.* New York.

Greenberg, Joseph H. 1966. *The Languages of Africa.* Bloomington, Ind.

———. 1987. *Language in the Americas.* Stanford.

Guha, Rangit, and Gayatri Chakravorty Spivak, eds. 1988. *Selected Subaltern Studies.* New York.

Guthrie, Malcolm. 1969–1970. *Comparative Bantu.* Farnham, U.K.

Harris, Joseph E., ed. 1982. *Global Dimensions of the African Diaspora.* Washington, D.C.

Harvey, David. 1989. *The Condition of Postmodernity.* Oxford.

Hawking, Stephen W. 1988. *A Brief History of Time: From the Big Bang to Black Holes.* New York.

Headrick, Daniel. 1981. *The Tools of Empire: Technology and European Imperialism in the Nineteenth Century.* New York.

———. 1988. *The Tentacles of Progress: Technology Transfer in the Age of Imperialism, 1850–1940.* New York.

Helms, Mary. 1988. *Ulysses' Sail: An Ethnographic Odyssey of Power, Knowledge, and Geographical Distance.* Princeton.

Henige, David P. 1982. *Oral Historiography.* London.

Herbert, Eugenia W. 1984. *Red Gold of Africa. Copper in Precolonial History and Culture.* Madison.

Hexter, J. H. 1971. *The History Primer.* New York.

Hicks, John D. 1969. *A Theory of Economic History.* Cambridge.

Hill, Robert A., ed. 1983–1996. *The Marcus Garvey and Universal Negro Improvement Association Papers.* 9 vols. Berkeley.

Hindess, Barry, and Paul Q. Hirst. 1975. *Pre-Capitalist Modes of Production.* London.

Hobsbawm, Eric J. 1969. *Industry and Empire: From 1750 to the Present Day.* Harmondsworth.

———. 1975. *The Age of Capital, 1848–1875.* New York.

———. 1987. *The Age of Empire, 1875–1914.* New York.

Hobsbawm, Eric J., and Terence Ranger, eds. 1983. *The Invention of Tradition.* Cambridge.

Hodgson, Marshall G. S. 1974. *The Venture of Islam: Conscience and History in a World Civilization.* 3 vols. New York.

Hogendorn, Jan, and Marion Johnson. 1986. *The Shell Money of the Slave Trade.* Cambridge.

Holm, John A. 1988–1989. *Pidgins and Creoles.* 2 vols. Cambridge.

Hopkins, A. G. 1973. *An Economic History of West Africa.* London.

Howard, Rhoda E. 1978. *Colonialism and Underdevelopment in Ghana.* New York.

Iliffe, John. 1987. *The African Poor: A History.* Cambridge.

Inikori, J. E. 1976a. "Measuring the Atlantic Slave Trade: An Assessment of Curtin and Anstey." *Journal of African History* 17, 197–233.

———. 1976b. "Measuring the Atlantic Slave Trade." *Journal of African History* 17, 607–27.

Israel, Jonathan I. 1989. *Dutch Primacy in World Trade, 1585–1740.* Oxford.

Jameson, Fredric. 1981. *The Political Unconscious: Narrative as a Socially Symbolic Act.* Ithaca.

Jewsiewicki, Bogumil. 1987. "The African Prism of Immanuel Wallerstein." *Radical History Review* 38, 50–68.

Johnson, Donald. 1987. "The American Educational Tradition: Hostile to a Humanistic World History?" *The History Teacher* 20, 519–44.

Johnson, W. McAllister, 1988. *Art History, its Use and Abuse.* Toronto.

Jones, E. L. 1981. *The European Miracle: Environments, Economies, and Geopolitics in the History of Europe and Asia.* Cambridge.

———. 1988. *Growth Recurring. Economic Change in World History.* Oxford.

Kapteijns, Lidwein. 1985. *Mahdist Faith and Sudanic Tradition: The History of the Masalit Sultanate, 1870–1930.* London.

Käsler, Dirk. [1979] 1988. *Max Weber: An Introduction to His Life and Work.* Trans. Philippa Hurd. Chicago.

Kearney, Hugh. 1971. *Science and Change 1500–1700.* New York.

Keddie, Nikki. 1972. *Sayyid Jamal ad-Din "al-Afghani": A Political Biography.* Berkeley.

Kennedy, Paul. 1987. *The Rise and Fall of the Great Powers: Economic Change and Military Conflict from 1500 to 2000.* New York.

Kiple, Kenneth F., and Virginia Himmelsteib King. 1981. *Another Dimension to the Black Diaspora: Diet, Disease, and Racism.* Cambridge.

Klein, Herbert S. 1967. *Slavery in the Americas: A Comparative Study of Virginia and Cuba.* Chicago.

Kulikoff, Allan. 1986. *Tobacco and Slaves: The Development of Southern Cultures in the Chesapeake, 1680–1800.* Chapel Hill, 1986.

Kuper, Adam. 1988. *The Invention of Primitive Society: Transformations of an Illusion.* London.

Lacan, Jacques. [1969] 1981. *Language of the Self.* Trans. Anthony Wilden. Baltimore.

Lamar, Howard, and Leonard Thompson, eds. 1981. *The Frontier in History: North America and Southern Africa Compared.* New Haven.

Lamb, H. H. 1982. *Climate, History and the Modern World.* London.

Landes, David S. 1969. *The Unbound Prometheus: Technological Change and Industrial Development in Western Europe from 1750 to the Present.* Cambridge.

Laslett, Peter. 1966. *The World We Have Lost.* New York.

———. 1972. *Household and Family in Past Time.* Cambridge.

Latham, A. J. H. 1979. *The International Economy and the Underdeveloped World, 1850–1914.* London.

Le Goff, Jacques. 1974. "Les mentalites: une histoire ambiguë." J. Le Goff and P. Nora, eds., *Faire de l'histoire* (Paris), I: 76–94.

Le Goff, Jacques, and P. Nora, eds. 1974. *Faire de l'histoire.* 2 vols. Paris.

Lerner, Gerda. 1986–1993. *Women and History.* 2 vols. New York.

Levenstein, Harvey. 1988. *Revolution at the Table: The Transformation of the American Diet.* New York.

Levtzion, Nehemiah, ed., trans. J. F. P. Hopkins. 1981. *Corpus of Early Arabic Sources for West African History.* Cambridge.

Lewis, Bernard. 1965. *The Emergence of Modern Turkey.* London.

———. 1973. *Islam in History: Ideas, Men and Events in the Middle East.* New York.

———. 1982. *The Muslim Discovery of Europe.* New York.

Liss, Peggy K. 1983. *Atlantic Empires: The Network of Trade and Revolution, 1713–1826.* Baltimore.

Louis, William Roger. 1967. *Great Britain and Germany's Lost Colonies, 1914–1919.* Oxford.

———. 1978. *Imperialism at Bay: The United States and the Decolonization of the British Empire, 1941–1945.* New York.

Lovejoy, Paul E. 1982. "The Volume of the Atlantic Slave Trade: A Synthesis." *Journal of African History* 23, 473–501.

M'Bokolo, Elikia. 1982. "Peste et société urbaine à Dakar: l'épidémie de 1914." *Cahiers d'Etudes Africaines,* No. 85–86, 13–46.

Mackenzie, John. 1986. *Propaganda and Empire: The Manipulation of British Public Opinion 1880–1960.* Manchester.

Maier, Charles S. 1988. *The Unmasterable Past: History, Holocaust, and German National Identity.* Cambridge, Mass.

Major-Poetzl, Pamela. 1983. *Michel Foucault's Archaeology of Western Culture: Toward a New Science of History.* Chapel Hill.

Mallory, J. P. 1989. *In Search of the Indo-Europeans: Language, Archaeology, and Myth.* London.

Mann, Michael. 1986. *The Origins of Social Power.* 2 vols. Cambridge.

Manning, Patrick. 1982. *Slavery, Colonialism and Economic Growth in Dahomey, 1640–1960.* Cambridge.

———. 1988. *Francophone Sub-Saharan Africa 1880–1985.* Cambridge.

Maquet, Jacques. [1962] 1972. *Civilizations of Black Africa.* Trans. Joan Rayfield. New York.

Marchand, Roland. 1985. *Advertising the American Dream: Making Way for Modernity, 1920–1940.* Berkeley.

McLellan, David. 1975. *Karl Marx.* New York.

McNeill, John R. 1985. *Atlantic Empires of France and Spain: Louisbourg and Havana, 1700–1763.* Chapel Hill.

McNeill, William H. 1967; later editions 1971, 1979, 1999. *A World History.* New York.

———. 1976. *Plagues and Peoples.* New York.

———. 1982. *The Pursuit of Power: Technology, Armed Force, and Society since A.D. 1000.* Chicago.

———. 1983. *The Great Frontier: Freedom and Hierarchy in Modern Times.* Princeton.

———. 1986. *Mythistory and Other Essays.* Chicago.

McNeill, William H. 1987. *History of the Human Community*. Englewood Cliffs, N.J. Previously *Ecumene: Story of Humanity*.

————. 1989. *Arnold J. Toynbee: A Life*. New York.

Menard, Russell R. 1985. *Economy and Society in Early Colonial Maryland*. New York.

Miller, Joseph C. 1988. *Way of Death: Merchant Capitalism and the Angolan Slave Trade, 1730–1830*. Madison.

Mintz, Sidney W. 1985. *Sweetness and Power: The Place of Sugar in Modern History*. New York.

Mitchell, B. R., ed. 1975. *European Historical Statistics*. New York.

Mokyr, Joel, ed. 1985. *The Economics of the Industrial Revolution*. Totowa, N.J.

Moore, Barrington. 1966. *Social Origins of Dictatorship and Democracy: Lord and Peasant in the Making of the Modern World*. Boston.

Mumford, Lewis. 1978. *The Transformations of Man*. Gloucester, Mass.

Munz, Peter 1983. "The Idea of 'New Science' in Vico and Marx." G. Tagliacozzo, ed., *Vico and Marx* (Atlantic Highlands, N.J., 1983), 1–19.

Murphy, Marjorie. 1984. "Telling Stories, Telling Tales: Literary theory, Ideology, and Narrative History." *Radical History Review*, No. 31, 33–38.

National Council for History Education. 1988. *Building a World History Curriculum: Guides for Implementing the History Curriculum Recommended by the Bradley Commission on History in the Schools*.

Ngoh, Victor Julius. 1989. *The World Since 1919: A Short History*. Yaounde.

Nisbet, Robert. 1980. *History of the Idea of Progress*. New York.

North, Douglass C. 1966. *Growth and Welfare in the American Past: A New Economic History*. Englewood Cliffs, N.J.

————. 1981. *Structure and Change in Economic History*. New York.

North, Douglass C., and Robert Paul Thomas. 1973. *The Rise of the Western World: A New Economic History*. Cambridge.

Novick, Peter. 1988. *That Noble Dream: The "Objectivity Question" and the American Historical Profession*. New York.

Nurse, Derek, and Thomas Spear. 1985. *The Swahili: Reconstructing the History and Language of an African Society, 800–1500*. Philadelphia.

Parker, William N. 1984–1991. *Europe, America, and the Wider World: Essays on the Economic History of Western Capitalism*. 2 vols. Cambridge.

Patterson, Orlando. 1982. *Slavery and Social Death: A Comparative Study*. Cambridge, Mass.

Perdue, Peter. 1987. *Exhausting the Earth: State and Peasant in Hunan, 1500–1850*. Cambridge, Mass.

Peterson, William. 1979. *Malthus*. Cambridge, Mass.

Phillips, J. R. S. 1988. *The Medieval Expansion of Europe*. Oxford.

Polanyi, Karl. 1966. *Dahomey and the Slave Trade, an Analysis of an Archaic Economy*. Ed. Abraham Rotstein. Seattle.

Pope-Hennessy, James. 1968. *Sins of the Fathers: A Study of the Atlantic Slave Traders, 1441–1807*. New York.

Preston, William, Edward S. Herman, and Herbert I. Schiller. 1989. *Hope and Folly, the United States and UNESCO 1945–1985*. Minneapolis.

Pryor, Frederic. 1977. *The Origins of the Economy: A Comparative Study of Distribution in Primitive and Peasant Economies*. New York.

Quale, G. Robina. 1988. *A History of Marriage Systems*. Westport, Conn.

Ransom, Roger L., and Richard Sutch. 1977. *One Kind of Freedom: The Economic Consequences of Emancipation*. New York.

Raychaudhuri, Tapan. 1988. *Europe Reconsidered: Perceptions of the West in Nineteenth-Century Bengal*. New York.

Reid, Anthony. 1988–1993. *Southeast Asia in the Age of Commerce, 1450–1680*. 2 vols. New Haven.

Reilly, Kevin. 1980. *The West and the World: A Topical History of Civilization*. New York.

Reilly, Kevin, ed. 1985. *World History: Selected Reading Lists and Course Outlines from American Colleges and Universities*. New York.

———. 1988. *Readings in World Civilizations*. 2 vols. New York.

Edwin Reischauer, 1970. *Japan: The Story of a Nation*. New York.

Renfrew, Colin. 1987. *Archaeology and Language: The Puzzle of Indo-European Origins*. London.

Roberts, J. M. 1976. *History of the World*. New York.

Rodney, Walter. 1972. *How Europe Underdeveloped Africa*. Kingston, Jamaica.

Rodzinski, Witold. 1979. *A History of China*. Oxford.

Rosenberg, Nathan, and L. E. Birdzell. 1986. *How the West Grew Rich*. New York.

Roszak, Theodore. 1969. *The Making of a Counter Culture: Reflections on the Technocratic Society and its Youthful Opposition*. Garden City, N.Y.

Rowbotham, Sheila. 1972. *Women, Resistance and Revolution: A History of Women and Revolution in the Modern World*. New York.

Rowlands, Michael, Mogens Larsen, and Kristian Kristiansen, eds. 1987. *Centre and Periphery in the Ancient World*. Cambridge.

Roxborough, Ian. 1979. *Theories of Underdevelopment*. London.

Ruete, Emily (Salme Said). 1989. *Memoirs of an Arabian Princess from Zanzibar*. New York.

Sabean, David. 1984. *Power in the Blood: Popular Culture and Village Discourse in Early Modern Germany*. Cambridge.

Sahlins, Marshall. *1981*. Historical Metaphors and Mythical Realities: Structure in the Early History of the Sandwich Islands Kingdom. *Ann Arbor*.

———. 1985. Islands of History. *Chicago*.

Said, Edward. 1978. *Orientalism*. New York.

Salaman, Redcliffe. 1985. *The History and Social Influence of the Potato*. Cambridge.

Sauer, Carl O. 1969. *Agricultural Origins and Dispersals: The Domestication of Animals and Foodstuffs*. Cambridge, Mass.

Shaffer, Lynda N. 1982. *Mao and the Workers: The Hunan Labor Movement, 1920–1923*. New York.

Shaw, Stanford. 1976. *History of the Ottoman Empire and Modern Turkey*. 2 vols. Cambridge.

Shorter, Edward. 1971. *The Historian and the Computer*. Englewood Cliffs, N.J.

Sigal, Phillip. 1988. *Judaism, the Evolution of a Faith*. Grand Rapids.

Skocpol, Theda. 1979. *States and Social Revolutions*. New York.

Smith, Joan, Immanuel Wallerstein, and Hans-Dieter Evers, eds. 1984. *Households and the World-Economy*. Beverly Hills.

Spüler, Bertold. [1969] 1972. *History of the Mongols, based on Eastern and Western Accounts of the Thirteenth and Fourteenth Centuries*. Trans. Helga and Stuart Drummond. Berkeley.

Stavrianos, Leften. 1971. *Man's Past and Present: A Global History*. Englewood Cliffs, N.J.

———. 1981. *Global Rift: The Third World Comes of Age*. New York.

———. 1989. *Lifelines from our Past: A New World History*. New York.

Stone, Lawrence. 1979. "The Revival of Narrative: Reflections on a New Old History." *Past and Present*, No. 85, 3–24.

Sweezy, Paul M. 1972. *Modern Capitalism and Other Essays*. New York.

Sweezy, Paul et al., introduced by Rodney Hilton. 1976. *The Transition from Feudalism to Capitalism*. London.

Tagliacozzo, Giorgio, and Hayden V. White, eds. 1969. *Giambattista Vico: An International Symposium.* Baltimore.

Taylor, John G. 1979. *From Modernization to Modes of Production: A Critique of the Sociologies of Development and Underdevelopment.* London.

Terkel, Studs. 1986. *Hard Times: An Oral History of the Great Depression.* New York.

Thernstrom, Stephan. 1969. *Poverty and Progress: Social Mobility in a Nineteenth Century City.* New York.

———. 1973. *The Other Bostonians: Poverty and Progress in the American Metropolis, 1870–1970.* Cambridge, Mass.

Thomas, Hugh. 1979. *A History of the World.* London.

Thompson, E. P. 1978. *Poverty of Theory and Other Essays.* New York.

Thompson, Robert Farris. 1983. *Flash of the Spirit: African and Afro-American Art and Philosophy.* New York.

Tilly, Charles. 1984. *Big Structures, Large Processes, Huge Comparisons.* New York.

Tilly, Louise, and Joan Scott. 1978. *Women, Work and Family.* New York.

Trigger, Bruce G. 1980. *Gordon Childe: Revolutions in Archaeology.* New York.

Tukey, John Wilder. 1977. *Exploratory Data Analysis.* Reading, Mass.

UNESCO. 1965. *History of Mankind: Cultural and Scientific Development.* Vol. 2. *The Ancient World,* by Luigi Pareti, assisted by Paolo Baezzi and Luciano Petech. New York.

———. 1966. *History of Mankind: Cultural and Scientific Development.* Vol. 6. *The Twentieth Century,* by Caroline F. Ware, K. M. Panikkar, and J. M. Romeen. New York.

———. 1975. *History of Mankind: Cultural and Scientific Development.* Vol. 3. *The Great Medieval Civilizations,* by Gaston Wiet, Vadime Elisseeff, Philippe Wolf, and Jean Nadou, with contributions by Jean Devisse, Betty Meggers, and Roger Green. New York.

United States, Bureau of the Census. 1975. *Historical Statistics of the United States: Colonial Times to 1970.* 2 vols. Washington.

Van Creveld, Martin. 1989. *Technology and War from 2000 B.C. to the Present.* New York.

Van Onselen, Charles. 1982. *New Babylon, New Nineveh: Studies in the Economic and Social History of the Witwatersrand.* 2 vols. London.

Vansina, Jan. 1966. *Kingdoms of the Savanna.* Madison.

———. 1984. *Art History in Africa: An Introduction to Method.* London.

Voll, John O. 1982. *Islam, Continuity and Change in the Modern World.* Boulder.

Von Bertalanffy, Ludwig. 1968. *General System Theory.* New York.

Von Laue, Theodore. 1987. *The World Revolution of Westernization: The Twentieth Century in Global Perspective.* New York.

Wallerstein, Immanuel. 1967. *Africa: The Politics of Unity; An Analysis of a Contemporary Social Movement.* New York.

———. 1974. *The Modern World-System: Capitalist Agriculture and the Origins of the European World-Economy in the Sixteenth Century.* New York.

———. 1979. *The Capitalist World-Economy.* Cambridge.

———. 1980. *The Modern World-System, II: Mercantilism and the Consolidation of the European World-Economy, 1600–1750.* New York.

———. 1983. *Historical Capitalism.* London.

———. 1989. *The Modern World-System, III: The Second Era of Great Expansion of the Capitalist World-Economy, 1730–1840s.* New York.

Wallerstein, Immanuel, and Paul Starr, eds. 1971. *The University Crisis Reader.* New York.

Watts, David. 1989. *The West Indies: Patterns of Development, Culture and Environmental Change since 1492.* New York.

Wells, C. M. 1984. *The Roman Empire.* Stanford.

Wheeler, Douglas L. 1969. "'Angola is Whose House?' Early Stirrings of Angolan Nationalism and Protest, 1822–1910." *African Historical Studies* 2, 1–22.

White, Hayden. 1973. *Metahistory: The Historical Imagination in Nineteenth-Century Europe.* Baltimore.

———. 1978. *Tropics of Discourse: Essays in Cultural Criticism.* Baltimore.

———. 1984. "The Question of Narrative in History." *History and Theory* 32, 1–33.

Wilks, Ivor. 1975. *Asante in the Nineteenth Century: The Structure and Evolution of a Political Order.* Cambridge.

Williams Raymond. [1976] 1983. *Keywords: A Vocabulary of Culture and Society.* Revised ed. New York.

Withey, Lynne. 1989. *Voyages of Discovery: Captain Cook and the Exploration of the Pacific.* Berkeley.

Wolf, Eric R. 1969. *Peasant Wars of the Twentieth Century.* New York.

———. 1982. *Europe and the People without History.* Berkeley.

Wood, Peter H. 1974. *Black Majority: Negroes in Colonial South Carolina from 1670 through the Stono Rebellion.* New York.

Worsley, Peter. 1968. *The Third World.* London.

———. 1984. *The Three Worlds: Culture and World Development.* Chicago.

Wright, Erik Olin. 1985. *Classes.* London

Wright, Gavin. 1978. *The Political Economy of the Cotton South.* New York.

Wrigley, E. A., and Roger S. Schofield. 1981. *Population History of England, 1581–1841.* Cambridge, Mass.

D. Works First Published in 1990 or Later

Abernethy, David. 2000. *The Dynamics of Global Dominance: European Overseas Empires, 1415–1980.* New Haven.

Acar, Feride, and Ayse Gunes-Ayata, eds. 2000. *Gender and Identity Construction: Women of Central Asia, the Caucasus and Turkey.* Leiden.

Adams, Robert M. 1996. *Paths of Fire: An Anthropologist's Inquiry into Western Technology.* Princeton.

Adamson, Lynda G. 1998. *Notable Women in World History: A Guide to Recommended Biographies and Autobiographies.* Westport, Conn.

Adas, Michael. 2001. "From Settler Colony to Global hegemon: Integrating the Exceptionalist Narrative of the American Experience into World History." *American Historical Review* 106, 1692–720.

Adas, Michael, ed. 1993. *Islamic and European Expansion: The Forging of a Global Order.* Philadelphia.

Adelman, Jeremy. 1999. *Republic of Capital: Buenos Aires and the Legal Transformation of the Atlantic World.* Stanford.

Adshead, S. A. M. 1992. *Salt and Civilization.* New York.

———. 1993. *Central Asia in World History.* New York.

———. 1997. *Material Culture in Europe and China, 1400–1800.* New York.

Alden, Dauril. 1992. "Changing Jesuit Perceptions of the Brasis during the Sixteenth Century." *Journal of World History* 3, 205–18.

———. 1996. *The Making of an Enterprise: The Jesuits in Portugal, its Empire and Beyond, 1540–1750.* Stanford.

Alencastro, Luiz Felipe de. 2000. *O Trato dos viventes: formação do Brasil no Atlântico sul.* São Paulo.

Allardyce, Gilbert. 1990. "Toward World History: American Historians and the Coming of the World History Course." *Journal of World History* 1, 23–76.

Altman, Ida. 2000. *Transatlantic Ties in the Spanish Empire: Brihuega, Spain and Puebla, Mexico, 1560–1620.* Stanford.

Amitai-Press, Reuven, and David O. Morgan, eds. 1999. *The Mongol Empire and Its Legacy.* Leiden.

Amsden, Alice. 2001. *The Rise of "The Rest": Challenges to the West from Late-Industrializing Economies.* New York.

Andrea, Alfred J., and James H. Overfield. 1990. *The Human Record: Sources of Global History.* Boston. (Subsequent editions 1994, 1998, 2001.)

Appadurai, Arjun. 1996. *Modernity at Large: Cultural Dimensions of Globalization.* Minneapolis.

Appleby, Joyce, Lynn Hunt, and Margaret Jacobs. 1994. *Telling the Truth about History.* New York.

Armitage, David. 2000. *The Ideological Origins of the British Empire.* Cambridge.

Arrighi, Giovanni. 1994. *The Long Twentieth Century: Money, Power, and the Origins of Our Times.* London.

Asimov M. S., and C. E. Bosworth, eds. 1998. *History of Civilizations of Central Asia,* Volume IV, *The Age of Achievement: A.D. 750 to the End of the Fifteenth Century.* Part One: *The Historical, Social and Economic Setting.* Paris.

Atwell, William S. 2001. "Volcanism and Short-Term Climatic Change in East Asian and World History, c. 1200–1699." *Journal of World History* 12, 29–98.

Baldick, Julian. 2000. *Animal and Shaman: Ancient Religions of Central Asia.* New York.

Barendse, R. J. 2000. "Trade and State in the Arabian Seas: A Survey from the Fifteenth to the Eighteenth Century." *Journal of World History* 11, 173–226.

———. 2002. *The Arabian Seas: The Indian Ocean World of the Seventeenth Century.* Armonk, N.Y.

Barkan, Elazar, and Marie-Denise Shelton. 1998. *Borders, Exiles, Diasporas.* Stanford.

Barnwell, P. S. 1992. *Emperor, Prefects, and Kings: The Roman West, 395–565.* Chapel Hill.

Barrow, Ian J. 2000. "Agency in the New World History." Neil L. Waters, ed., *Beyond the Area Studies Wars: Toward a New International Studies* (Hanover, NH), 190–212.

Bartholomew, James R. 1993. "Modern Science in Japan: Comparative Perspectives." *Journal of World History* 4, 101–16.

Bartlett, Kenneth R. 1995–1996. "Burckhardt's Humanist Myopia: Machiavelli, Guicciardini and the Wider World." *Scripta Mediterranea* 16–17, 17–30.

Bartlett, Robert. 1993. *The Making of Europe: Conquest, Colonization, and Cultural Change, 950–1350.* Princeton.

Basch, Linda, Nina Glick Schiller, and Cristina Szanton Blanc. 1994. *Nations Unbound: Transnational Projects, Postcolonial Predicaments, and Deterritorialized Nation-States.* Amsterdam.

Bates, Robert H., V. Y. Mudimbe, and Jean O'Barr, eds. 1993. *Africa and the Disciplines: The Contributions of Research in Africa to the Social Sciences and Humanities.* Chicago.

Bayly, Christopher A. 1996. *Empire and Information: Intelligence Gathering and Social Communication in India, 1780–1870.* Cambridge.

———. 2002. "'Archaic' and 'Modern' Globalization in the Eurasian and African Arena, c. 1750–1850." A. G. Hopkins, ed., *Globalization in World History* (London, 2002), 47–73.

Becker, Gary S. 1991. *A Treatise on the Family.* Cambridge, Mass.

Begley, Vimala, and Richard Daniel de Pubma, eds. 1991. *Rome and India: The Ancient Sea Trade.* Madison.

Bellwood, Peter. 1991. "The Austronesian Expansion and the Origin of Languages." *Scientific American* 265 (July 1991), 88–93.

———. 1997. *The Prehistory of the Indo-Malaysian Archipelago.* Revised ed. Honolulu.

Bellwood, Peter, Darrell Tyron, and James J. Fox, eds. 1995. *The Austronesians.* Canberra.

Bendix, Regina. 1997. In Search of Authenticity: The Formation of Folklore Studies. Madison.

Bentley, Jerry H. 1993. *Old World Encounters: Cross-Cultural Contacts and Exchanges in Pre-Modern times.* New York.

———. 1995. *Shapes of World History in Twentieth-Century Scholarship.* Washington.

———. 1996. "Cross-Cultural Interaction and Periodization in World History." *American Historical Review* 101, 749–70.

Bentley, Jerry H., and Herbert F. Ziegler. 1999. *Traditions and Encounters: A Global Perspective on the Past.* Boston.

Benton, Lauren. 2002. *Law and Colonial Cultures: Legal Regimes in World History, 1400–1900.* New York.

Berger, Iris, and E. Frances White. 1999. *Women in Sub-Saharan Africa: Restoring Women to History.* Bloomington, Ind.

Berman, Daniel, and Robert Rittner. 1998. *The Industrial Revolution: A Global Event.* Los Angeles. (Teaching unit.)

Blaut, James M. 1993. *The Colonizer's Model of the World: Geographical Diffusionism and Eurocentric History.* New York.

———. 2000. *Eight Eurocentric Historians.* New York.

Blier, Suzanne Preston. 1993. "Truth and Seeing: Magic, Custom, and Fetish in Art History." Robert H. Bates, V. Y. Mudimbe, and Jean O'Barr, eds., *Africa and the Disciplines: The Contributions of Research in Africa to the Social Sciences and Humanities* (Chicago, 1993), 139–66.

Blum, Stephen, Philip V. Bohlman, and Daniel Neuman, eds. 1991. *Ethnomusicology and Modern Music History.* Urbana.

Bodde, Derek. 1991. *Chinese Thought, Society, and Science: The Intellectual and Social Background of Science and Technology in Pre-Modern China.* Honolulu.

Bogucki, Peter. 1999. *The Origins of Human Society.* Oxford.

Boucher, Philip P. 1992. *Cannibal Encounters: Europeans and Island Caribs, 1492–1763.* Baltimore.

Boyce, Mary. 1992. *Zoroastrianism: Its Antiquity and Constant Vigour.* Costa Mesa, CA.

Brody, Gene H., and Irving E. Siegel, eds. 1990. *Methods of Family Research.* 2 vols. Hillsdale, N.J.

Brooks, George. 1993. *Landlords and Strangers: Ecology, Society, and Trade in Western Africa, 1000–1630.* Boulder.

Brooks, Pamela E. 2000. "Buses, Boycotts and Passes: Black Women's Resistance in Montgomery, Alabama and Johannesburg, South Africa from Colonization to 1960." Ph.D. dissertation, Northeastern University.

Brower, Daniel R., and Edward J. Lazzerini, eds. 1997. *Russia's Orient: Imperial Borderlands and Peoples, 1700–1917.* Bloomington, Ind.

Bulliet, Richard. 1990. *The Camel and the Wheel.* New York.

Bulliet, Richard W., Pamela Kyle Crossley, Daniel R. Headrick, Steven W. Hirsch, Lyman L. Johnson, and David Northrup. 1998. *The Earth and Its Peoples: A Global History,* 2nd ed. Boston.

Burke, James, and Robert Ornstein. 1995. *The Axemaker's Gift: A Double-Edged History of Human Culture.* New York.

Burke, Peter. 1992. *History and Social Theory.* Ithaca.

Burstein, Stanley. 1995. *Graeco-Africana: Studies in the History of Greek Relations with Egypt and Nubia*. New Rochelle, N.Y.

Burstein, Stanley, ed. 1998. *Ancient African Civilizations: Kush and Axum*. Princeton.

Burton, Antoinette. 1994. *Burdens of History: British Feminists, Indian Women and Imperial Culture, 1865–1915*. Chapel Hill.

Bush, Barbara. 1990. *Slave Women in Caribbean Society, 1650–1838*. Bloomington, Ind.

Buzan, Barry. 1991. *People, States, and Fear*. New York.

Cain, P. J., and A. G. Hopkins. 1993. *British Imperialism, 1688–1914*. 2 vols. London.

Calder, Alex, Jonathan Lamb, and Bridget Orr, eds. 1999. *Voyages and Beaches: Pacific Encounters, 1769–1840*. Honolulu.

Calloway, Colin G. 1997. *New Worlds for All: Indians, Europeans, and the Remaking of Early America*. Baltimore.

Campbell, I. C. 1995. "The Lateen Sail in World History." *Journal of World History* 6, 1–24.

———. 1997. "Culture Contact and Polynesian Identity in the European Age." *Journal of World History* 8, 29–56.

Cañizares-Esguerra, Jorge. 1999. "New World, New Stars: Patriotic Astrology and the Invention of Indian and Creole Bodies in Colonial Spanish America, 1600–1650." *American Historical Review* 104, 33–68.

———. 2001. *How to Write the History of the New World: Historiographies, Epistemologies, and Identities in the Eighteenth-Century Atlantic World*. Stanford, 2001.

Canning, Kathleen. 1992. "Gender and the Politics of Class Formation: Rethinking German Labor History." *American Historical Review* 97, 736–768.

Canny, Nicholas. 1994. *Europeans on the Move: Studies on European Migration, 1500–1800*. Oxford.

Caras, Roger A. 1996. *A Perfect Harmony: The Intertwining Lives of Animals and Humans throughout History*. New York.

Carney, Judith. 2001. *Black Rice: The African Origins of Rice Cultivation in the Americas*. Cambridge.

Carpenter, Kenneth J. 1994. *Protein and Energy: A Study of Changing Ideas in Nutrition*. Cambridge.

Catanach, I. J. 2001. "The 'Globalization' of Disease? India and the Plague." *Journal of World History* 12, 131–54.

Çelik, Zeynep. 1992. *Displaying the Orient: Architecture of Islam at Nineteenth-Century World's Fairs*. Berkeley.

Chakrabarty, Dipesh. 2000. *Provincializing Europe*. Princeton.

Chaplin, Joyce E. 2001. *Subject Matter: Technology, the Body, and Science on the Anglo-American Frontier, 1500–1676*. Cambridge, Mass.

Chappell, David A. 1993a. "Ethnogenesis and Frontiers." *Journal of World History* 4, 267–276.

———. 1993b. "Frontier Ethnogenesis: The Case of New Caledonia." *Journal of World History* 4, 307–24.

———. 1997. *Double Ghosts: Oceanian Voyagers on Euroamerican Ships*. Armonk, N.Y.

Chase-Dunn, Christopher. 1998. *Global Formation: Structures of the World-Economy*. 2nd ed. Oxford.

Chase-Dunn, Christopher, and Thomas D. Hall. 1997. *Rise and Demise: Comparing World Systems*. Boulder.

Chase-Dunn, Christopher, and Thomas D. Hall, eds. 1991. *Core/Periphery Relations in Precapitalist Worlds*. Boulder.

Chase-Dunn, Christopher, E. Susan Manning, and Thomas D. Hall. 2000. "Rise and Fall, East-West Synchronicity and Indic Exceptionalism Revisited." *Social Science History* 24, 727–54.

Chaudhuri, K. N. 1990. *Asia before Europe: Economy and Civilisation of the Indian Ocean from the Rise of Islam to 1750.* Cambridge.

Chaudhuri, Nupur, and Margaret Strobel, eds. 1992. *Western Women and Imperialism: Complicity and Resistance.* Bloomington, Ind.

Cheng, Yinghong. 2001. "Creating the New Man—Communist Experiments in China and Cuba: A World History Perspective." Ph.D. dissertation, Northeastern University.

Chomsky, Noam. 1994. *World Orders Old and New.* New York.

Christian, David. 1991. "The Case for 'Big History'." *Journal of World History* 2, 223–38.

———. 1998. *Inner Eurasia from Prehistory to the Mongol Empire.* Vol. 1 of *A History of Russia, Central Asia and Mongolia.* Oxford, 1998.

———. 2000. "Silk Roads or Steppe Roads? The Silk Roads in World History." *Journal of World History* 11, 1–26.

Clancy-Smith, Julia. 1994. *Rebel and Saint: Muslim Notables, Populist Protest, Colonial Encounters (Algeria and Tunisia, 1800–1904).* Berkeley.

Clancy-Smith, Julia, and Frances Gouda, eds. 1998. *Domesticating the Empire: Race, Gender, and Family Life in French and Dutch Colonialism.* Charlottesville.

Clark, Hugh R. 1995. "Muslims and Hindus in the Culture and Morphology of Quanzhou from the Tenth to the Thirteenth Century." *Journal of World History* 6, 49–74.

Clark, Robert P. 1997. *The Global Imperative: An Interpretive History of the Spread of Humankind.* Boulder.

Coclanis, Peter. 2002. "*Drang Nach Osten:* Bernard Bailyn, the World-Island, and the Idea of Atlantic History." *Journal of World History* 13, 169–82.

Cohen, Robin. 1997. *Global Diasporas: An Introduction.* London.

Cohen, Robin, ed. 1995. *The Cambridge Survey of World Migration.* Cambridge.

Cohen, Robin, and Steven Vertovec, eds. 1999. *Migration, Diasporas and Transnationalism.* Cheltenham, U.K.

Comaroff, John and Jean. 1992. *Ethnography and the Historical Imagination.* Boulder.

Comaroff, Jean and John, eds. 1993. *Modernity and Its Malcontents: Ritual and Power in Postcolonial Africa.* Chicago.

Coope, Jessica A. 1993. "Religious and Cultural Conversion to Islam in Ninth-Century Umayyad Córdoba." *Journal of World History* 4, 47–68.

Cooper, Frederick. 1996. *Decolonization and African Society: The Labor Question in French and British Africa.* Cambridge.

Cooper, Frederick, Thomas C. Holt, and Rebecca J. Scott. 2000. *Beyond Slavery: Explorations of Race, Labor, and Citizenship in Postemancipation Societies.* Chapel Hill.

Cooper, Frederick, Florencia E. Mallon, Steve J. Stern, Allen F. Isaacman, and William Roseberry. 1993. *Confronting Historical Paradigms: Peasants, Labor, and the Capitalist World System in Africa and Latin America.* Madison.

Cooper, Frederick, and Ann Laura Stoler, eds. 1997. *Tensions of Empire: Colonial Cultures in a Bourgeois World.* Berkeley.

Costello, Paul. 1993. *World Historians and their Goals: Twentieth-Century Answers to Modernism.* DeKalb, IL.

Croizier, Ralph. 1990. "World History in the People's Republic of China." *Journal of World History* 1, 151–69.

Crosby, Alfred W. 1991. "Infectuous Disease and the Demography of the Atlantic Peoples." *Journal of World History* 2, 119–34.

———. 1994. *Germs, Seeds, and Animals: Studies in Ecological History.* Armonk, N.Y.

———. 1995. "The Past and Present in Environmental History." *American Historical Review* 100, 1,177–89.

———. 1997. *The Measure of Reality: Quantification and Western Society, 1250–1600.* Cambridge.

Crosby, Alfred W. 2000. *Throwing Fire: Projectile Technology Through History.* New York.

Curtin, Philip D. 1990a. *The Rise and Fall of the Plantation Complex: Essays in Atlantic History.* Cambridge.

———. 1990b. "The Environment beyond Europe and the European Theory of Empire." *Journal of World History* 1, 131–50.

———. 1991. "Graduate Teaching in World History." *Journal of World History* 2, 81–89. Reprinted in Dunn, *New World History,* 534–40.

———. 1998. *Disease and Empire: The Health of European Troops in the Conquest of Africa.* Cambridge.

———. 2000. *The World and the West: The European Challenge and the Overseas Response in the Age of Empire.* New York.

Curtin, Philip D., ed. 2001. *Migration and Mortality in Africa and the Atlantic World, 1700–1900.* London.

Cwiertka, Katarzyna. 1999. *The Making of Modern Culinary Tradition in Japan.* Leiden.

Dathorne, O. R. 1996. *Asian Voyages: Two Thousand Years of Constructing the Other.* Westport, Conn.

Davis, Mike. 1990. *City of Quartz.* London.

———. 2001. *Late Victorian Holocausts: El Niño Famines and the Making of the Third World.* London.

De Pauw, Linda Grant. 1998. *Battle Cries and Lullabies: Women in War from Prehistory to the Present.* Norman.

De Vries, Jan. 1994. "The Industrious Revolution and the Industrial Revolution." *Journal of Economic History* 54, 249–70.

De Vries, Jan, and Ad van der Woude. 1997. *The First Modern Economy: Success, Failure, and Perseverance of the Dutch Economy, 1500–1815.* Cambridge.

Devens, Carol. 1992a. *Countering, Colonization: Native American Women and Great Lakes Missions, 1630–1900.* Berkeley.

———. 1992b. " 'If We Get the Girls, We Get the Race': Missionary Education of Native American Girls." *Journal of World History* 3, 219–38.

Diamond, Jared. 1993. The Third Chimpanzee: The Evolution and Future of the Human Animal. New York.

———. 1997. *Guns, Germs, and Steel: The Fates of Human Societies.* New York.

Diamond, Larry, ed. 1992. *The Democratic Revolution: Struggles for Freedom and Pluralism in the Developing World.* New York.

Diène, Doudou, ed. 2001. *From Chains to Bonds: The Slave Trade Revisited.* Paris.

Dirlik, Arif, ed. 1993. *What is a Rim? Critical Perspectives on the Pacific Region Idea.* Boulder.

Douglass, Susan L. 2000. *Teaching About Religion in National and State Social Studies Standards.* Fountain Valley, CA.

Drayton, Richard. 2000. *Nature's Government: Science, Imperial Britain, and the "Improvement" of the World.* New Haven.

Duncan, David Ewing. 1998. *Calendar: Humanity's Epic Struggle to Determine a True and Accurate Year.* New York.

Dunn, John. 1992. *Democracy: The Unfinished Journey, 508 B.C. to A.D. 1993.* Oxford.

Dunn, Ross E. 1990. *Links Across Time and Place: A World History.* New York.

Dunn, Ross E., ed. 2000. *The New World History.* New York.

Dunn, Ross, and David Vigilante, eds. 1996. *Bring History Alive! A Sourcebook for Teaching World History.* Los Angeles.

Durham, William. 1991. *Coevolution: Genes, Culture, and Human Diversity.* Stanford.

Earle, Timothy. 1997. *How Chiefs Came to Power: The Political Economy in Prehistory.* Stanford.

Eaton, Richard. 1993. *The Rise of Islam and the Bengal Frontier, 1204–1760.* Berkeley.

———. 2000. *Essays on Islam and Indian History.* New Delhi.

Eckhardt, William. 1992. *Civilizations, Empires, and Wars: A Quantitative History of War.* Jefferson, NC.

Ehret, Christopher. 1998. *An African Classical Age: Eastern and Southern Africa in World History, 1000 B.C. to 400 A.D.* Charlottesville.

———. 2002. *The Civilizations of Africa: A History to 1800.* Charlottesville.

Ehrlich, Paul R., and A. H. Ehrlich. 1990. *The Population Explosion.* New York.

Eickelman, Dale F., and James Piscatori, eds. 1990. *Muslim Travellers: Pilgrimage, Migration, and the Religious Imagination.* Berkeley.

Eliade, Mircea, and Ioan P. Couliano. 1991. *The Eliade Guide to World Religions.* New York.

Eltis, David, Stephen D. Behrendt, David Richardson, and Herbert S. Klein. 1999. *The Trans-Atlantic Slave Trade: A Database on CD-ROM.* New York.

Embree, Ainslie T., and Carol Gluck, eds. 1997. *Asia in Western and World History: A Guide for Teaching.* Armonk, N.Y.

Enloe, Cynthia. 1990. *Bananas, Beaches, and Basis: Making Feminist Sense of International Politics.* Berkeley.

Erturk, Korkut A., ed. 1999. *Rethinking Central Asia: Non-Eurocentric Studies in History, Social Structure and Identity.* Ithaca.

Fagan, Brian M. 1998. *From Black Land to Fifth Sun: The Science of Sacred Sites.* Reading, Mass.

———. 2000a. *Floods, Famines, and Emperors: El Nino and the Fate of Civilizations.* New York.

———. 2000b. *The Little Ice Age: How Climate Made History 1300–1850.* New York.

Fernández-Armesto, Felipe. 1995. *Millennium: A History of the Last Thousand Years.* New York.

———. 2001. *Civilizations: Culture, Ambition, and the Transformation of Nature.* New York.

Fetter, Bruce, ed. 1990. *Demography from Scanty Evidence: Central Africa in the Colonial Era.* Boulder.

Findley, Carter, and John Rothney, 1994. *Twentieth-Century World.* 3rd ed. Boston.

Finlay, Robert. 1998. "The Pilgrim Art: The Culture of Porcelain in World History." *Journal of World History* 9, 141–88.

———. 2000. "China, the West, and World History in Joseph Needham's Science and Civilisation in China." *Journal of World History* 11, 265–304.

Finney Ben, with Marlene Among. 1994. *Voyage of Rediscovery: A Cultural Odyssey through Polynesia.* Berkeley.

Fisher, Gayle V. 1992. *Journal of Women's History Guide to Periodical Literature.* Bloomington, Ind.

Fletcher, Joseph. 1995. *Studies on Chinese and Islamic Inner Asia.* Ed. Beatrice Forbes Manz. Aldershot, U.K.

Fletcher, Richard. 1998. *The Barbarian Conversion: From Paganism to Christianity.* New York.

Flint, Valerie I. J. 1991. *The Rise of Magic in Early Medieval Europe.* Princeton.

Flynn, Dennis O., and Arturo Giráldez. 1995. "Born with a 'Silver Spoon': The Origin of World Trade in 1571." *Journal of World History* 6, 201–21.

———. 2002. "Cycles of Silver: Global Economic Unity through the Mid-Eighteenth Century." *Journal of World History* 13, 391–427.

Foltz, Richard C. 1998. *Mughal India and Central Asia.* Karachi.

———. 1999. *Religions of the Silk Road: Overland Trade and Cultural Exchange from Antiquity to the Fifteenth Century.* New York.

Forbes, Jack D. 1993. *Africans and Native Americans: The Language of Race and the Evolution of Red-Black Peoples.* Urbana.

Francis, Mark. 1998. "The 'Civilizing' of Indigenous People in Nineteenth-Century Canada." *Journal of World History* 9, 51–88.

Frank, Andre Gunder. 1991. "A Plea for World System History." *Journal of World History* 2, 1–28.

———. 1992. *The Centrality of Central Asia.* Amsterdam

———. 1998a. *ReOrient: Global Economy in the Asian Age.* Berkeley.

———. 1998b. "Materialistically Yours: The Dynamic Society of Graeme Snooks." *Journal of World History* 9, 107–16.

Frank, Andre Gunder, and Barry K. Gills, eds. 1993. *The World System: Five Hundred Years or Five Thousand?* New York.

Friedman, Jonathan. 1994. *Cultural Identity and Global Process.* New York.

Fromkin, David. 1998. *The Way of the World: From the Dawn of Civilizations to the Eve of the Twenty-first Century.* New York.

Frye, Richard N. 1996. *The Heritage of Central Asia: From Antiquity to the Turkish Expansion.* Princeton.

Furbank, P. N. 1992. *Diderot: A Critical Biography.* New York.

Gandy, Matthew, and Alimuddin Zumla, eds. 2001. *Return of the White Plague: Global Poverty and the New Tuberculosis.* London.

Garbutt, Nick. 1999. *Mammals of Madagascar.* New Haven.

Geriffi, Gary, and Miguel Korzeniewicz, eds. 1994. *Commodity Chains and Global Capitalism.* Westport, Conn.

Geyer, Michael and Charles Bright. 1995. "World History in a Global Age." *American Historical Review* 100, 1,034–60.

Gillis, John R., ed. 1994. *Commemorations: The Politics of National Identity.* Princeton.

Gilroy, Paul. 1993. *The Black Atlantic: Modernity and Double Consciousness.* Cambridge, Mass.

Goldstone, Jack A. 1991. *Revolution and Rebellion in the Early Modern World.* Berkeley.

Gomez, Michael A. 1998. *Exchanging our Country Marks: The Transformation of African Identities in the Colonial and Antebellum South.* Chapel Hill.

Goonan, Jessica. 1999. *Africa: Cultural and Geographic Diversity.* Boston. (Teaching unit.)

Goudsblom, Johan. 1992a. *Fire and Civilization.* London.

———. 1992b. "The Civilizing Process and the Domestication of Fire." *Journal of World History* 3, 1–12.

Goudsblom, Johan, Eric Jones, and Stephen Mennell, eds. 1996. *The Course of Human History: Economic Growth, Social Process, and Civilization.* Armonk, N.Y.

Gran, Peter. 1996. *Beyond Eurocentrism: A New View of Modern World History.* Syracuse.

Green, William A. 1992. "Periodization in European and World History." *Journal of World History* 3, 13–54.

Greene, K. 1999. "V. Gordon Childe and the Vocabulary of Revolutionary Change." *Antiquity* 73, 97–109.

Greenberg, Joseph H. *Indo-European and its Closest Relatives: The Eurasiatic Language Family.* Vol. 1, *Grammar.* Stanford.

Greenberg, Joseph, and Merritt Ruhlen. 1992. "Linguistic Origins of Native Americans." *Scientific American* 267 (November 1992), 94–99.

Greenstein, Daniel I. 1994. *A Historian's Guide to Computing.* Oxford.

Gress, David. 1998. *From Plato to NATO: The Idea of the West and its Opponents.* New York.

Guarneri, Carl J. 1997. *America Compared: American History in International Perspective.* 2 vols. Boston.

Guha, Ranajit. 1998. *Dominance without Hegemony: History and Power in Colonial India.* Cambridge, Mass.

Hall, Gwendolyn Midlo. 1992. *Africans in Colonial Louisiana: The Development of Afro-Creole Culture in the Eighteenth Century.* Baton Rouge.

Haraven, Tamara K. 1991. "The History of the Family and the Complexity of Social Change." *American Historical Review* 96, 95–124.

Hardt, Michael, and Antonio Negri. 2000. *Empire.* Cambridge, Mass.

Harris, Sheldon H. 1994. *Factories of Death: Japanese Biological Warfare, 1932–1945.* London.

Hartog, Leo de. 1996. *Russia and the Mongol Yoke: The History of the Russian Principalities and the Golden Horde, 1221–1502.* London.

Harvey, Charles, and Jon Press. 1996. *Databases in Historical Research: Theory, Methods and Applications.* New York.

Hatton, Timothy J., and Jeffrey G. Williamson, eds. 1994. *Migration and the International Labor Market 1850–1939.* London.

Headley, John M. 1997. "The Sixteenth-Century Venetian Celebration of the Earth's Total Habitability: The Issue of the Fully Habitable World for Renaissance Europe." *Journal of World History* 8, 1–28.

Headrick, Daniel R. 1991. *The Invisible Weapon: Telecommunications and International Politics, 1851–1945.* New York.

———. 1996. "Botany, Chemistry, and Tropical Development." *Journal of World History* 7, 1–20.

———. 2000. *When Information Came of Age: Technologies of Knowledge in the Age of Reason and Revolution, 1700–1850.* New York.

Heine Bernd, and Derek Nurse, eds. 2000. *African Languages: An Introduction.* Cambridge.

Hillman, David, and Carla Mazzio. 1997. The Body in Parts: Fantasies of Corporeality in Early-Modern Europe. *New York.*

Hobsbawm, Eric J. 1990. *Nations and Nationalism since 1780: Programme, Myth, Reality.* Cambridge.

———. 1994. *The Age of Extremes: A History of the World, 1914–1991.* New York.

Hodgson, Marshall G. S. 1993. *Rethinking World History: Essays on Europe, Islam, and World History.* Edmund Burke III, ed. Cambridge.

Holt, Thomas C. 1992. *The Problem of Freedom: Race, Labor, and Politics in Jamaica and Britain, 1832–1938.* Baltimore.

Hopkins, A. G. 2002a. "Introduction: Globalization—An Agenda for Historians." Hopkins, ed., *Globalization in World History,* 1–10. London.

———. 2002b. "Globalization with and without Empires: From Bali to Labrador." Hopkins, ed., *Globalization in World History* (London), 220–42.

Hopkins, A. G., ed. 2002. *Globalization in World History.* London.

Hourani, Albert. 1991. *A History of the Arab Peoples.* Cambridge, Mass.

Hovannisian, Richard G. 1997. *The Armenian People from Ancient to Modern Times.* 2 vols. New York.

Hubel, Teresa. 1996. *Whose India? The Independence Struggle in British and Indian Fiction and History.* Durham.

Hudson, Mark J. 1999. *The Ruins of Identity: Ethnogenesis in the Japanese Islands.* Honolulu.

Huff, Toby E. 1993. *The Rise of Early Modern Science: Islam, China, and the West.* Cambridge.

Hughes, J. Donald, ed. 1999. *The Face of the Earth: Environment and World History.* Armonk, N.Y.

Hung, Ho-fung. 2001. "Imperial China and Capitalist Europe in the Eighteenth-Century Global Economy." *Review* 24, 473–513.

Huntington, Samuel P. 1992. *The Third Wave: Democratization in the Late Twentieth Century.* Norman.

———. 1996. *The Clash of Civilizations and the Remaking of World Order.* New York.

Hutchinson, John F. 1996. Champions of Charity: War and the Rise of the Red Cross. Boulder, 1996.

Iggers, Georg. G. 1997. *Historiography in the Twentieth Century: From Scientific Objectivity to the Postmodern Challenge.* Hanover, NH.

Iriye, Akira. 1992. *China and Japan in the Global Setting.* Cambridge, Mass.

———. 1997. *Cultural Internationalism and World Order.* Baltimore.

Isaacman, Allen, and Richard Roberts. 1995. *Cotton, Colonialism, and Social History in Sub-Saharan Africa.* Portsmouth, NH.

Israel, Jonathan I. 2001. *Radical Enlightenment: Philosophy and the Making of Modernity, 1650–1750.* Oxford.

James, H. Parker. 2001. "Up on Stilts: The Stilt House in World History." Ph.D. dissertation, Tufts University.

Johnson, Jean Elliott, and Donald James Johnson. 1999. *Emperor Ashoka of India: What Makes a Ruler Legitimate?* Los Angeles. (Teaching unit.)

Johnston, Deborah Smith. 2002. "Rethinking World History: Conceptual Frameworks for the World History Survey." Ph.D. dissertation, Northeastern University.

Jolly, Karen Louise, ed. 1997. *Tradition and Diversity: Christianity in a World Context to 1500.* Armonk, N.Y.

Jones, Eric, Lionel Frost, and Colin White. 1993. *Coming Full Circle: An Economic History of the Pacific Rim.* Boulder.

Kalivas, David M. 2000. "A World History Worldview: Owen Lattimore, a Life Lived in Interesting Times, 1900–1950." Ph.D. dissertation, Northeastern University.

Karttunen, Frances. 1992. "After the Conquest: The Survival of Indigenous Patterns of Life and Belief." *Journal of World History* 3, 239–56.

———. 1994. *Between Worlds: Interpreters, Guides, and Survivors.* New Brunswick, N.J.

Karttunen, Frances, and Alfred W. Crosby. 1995. "Language Death, Language Genesis, and World History." *Journal of World History* 6, 157–74.

Keddie, Nikki R. 1990. "The Past and Present of Women in the Muslim World." *Journal of World History* 1, 77–108.

Keita, Maghan. 2000. *Race and the Writing of History: Riddling the Sphinx.* New York.

Kelley, Donald R., ed. 1997. *History and the Disciplines: The Reclassification of Knowledge in Early Modern Times.* Rochester.

Keylor, William. 1995. *The Twentieth Century World.* Oxford.

Keys, David. 1999. *Catastrophe: An Investigation into the Origins of the Modern World.* New York.

King, Anthony, ed. 1991. *Culture, Globalization and the World-System: Contemporary Conditions for the Representation of Identity.* Minneapolis.

Kiple, Kenneth, and Krlemhild Conee Ornelas, eds. 2000. *The Cambridge World History of Food.* Cambridge.

Kirch, Patrick Vinton. 1997. *The Lapita Peoples: Ancestors of the Oceanic World.* Cambridge, Mass.

Knutsen, Torbjorn. 1999. *The Rise and Fall of World Orders.* Manchester.

Knysh, Alexander. 2000. *Islamic Mysticism: A Short History.* Leiden.

Kuklick, Henrika. 1991. *The Savage Within: The Social History of British Anthropology, 1885–1945.* Cambridge.

Kumar, Deepak. 1995. *Science and the Raj, 1857–1905.* New York.

Kuper, Adam. 1994. *The Chosen Primate: Human Nature and Cultural Diversity.* Cambridge, Mass.

———. 1999. *Culture: The Anthropologists' Account.* Cambridge, Mass.

Kupperman, Karen Ordahl. 2000. *Indians and English: Facing Off in Early America.* Ithaca.

Kurlansky, Mark. 1997. *Cod: A Biography of the Fish that Changed the World.* New York.

———. 2002. *Salt: A World History.* New York.

Lance, James, and Richard Roberts. 1991. " 'The World Outside the West' Course Sequence at Stanford University." *Perspectives* (March) 18, 22–24.

Landes, David S. 1998. *The Wealth and Poverty of Nations: Why Some Are So Rich and Others So Poor.* New York.

Laver, James. 1995. *Costume and Fashion: A Concise History.* New York.

Lazreg, Marnia. 1994. *The Eloquence of Silence: Algerian Women in Question.* London.

Lee, James Z., and Wang Feng. 1999. *One Quarter of Humanity: Malthusian Mythology and Chinese Realities, 1700–2000.* Cambridge, Mass.

Lemarchand, René. 1994. *Burundi: Ethnocide as Discourse and Practice.* New York.

Levene, Mark. 2000. "Why is the Twentieth Century the Century of Genocide?" *Journal of World History* 11, 305–36.

Lewis, Bernard. 2002. *What Went Wrong? Western Impact and Middle Eastern Response.* New York.

Lewis, Martin W., and Karen E. Wigen. 1997. *The Myth of Continents: A Critique of Metageography.* Berkeley.

Lieberman, Victor, ed. 1999. *Beyond Binary Histories: Re-imagining Eurasia to c. 1830.* Ann Arbor.

Linebaugh, Peter, and Marcus Rediker. 2000. *The Many-Headed Hydra: Sailors, Slaves, Commoners, and the Hidden History of the Revolutionary Atlantic.* Boston.

Lipman, Jonathan. 1997. *Familiar Strangers: A History of Muslims in Northwest China.* Seattle.

Liu Xinru. 1995a. *Ancient India and Ancient China: Trade and Religious Exchanges AD 1–600.* Oxford.

———. 1995b. "Silks and Religions in Eurasia, c. A.D. 600–1200." *Journal of World History* 6, 25–48.

———. 1998. *The Silk Road: Overland Trade and Cultural Interactions in Eurasia.* Washington.

———. 2001. "Migration and Settlement of the Yuezhi-Kushan: Interaction and Interdependence of Nomadic and Sedentary Societies." *Journal of World History* 12, 261–92.

Lockard, Craig A. 1994. "The Contribution of Philip Curtin and the 'Wisconsin School' to the Study and Promotion of Comparative World History." *Journal of Third World Studies* 11, 180–82, 199–211, 219–23.

Louis, William Roger, ed. 1998–1999. *The Oxford History of the British Empire.* 5 vols. Oxford.

Lovejoy, Paul E., and Jan S. Hogendorn. 1993. *Slow Death for Slavery: The Course of Abolition in Northern Nigeria, 1897–1936.* Cambridge.

Lovell, W. George, and Christopher H. Lutz. 1995. *Demography and Empire: A Guide to the Population History of Spanish Central America, 1500–1821.* Boulder, Colo.

Ludden, David. 2002. "Modern Inequality and Early Modernity: A Comment on the AHR Articles by R. Bin Wong and Kenneth Pomeranz." *American Historical Review* 107, 470–80.

Mackenzie, John, ed. 1992. *Imperialism and Popular Culture.* Manchester.

MacLeod, Roy. 1993. "Passages in Imperial Science: From Empire to Commonwealth" *Journal of World History* 4, 117–50.

Maddison, Angus. 2001. *The World Economy: A Millennial Perspective.* Paris.

Maier, Charles S. 1997. *Dissolution: The Crisis of Communism and the End of East Germany.* Princeton.

Mair, Victor, ed. 1998. *The Bronze Age and Early Iron Age Peoples of Eastern Central Asia.* 2 vols. Philadelphia.

Mallory, J. P., and Victor Mair. 2000. *The Tarim Mummies: Ancient China and the Mystery of the Earliest Peoples from the West.* London.

Mani, Lata. 1998. *Contentious Traditions: The Debate on Sati in Colonial India.* Berkeley.

Manning, Patrick. 1990. *Slavery and African Life: Occidental, Oriental and African Slave Trades.* Cambridge.

———. 1990–1991. "African Economic Growth and the Public Sector: Lessons from Historical Statistics of Cameroon." *African Economic History* 19, 135–70.

———. 1992. "Methodology and World History in a Ph.D. Program." *Perspectives* (April) 23, 25.

———. 1994. "Cultural History: Paths in Academic Forests." Robert W. Harms, Joseph C. Miller, David S. Newbury, and Michele D. Wagner, eds., *Paths Toward the Past: African Historical Essays in Honor of Jan Vansina* (Madison), 439–54.

———. 1996a. "Introduction." Manning, *Slave Trades, 1500–1800,* xv–xxxiv.

———. 1996b. "The Problem of Interactions in World History." *American Historical Review* 101, 771–82.

———. 1999a. "The Monograph in World History: Philip Curtin's Comparative Approach." *World History Bulletin* 15 (Spring), 12–17.

———. 1999b. "Doctoral Training in World History: The Northeastern University Experience." *Perspectives* 37 (March), 35–38. Reprinted in Dunn, *New World History,* 541–47.

———. 2002. "Asia and Europe in the World Economy: Introduction." *American Historical Review* 107, 419–24.

Manning, Patrick, ed. 1996. *Slave Trades, 1500–1800: Globalization of Forced Labour.* Aldershot, U.K.

Manning, Patrick, et al. 2000. *Migration in Modern World History, 1500–2000.* Belmont, Calif. CD-ROM.

Markovits, Claude. 2000. *The Global World of Indian Merchants, 1750–1947: Traders of Sind from Bukhara to Panama.* Cambridge.

Marks, Robert. 1998. *Tigers, Rice, Silk, and Silt: Environment and Economy in Late Imperial South China.* Cambridge.

Marks, Steven G. 1991. *Road to Power: The Trans-Siberian Railroad and the Colonization of Asian Russia, 1850–1917.* Ithaca.

Martin, Dorothea A. L. 1991. *The Making of a Sino-Marxist World View: Perceptions and Interpretations of World History in the People's Republic of China.* Armonk, N.Y.

Martin, Eric L. 2001. "Anti-colonial Worldviews: An Intellectual World History of the Twentieth Century." Ph.D. dissertation, Northeastern University.

Martin, William G., ed. 1990. *Semiperipheral States in the World-Economy.* New York.

Martin, William G., and Michael O. West, eds. 1999. *Out of One, Many Africas: Reconstructing the Study and Meaning of Africa.* Urbana.

Marwick, Arthur. 1998. *The Sixties: Cultural Revolution in Britain, France, Italy and the United States, c. 1958–c. 1974.* Oxford.

Mazlish, Bruce. 1993. "An Introduction to Global History," in Mazlish and Buultjens, *Conceptualizing Global History,* 1–24.

————. 1998a. *The Uncertain Sciences*. New Haven.

————. 1998b. "Comparing Global to World History." *Journal of Interdisciplinary History* 28, 385–95.

Mazlish, Bruce, and Ralph Buultjens, eds. 1993. *Conceptualizing Global History*. Boulder.

Mazrui Ali A., and Alamin M. Mazrui. 1998. *The Power of Babel: Language and Governance in the African Experience*. London.

Mazumdar, Sucheta. 1998. *Sugar and Society in China: Peasants, Technology and the World Market*. Cambridge, Mass.

McBrearty, Sally, and Alison S. Brooks. 2000. "The Revolution that Wasn't: A New Interpretation of the Origin of Modern Human Behavior." *Journal of Human Evolution* 39, 453–563.

McCann, James. 1999. *Green Land, Brown Land, Black Land: An Environmental History of Africa, 1800–1990*. Portsmouth, NH.

McCarthy, Justin. 1996. *Death and Exile: The Ethnic Cleansing of Ottoman Muslims, 1821–1922*. Princeton.

McClellan James E., III, and Harold Dorn. 1999. *Science and Technology in world History: An Introduction*. Baltimore.

McKay, John P., Bennett D. Hill, and John Buckler. 1992. *A History of World Societies*. 3rd ed. Boston.

McKeown, Adam. 1997. "Chinese Migrants among Ghosts: Chicago, Peru, and Hawaii in the Early Twentieth Century." Ph.D. dissertation, University of Chicago.

————. 2001. *Chinese Migrant Networks and Cultural Change: Peru, Chicago, Hawaii, 1900–1936*. Chicago.

McNay, Lois. 1994. *Foucault, A Critical Introduction*. New York.

McNeill, John R. 1992. *The Mountains of the Mediterranean World: An Environmental History*. Cambridge.

————. 1994. "Of Rats and Men: A Synoptic Environmental History of the Island Pacific," *Journal of World History* 5, 299–350.

————. 2000. *Something New Under the Sun: An Environmental History of the Twentieth-Century World*. New York.

McNeill, John R., ed. 2001. *Environmental History in the Pacific*. Aldershot, U.K.

McNeill, William H. 1990a. "*The Rise of the West* after Twenty-Five Years." *Journal of World History* 1, 1–21.

————. 1990b. *Keeping Together in Time: Dance and Drill in Human History*. Cambridge, Mass.

————. 1990c. *The Age of Gunpowder Empires, 1450–1800*. Washington.

Mernissi, Fatima. 1993. *The Forgotten Queens of Islam*. Trans. Mary Jo Lakeland. Minneapolis.

Meyer, Howard N. 2001. *The World Court in Action: Judging among the Nations*. Lanham, MD.

Michaels, Athanasios S. 2000. "Masculinity and Imperialism in England: A Patriotic Construct from the Indian Mutiny to the South African War, 1857–1902." M.A. thesis, Northeastern University.

Mih, Walter C. 2000. *The Fascinating Life and Theory of Albert Einstein*. Commack, N.Y.

Miller, Joseph C. 1993. *Slavery and Slaving in World History: A Bibliography, 1900–1991*. Millwood, NH.

Miller, Susan Gilson, ed. and trans. 1992. *Disorienting Encounters: Travels of a Moroccan Scholar in France in 1845–1846*. Berkeley.

Mitchell, B. R., ed. 1992. *International Historical Statistics: Europe, 1750–1988*. New York.

————. 1993. *International Historical Statistics: The Americas 1750–1988*. New York.

Mitchell, B. R. 1995. *International Historical Statistics: Africa, Asia and Oceania, 1750–1988.* New York.

Mokyr, Joel. 1990a. *The Lever of Riches, Technological Creativity and Economic Progress.* New York.

———. 1990b. *Twenty-five Centuries of Technological Change: An Historical Survey.* New York.

Morris-Suzuki, Tessa. 1998. *Re-inventing Japan: Time, Space, Nation.* Armonk, N.Y.

Muldoon, James. 1991. "Solórzano's De indiarum iure: Applying a Medieval Theory of World Order in the Seventeenth Century." *Journal of World History* 2, 29–46.

Murphy, Craig N. 1994. *International Organization and Industrial Change: Global Governance since 1850.* Cambridge.

Murphy, Marjorie. 1991. *Blackboard Unions: The AFT and the NEA, 1900–1980.* Ithaca.

Nash, Gary B., Charlotte Crabtree, and Ross E. Dunn. 1997. *History on Trial: Culture Wars and the Teaching of the Past.* New York.

National Center for History in the Schools. 1994. *National Standards for World History: Exploring Paths to the Present.* Grades 5–12, expanded edition. Los Angeles.

National Council for Social Studies. 1994. *Expectations of Excellence: Curriculum Standards for Social Studies.*

Navarro, Marysa, and Virginia Sánchez Korrol, with Kecia Ali. 1999. *Women in Latin America and the Caribbean.* Bloomington, Ind.

Needham, Joseph, Robin D. S. Yates, Krzysztof Gawlikowsky, Edward McEwen, and Wang Ling. 1994. *Science and Civilisation in China.* Volume 5, part 6, section 30. *Military Technology: Missiles and Sieges.* Cambridge.

Nicholson, Philip Yale. 1999. *Who Do We Think We Are? Race and Nation in the Modern World.* Armonk, N.Y.

Nizza da Silva, Maria Beatriz, ed. 1998. *Families in the Expansion of Europe.* Aldershot, U.K.

North, Douglass C. 1990. *Institutions, Institutional Change and Economic Performance.* Cambridge.

Northrup, David. 1995. *Indentured Labor in the Age of Imperialism, 1831–1922.* New York.

———. 2002. *Africa's Discovery of Europe, 1450 to 1850.* New York.

Nordquist, Joan, compiler. 1992. *The Multicultural Education Debate in the University: A Bibliography.* Contemporary Social Issues: A Bibliographical Series, No. 25. Santa Cruz.

Norton, Mary Beth, and Pamela Gerardi, eds. 1995. *The American Historical Association's Guide to the Historical Literature.* 2 vols. New York.

Nugent, Walter. 1992. *Crossings: The Great Transatlantic Migrations, 1870–1914.* Bloomington, Ind.

O'Connor, James. 1998. *Natural Causes: Essays in Ecological Marxism.* New York.

Obeyesekere, Gananath. 1992. *The Apotheosis of Captain Cook: European Mythmaking in the Pacific.* Princeton.

Ohnuki-Tierney, Emiko. 1993. *Rice as Self: Japanese Identities Through Time.* Princeton.

O'Rourke, Kevin H., and Jeffrey G. Williamson. 2000. *Globalization and History: The Evolution of a Nineteenth-Century Atlantic Economy.* Cambridge, Mass.

Pacey, Arnold. 1990. *Technology in World Civilization: A Thousand-Year History.* Cambridge, Mass.

Pagden, Anthony. 1995. *Lords of All the World: Ideologies of Empire in Spain, Britain and France, c. 1500–c. 1800.* New Haven.

Palat, Ravi Arvind. 2000. "Fragmented Visions: Excavating the Future of Area Studies in a Post-American World." Waters, *Beyond the Area Studies Wars,* 64–108.

Palmer, Colin A. 1995. "From Africa to the Americas: Ethnicity in the Early Black Communities of the Americas." *Journal of World History* 6, 223–36.

Palmer, Kristin. 1999. *Mecca: Islam's Mosque*. Boston. (Teaching unit.)

Pan, Lyn. 1990. *Sons of the Yellow Emperor: A History of the Chinese Diaspora*. Boston.

Parrinder, Geoffrey. 1996. *Sexual Morality in the World's Religions*. Rockport, Mass.

Parthasarathi, Prasannan. 1998. "Rethinking Wages and Competitiveness in the Eighteenth Century: Britain and South Asia." *Past and Present*, No. 158, 79–109.

———. 2001. *The Transition to a Colonial Economy: Weavers, Merchants and Kings in South India, 1720–1800*. New York.

Partner, Peter. 1998. *God of Battles: Holy Wars of Christianity and Islam*. Princeton.

Pearson, Michael. N. 1998. *Port Cities and Intruders: The Swahili Coast, India, and Portugal in the Early Modern Era*. Baltimore.

Pearson, Michael. N., ed. 1996. *Spices in the Indian Ocean World*. London.

Pederson, Gorm, and Ida Nicolaisen. 1995. *Afghan Nomads in Transition: A Century of Change Among the Zala Khan Khel*. New York.

Pérez-Malláina, Pablo. 1998. *Spain's Men of the Sea: Daily Life on the Indies Fleets in the Sixteenth Century*. Trans. Carla Rahn Phillips. Baltimore.

Pickering, Mary. 1993. *Auguste Comte: An Intellectual Biography*. Vol. 1. Cambridge.

Pierson, Ruth Roach, and Nupur Chaudhuri, eds. 1998. *Nation, Empire, Colony: Historicizing Gender and Race*. Bloomington, Ind.

Pipes, Richard. 1999. *Property and Freedom*. New York.

———. 2001. *Communism: A History*. New York.

Pomeranz, Kenneth. 2000. *The Great Divergence: China, Europe, and the Making of the Modern World Economy*. Princeton.

———. 2002. "Political Economy and Ecology on the Eve of Industrialization: Europe, China, and the Global Conjuncture." *American Historical Review* 107, 425–46.

Pomeranz, Kenneth, and Steven Topik. 1999. *The World that Trade Created: Society, Culture, and the World Economy, 1400–the Present*. Armonk, N.Y.

Pomper, Philip, Richard H. Elphick and Richard T. Vann, eds. 1998. *World History: Ideologies, Structures, and Identities*. Oxford. Based on "World Historians and Their Critics," theme issue 34, *History & Theory* (1995).

Ponting, Clive. 1991. *A Green History of the World: The Environment and the Collapse of Great Civilizations*. New York.

Powell, Richard J. 1997. *Black Art and Culture in the 20th Century*. New York.

Power, Daniel, and Naomi Standen, eds. 2000. *Frontiers in Question: Eurasian Borderlands, 700–1700*. New York.

Prakash, Gyan, ed. 1995. *After Colonialism: Imperial Histories and Postcolonial Displacements*. Princeton.

Pratt, Mary Louise. 1992. *Imperial Eyes: Studies in Travel Writing and Transculturation*. New York.

Prazniak, Roxann. 1996. *Dialogues across Civilizations: Sketches in World History from the Chinese and European Experiences*. Boulder.

Quale, G. Robina. 1992. *Families in Context: A World History of Population*. New York.

Ralston, David B. 1990. *Importing the European Army: The Introduction of European Military Techniques and Institutions into the Extra-European World, 1600–1914*. Chicago.

Ramusack, Barbara, and Sharon L. Sievers. 1999. *Women in Asia: Restoring Women to History*. Bloomington, Ind.

Ranger, Terence. 1998. "Europeans in Black Africa." *Journal of World History* 9, 255–68.

Rehbock, Philip F. 2001. "Globalizing the History of Science." *Journal of World History* 12, 183–92.

Reid, Anthony, ed. 1993. *Southeast Asia in the Early Modern Era: Trade, Power, and Belief*. New York.

Reiff, Janice L. 1991. *Structuring the Past: The Use of Computers in History.* Washington.

Reynolds, Clark G. 1998. *Navies in History.* Annapolis.

Ricklefs, Merle C. 1998. *The Seen and Unseen Worlds in Java, 1726–1749: History, Literature and Islam in the Court of Pakubuwana II.* Honolulu.

Risso, Patricia. 1995. *Merchants and Faith: Muslim Commerce and Culture in the Indian Ocean.* Boulder.

————. 2001. "Cross-Cultural Perceptions of Piracy: Maritime Violence in the Western Indian Ocean and Persian Gulf Region during a Long Eighteenth Century." *Journal of World History* 12, 292–320.

Roberts, J. M. 1999. *Twentieth Century: The History of the World, 1901 to 2000.* New York.

Robertson, Roland. 1992. *Globalization: Social Theory and Global Culture.* London.

Roediger, David. 1999. *Wages of Whiteness: Race and the Making of the American Working Class.* London.

Root, Deborah. 1996. Cannibal Culture: Art Appropriation, and the Commodification of Difference. *Boulder.*

Ross, Dorothy. 1991. *The Origins of American Social Science.* Cambridge.

Roupp, Heidi, ed. 2000. *Jump Start Manual for Teaching World History.* Aspen, Colo.

Russell, James C. 1994. *The Germanization of Early Medieval Christianity: A Sociohistorical Approach to Religious Transformation.* New York.

Russell-Wood, A. J. R. 1993. *A World on the Move: The Portuguese in Africa, Asia, and America, 1415–1808.* New York.

Russell-Wood, A. J. R., general editor. 1995–2000. *An Expanding World: The European Impact on World History, 1450–1800.* 31 vols. Aldershot, U.K.

Said, Edward. 1993. *Culture and Imperialism.* New York.

Samman, Khaldoun. 2001. "The Limits of the Classical Comparative Method." *Review* 24, 533–73.

Sanderson, Stephen K., ed. 1995. *Civilizations and World Systems: Studying World-Historical Change.* Walnut Creek.

Schieffelin Edward L., and Robert Crittenden. 1991. *Like People You See in a Dream: First Contact in Six Papuan Societies.* Stanford.

Schivelbusch, Wolfgang. 1992. *Tastes of Paradise: A Social History of Spices, Stimulants, and Intoxicants.* New York.

Schmidt, Peter R., and Roderick J. McIntosh, eds. 1996. *Plundering Africa's Past.* Bloomington, Ind.

Scott, Joan Wallach, ed. 1996. *Feminism and History.* New York.

Seccombe, Wally. 1992. *A Millennium of Family Change: Feudalism to Capitalism in Northwestern Europe.* London.

————. 1993. *Weathering the Storm: Working-Class Families from the Industrial Revolution to the Fertility Decline.* London.

Seed, Patricia. 1995. *Ceremonies of Possession in Europe's Conquest of the New World, 1492–1640.* New York.

————. 2001. *American Pentimento: The Invention of Indians and the Pursuit of Riches.* Minneapolis.

Segal, Daniel. 2000. " 'Western Civ' and the Staging of History in American Higher Education." *American Historical Review* 105, 770–805.

Shaffer, Lynda N. 1992. *Native Americans before 1492: The Moundbuilding Centers of the Eastern Woodlands.* Armonk, N.Y.

————. 1994. "Southernization." *Journal of World History* 5, 1–22. Reprinted in Ross E. Dunn, *The New World History: A Teacher's Companion* (Boston, 2000), 175–91.

————. 1996. *Maritime Southeast Asia to 1500.* Armonk, N.Y.

Sheffer, Gabriel, ed. 1986. *Modern Diasporas and International Politics.* New York.

Shlomowitz, Ralph, and Lance Brennan. 1994. "Epidemiology and Indian Labor Migration at Home and Abroad." *Journal of World History* 5, 47–70.

Shoemaker, Robert, and Mary Vincent, eds. 1998. *Gender and History in Western Europe.* London.

Simmons, Aan G., and Ian G. Simmons. 1996. *Changing the Face of the Earth: Culture, Environment, History.* Oxford.

Sinha, Mrinalini. 1995. *Colonial Masculinity: the "Manly" Englishman and "Effeminate" Bengali in Nineteenth Century India.* New York.

Sinn, Elizabeth, ed. 1998. *The Last Half Century of the Chinese Overseas.* Hong Kong.

Smith, Alan K. 1991. *Creating a World Economy: Merchant Capital, Colonialism, and World Trade, 1400–1825.* Boulder.

Smith, Bonnie G. 1998. *The Gender of History: Men, Women, and Historical Practice.* Cambridge, Mass.

Smith, Daniel Scott. 1995. "Recent Change and the Periodization of American Family History," *Journal of Family History* 20, 329–46.

Smith, Page. 1990. *Killing the Spirit: Higher Education in America.* New York.

Smith, Roger C. 1993. *Vanguard of Empire: Ships of Exploration in the Age of Columbus.* New York.

Snooks, Graeme. 1996. *The Dynamic Society: Exploring the Sources of Global Change.* London.

———. 1998a. *The Laws of History.* London.

———. 1998b. *The Ephemeral Civilization.* London.

———. 1998c. *Longrun Dynamics: A General Economic and Political Theory.* London.

———. 1999. *Global Transition: A General Theory of Economic Development.* London.

Sommers, Jeffrey W. 2001. "The Entropy of Order: Democracy and Governability in the Age of Liberalism." Ph.D. dissertation, Northeastern University.

Southall, Aidan. 1998. *The City in Time and Space: From Birth to Apocalypse.* Cambridge.

Sowell, Thomas. 1996. *Migrations and Cultures: A Worldview.* New York.

Spier, Fred. 1996. *The Structure of Big History: From the Big Bang until Today.* Amsterdam.

Spodek, Howard. 2001. *The World's History.* 2nd ed. Upper Saddle River, N.J.

Stearns, Peter. 1993a. *Meaning Over Memory: Recasting the Teaching of History and Culture.* Chapel Hill.

———. 1993b. *The Industrial Revolution in World History.* Boulder.

———. 2000. *Gender in World History.* New York.

Stearns, Peter N., Michael Adas, Stuart B. Schwartz, and Marc Jason Gilbert. 2000. *World Civilizations, the Global Experience.* 3rd ed. New York.

Stearns, Peter, ed., William L. Langer, compiler. 2001. *The Encyclopedia of World History.* 6th edition. Boston.

Stein, Gil J. 1999. *Rethinking World-Systems: Diasporas, Colonies, and Interaction in Uruk Mesopotamia.* Tucson.

Stokes, Gale. 2001. "The Fates of Human Societies: A Review of Recent Macrohistories." *American Historical Review* 106, 508–25.

Stoler, Ann Laura. 1995. *Race and the Education of Desire: Foucault's History of Sexuality and the Colonial Order of Things.* Durham.

Storey, William K. 1991. "Big Cats and Imperialism: Lion and Tiger Hunting in Kenya and Northern India, 1898–1930." *Journal of World History* 2, 135–74.

———. 1999. *Writing History: A Guide for Students.* New York.

Storey, William K., ed. 1996. *Scientific Aspects of European Expansion.* Aldershot, U.K.

Strayer, Robert. 1998. *Why Did the Soviet Union Collapse? Understanding Historical Change.* Armonk, N.Y.

Strobel, Margaret. 1991. *European Women and the Second British Empire*. Bloomington, Ind.

Subrahmanyam, Sanjay. 1990. *The Political Economy of Commerce. Southern India 1500–1650*. Cambridge.

———. 1997. *The Career and Legend of Vasco Da Gama*. Cambridge.

———. 2001. *Penumbral Visions: Making Polities in Early Modern South India*. Ann Arbor.

Swedberg, Sarah. 1999. "The Cranch Family, Communication, and Identity Formation in the Early Republic." Ph.D. dissertation, Northeastern University.

Tapper, Perry M. 1991. "Who Are We? Tales of National Identity." M.A. thesis, Northeastern University.

Thomas, Hugh. 1997. *The Slave Trade. The Story of the Atlantic Slave Trade: 1440–1870*. New York.

Thompson, Jason. 1994. "Osman Effendi: A Scottish Convert to Islam in Early Nineteenth-Century Egypt." *Journal of World History* 5, 99–124.

Thompson, William R. 2000. *The Emergence of the Global Political Economy*. London.

Thongchai Winichakul. 1994. *Siam Mapped: A History of the Geo-Body of a Nation*. Honolulu.

Thornton, John K. 1991. "African Dimensions of the Stono Rebellion," *American Historical Review* 96, 1,101–13.

Thornton, John. 1992. *Africa and Africans in the Making of the Atlantic World, 1400–1680*. New York.

Tilly, Charles. 1990. *Coercion, Capital, and European States, AD 990–1990*. Oxford.

———. 1998. *Durable Inequality*. Berkeley.

Tracy, James D., ed. 1990. *The Rise of Merchant Empires: Long-Distance Trade in the Early Modern World, 1350–1750*. Cambridge.

———. 1991. *The Political Economy of Merchant Empires: State Power and World Trade, 1350–1750*. Cambridge.

Trigger, Bruce G. 1993. *Early Civilizations: Ancient Egypt in Context*. Cairo.

Turner, Mary, ed. 1995. *From Chattel Slaves to Wage Slaves: The Dynamics of Labour Bargaining in the Americas*. Bloomington, Ind.

Tyrrell, Ian. 1999. *True Gardens of the Gods: Californian–Australian Environmental Reform, 1860–1930*. Berkeley.

Van Krieken, Robert. 1998. *Norbert Elias*. London.

Vansina, Jan. 1994. *Living with Africa*. Madison.

———. 1990. *Paths in the Rainforest: Toward a History of Political Tradition in Equatorial Africa*. Madison.

Vasey, Daniel E. 1992. *An Ecological History of Agriculture, 10,000 B.C.–A.D. 10,000*. Ames.

Vidal, Gore. 2000. *The Decline and Fall of the American Empire*. Monroe, Maine.

Viola, Herman, and Carolyn Margolis, eds. 1990. *Seeds of Change*. Washington, D.C.

Vivante, Bella. 1999. *Women's Roles in Ancient Civilizations: A Reference Guide*. Westport, Conn.

Vlastos, Stephen. 1997. *Mirror of Modernity: Invented Traditions of Modern Japan*. Berkeley.

Voll, John. 1994. "Islam as a Special World System." *Journal of World History* 5, 213–26.

Von Glahn, Richard. 1996. *Fountain of Fortune: Money and Monetary Policy in China, 1000–1700*. Berkeley.

Wallerstein, Immanuel. 1990. "*Culture as the Ideological Battleground of the Modern World-System*." Theory, Culture and Society 7, 31–55.

———. 2001. *The End of the World as We Know it: Social Science for the Twenty-first Century*. Minneapolis.

Wallerstein, Immanuel et al. 1996. *Open the Social Sciences: Report of the Gulbenkian Commission on the Restructuring of the Social Sciences.* Stanford.

Wang Gungwu, ed. 1997. *Global History and Migrations.* Boulder.

Warner, R. Stephen, and Judith G. Wittner. 1998. *Gatherings in Diaspora: Religious Communities and the New Immigration.* Philadelphia.

Wasserstrom, Jeffrey N., Lynn Hunt, and Marilyn B. Young, eds. 2000. *Human Rights and Revolutions.* Boston.

Waters, Neil L., ed. 2000. *Beyond the Area Studies Wars: Toward a New International Studies.* Hanover, NH.

Watts, Sheldon. 1999. *Epidemics and History: Disease, Power and Imperialism.* New Haven.

———. 2001. "From Rapid Change to Stasis: Official Responses to Cholera in British-Ruled India and Egypt, 1860 to c. 1921." *Journal of World History* 12, 321–74.

Wells, Peter. 1999. *The Barbarians Speak: How the Conquered Peoples Shaped Roman Europe.* Princeton.

Wheatley, Helen. 1992. "From Traveler to Notable: Lady Duff Gordon in Upper Egypt, 1862–1869." *Journal of World History* 3, 81–104.

White, Donald Wallace. 1996. *The American Century: The Rise and Decline of the United States as a World Power.* New Haven.

White, Philip L., ed. "Doctoral Training in World History: What Changes in Ph.D. Programs Will it Require?" *World History Bulletin* 17 (Spring 2002), 8–17.

Wildenthal, Lora Joyce. 1994. *Colonizers and Citizens: Bourgeois Women and the Woman Question in the German Colonial Movement, 1886–1914.* Ann Arbor.

Wills, John E., Jr. 2001. *1688: A Global History.* New York.

Wilson, Samuel M. 1999. *The Emperor's Giraffe, and Other Stories of Cultures in Contact.* Boulder.

Wong, R. Bin. 1997. *China Transformed: Historical Change and the Limits of European Experience.* Ithaca.

———. 2002. "The Search for European Differences and Domination in the Early Modern World: A View from Asia." *American Historical Review* 107 (2002), 447–69.

Worster, Donald. 1993. *The Wealth of Nature: Environmental History and the Ecological Imagination.* New York.

Wriggins, Sally Hovey. 1996. *Xuanzang: A Buddhist Pilgrim on the Silk Road.* Boulder.

Wright, Donald R. 1997. *The World and a Very Small Place in Africa.* Armonk, N.Y.

Wright, Tim, ed. 1997. *Migration and Ethnicity in Chinese History: Hakkas, Pengmin, and Their Neighbors.* Stanford.

Zamoyski, Adam. 2001. *Holy Madness: Romantics, Patriots, and Revolutionaries, 1776–1871.* New York.

Znamenski, Andre. 1999. *Shamanism and Christianity—Native Encounters with Russian Orthodox Missions, 1820–1917.* Westport, Conn.

Zohary, Daniel, and Maria Hopf. 1993. *Domestication of Plants in the Old World.* Oxford.

Zuckerman, Larry. 1998. *The Potato: How the Humble Spud Rescued the Western World.* Boston.

Index

Printed in the United States
135123LV00001B/64/A